Breakthrough Research and Technologies in Physics for a Sustainable Future

nova science publishers
www.novapublishers.com
New York

Physics Research and Technology

More information about this series can be found at
https://novapublishers.com/product-category/series/physics-research-and-technology/

Environmental Science, Engineering and Technology

More information about this series can be found at
https://novapublishers.com/product-category/series/environmental-science-engineering-and-technology/

Koyel Ganguly
Soumyadipta Pal
Kamakhya Prasad Ghatak
Editors

Breakthrough Research and Technologies in Physics for a Sustainable Future

nova science publishers
www.novapublishers.com
New York

DOI: https://doi.org/10.52305/UYZK8757

Library of Congress Cataloging-in-Publication Data

ISBN: 979-8-90134-111-7 (Hardcover)
ISBN: 979-8-90134-178-0 (eBook)

Published by Nova Science Publishers, Inc. † New York

Contents

Preface

In an era where the global community faces unprecedented environmental, energy, and technological challenges, the field of physics continues to be at the forefront of innovation, offering transformative solutions for a more sustainable future. "Breakthrough Research and Technologies in Physics for a Sustainable Future" brings together a diverse collection of pioneering studies and technological advancements that illustrate the powerful intersection of theoretical insights, experimental discoveries, and applied physics in addressing sustainability concerns.

This book explores the rich landscape of physics-driven breakthroughs aimed at ensuring long-term environmental balance, energy efficiency, and technological sustainability. Covering a wide range of topics, including the development of smart materials, renewable energy innovations, quantum technologies, and eco-friendly nanotechnology applications, each chapter presents both foundational research and novel approaches to practical implementation. These contributions reflect not only the ingenuity of researchers but also a shared vision for leveraging scientific knowledge to meet the needs of a rapidly evolving world.

The chapters included in this volume are authored by experts from academic and research institutions who have demonstrated a commitment to pushing the boundaries of physics in service of societal and ecological well-being. Whether through the design of materials that reduce carbon footprints, the integration of artificial intelligence in energy systems, or the advancement of quantum sensors for environmental monitoring, the work presented here is both intellectually rigorous and profoundly relevant.

We hope this compilation serves as a valuable resource for researchers, practitioners, educators, and policy-makers who are engaged in the pursuit of sustainable development through science and technology. It is our belief that physics, as both a fundamental and applied discipline, holds the key to many of the solutions our planet urgently requires. May this book inspire further inquiry, collaboration, and innovation in the ongoing journey toward a sustainable and equitable future.

We extend our sincere gratitude to all the authors and reviewers who contributed their time, expertise, and passion to this project. Their efforts have culminated in a work that underscores the transformative power of physics and its pivotal role in shaping a better tomorrow. Our sincere thanks to NOVA publishers for allowing us to publish this book. We also thank our administrative authorities for giving us this opportunity to be the editors of this book.

Koyel Ganguly
Soumyadipta Pal
Kamakhya Prasad Ghatak
Editors

Institute of Engineering & Management
Salt Lake Electronics Complex
School of University of Engineering and Management
Kolkata, India

Chapter 1

The Challenge in the Third World Is Unseen Toxicology, Which Hinders Awareness About Hidden Health Threats

Abhijit Kargupta
Goutam Ghoshal
and Kakoli Dutta*, PhD
Department of Basic Science and Humanities, Institute of Engineering & Management, Management House, University of Engineering and Management, Kolkata, India

Abstract

Toxicology investigates how chemical, biological, and physical agents affect living organisms, primarily focusing on humans. All nine US-approved dyes have been identified as causes of health disorders. Red pigment is another color that causes disease in animals, suggesting carcinogenicity in other dyes as well. Contamination with Benzedrine or other carcinogens has been found in three red dyes. Different dyes have been identified as having adverse effects in humans and animals. Research has shown the impact of Genotoxicity on yellow pigments. Toxicological studies conducted on other dyes also indicate risks and health hazards.

It is recommended that dyes used for coloring different food materials be avoided or disallowed and substituted with safer alternatives. Health and safety authorities concerned should conduct better and independent toxicological studies, implement greater care in approving or permitting

* Corresponding Author's Email: kakoli.dutta@iem.edu.in

In: Breakthrough Research and Technologies in Physics …
Editors: Koyel Ganguly, Soumyadipta Pal and Kamakhya Prasad Ghatak
ISBN: 979-8-90134-111-7

these dyes, and consider allowing only well-tested, safe alternatives in the future.

Keywords: toxicology, dyes, hypersensitivity, genotoxicity, multi-faced approach, regulatory frameworks

Introduction

Artificial food dyes, made of petroleum-based substances, are frequently used to enhance the appearance and appeal of different food products. However, use of these food additives may lead to serious health concerns like hyperactivity, some allergic reactions, neurobehavioral changes (in kids) and even cancers. Therefore, in recent times, the potential health risks associated with the frequent use of these dyes have garnered considerable attention from researchers as well as food safety regulators.

The United States Food and Drug Administration (FDA) approved nine synthetic colour additives as artificial food dyes (derived from petroleum), which can be applied for the enhancement of the color of food products. However, there are still some concerns regarding the potential risks associated with the consumption of these dyes as food additives.

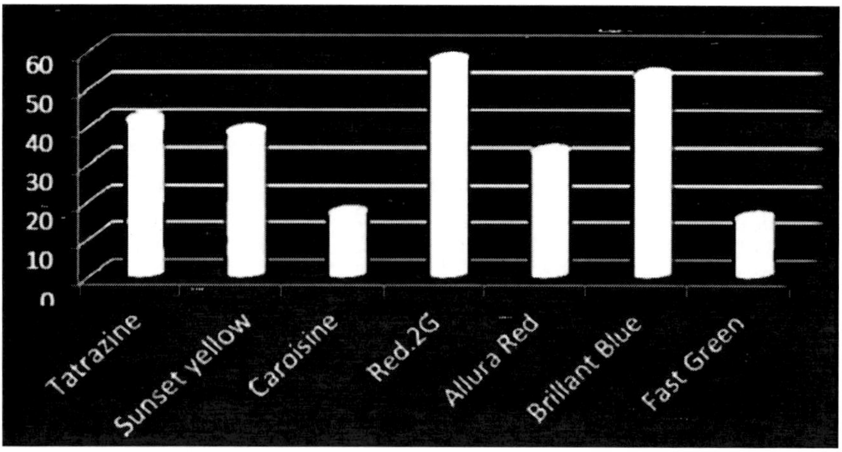

Graph 1. Percentage distribution of food color additives: Comparison of food color additive levels in desserts like jelly with gmp-permitted standards (mg/kg).

Table 1.1. Concentration range of food colorings additives in different foods and products

Food Category	Tartrazine	Sunset Yellow	Carmoisine	Erythrosine	Red 2G	Allura Red	Indigo Carmine	Brilliant Blue	Fast Green	Black PN
Chocolates (A)	-	0-217.0	-	-	-	-	-	-	-	-
Ice Cream (B)	0-0.19	0-4.98	-	-	-	0-2.8	-	-	-	-
Juices/Drinks (C)	0-26.26	0-224	0-114	-	0-109.36	-	-	0-5.12	-	-
Candy (D)	0-19.67	0-47	0-12.3	0-1 1	0-49.7	0-76.26	0-59.3	0-193.21	0-5.58	-
Jelly (E)	0-5.8	0-14.3	0-5 14	0-0 2	0-26.23	0-42.8	-	0-33.24	-	-
Gums (F)	-	0-8.2	0-15.89	-	-	-	-	0-0.57	0-11.92	0-175.3

In this regard, many countries have also established different sets of rules and regulations to monitor and control the use of artificial color ingredients in food items to ensure the quality and purity of these additives. Even though there is a lack of conclusive evidence, there is a consensus that the unregulated use of food additives, derived from petroleum-based compounds, which are commonly employed to enhance the visual appeal of food products, can pose a serious risk to human health (Bell, 2023) [1]. Therefore, it is important to study these nine FDA-approved food additive dyes in detail to get a deeper insight into their long-term adverse effects on human health.

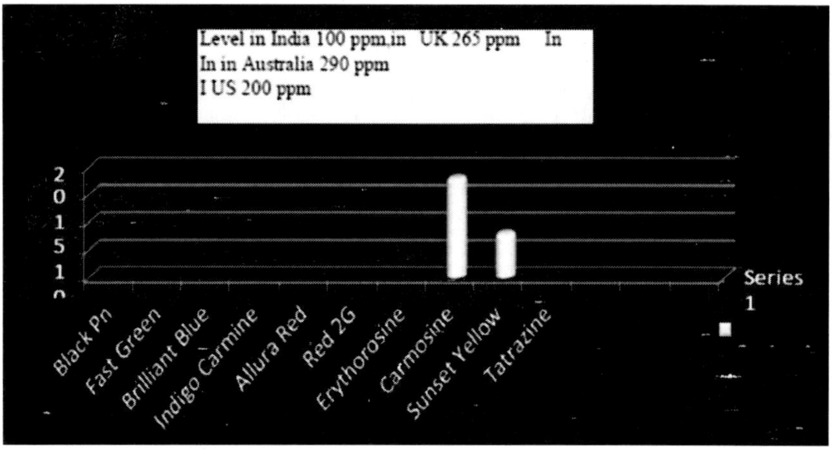

Graph 2. Percentage distribution of food color additives: comparison of food color additive levels in flavoured milk with GMP-permitted standards (mg/kg).

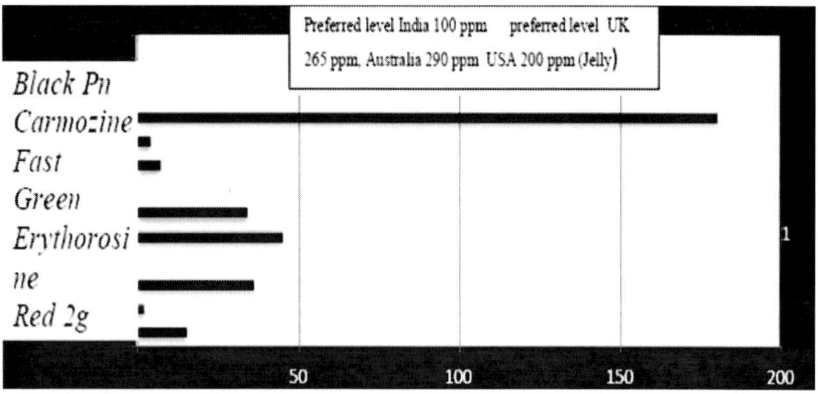

Graph 3. Percentage distribution of food color additives: Comparison of food color additive levels in carmosine (azorubine) with GMP-permitted standards (mg/kg).

Additive Levels in Fruit Juices (Artificial Blends) with GMP-Permitted Standards

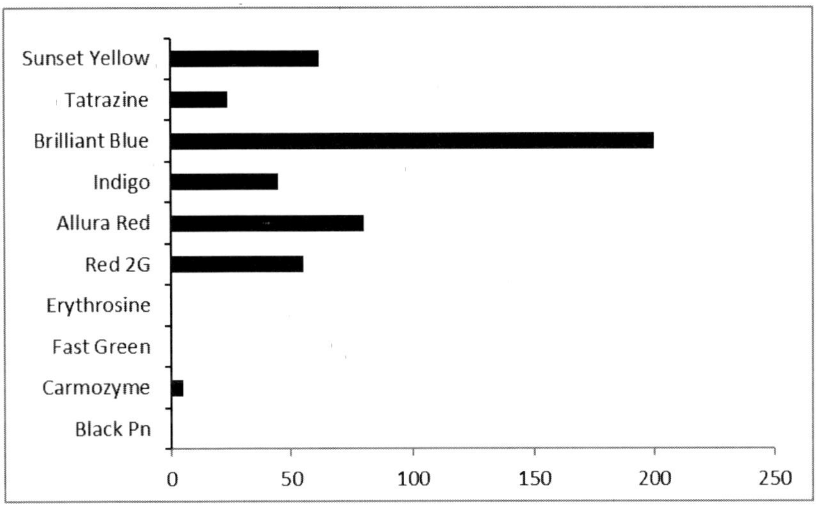

Graph 4. Concentration of food color additives in candy compared to GMP-permitted levels (mg/kg)

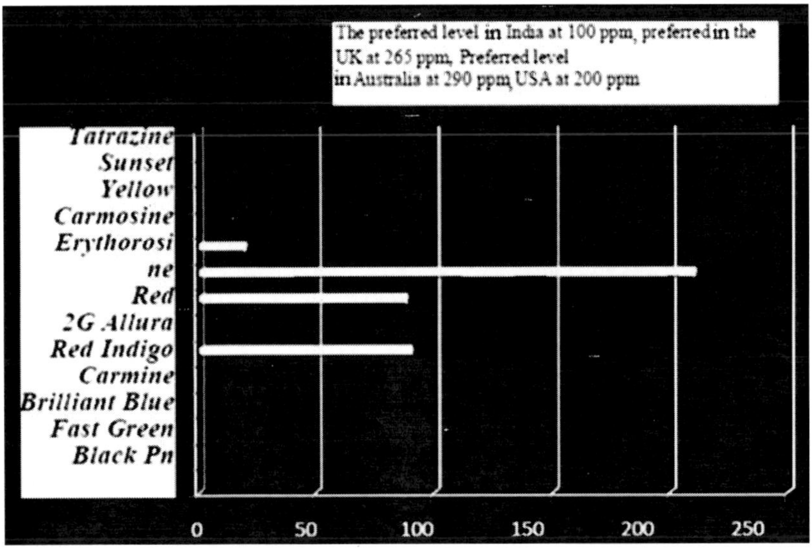

Graph 5. Comparison of food color additive concentrations in juices and beverages with GMP-permitted levels (mg/kg).

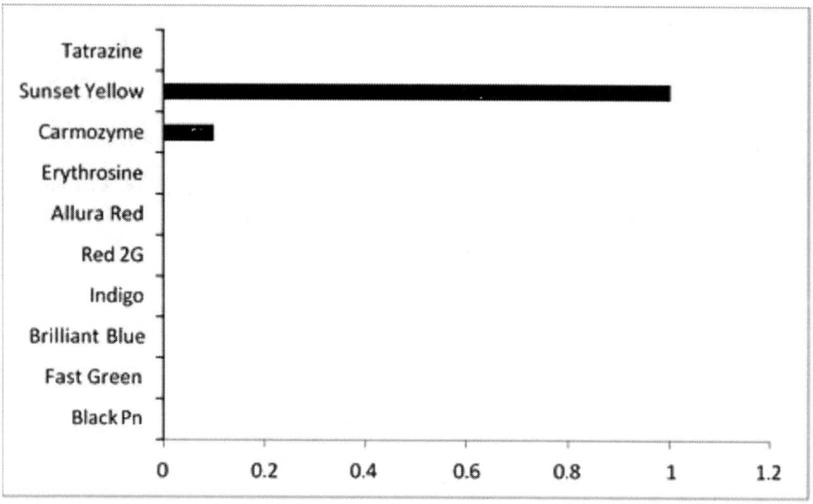

Graph 6. Comparison of food color additive concentrations in ice cream with GMP-permitted levels (mg/kg).

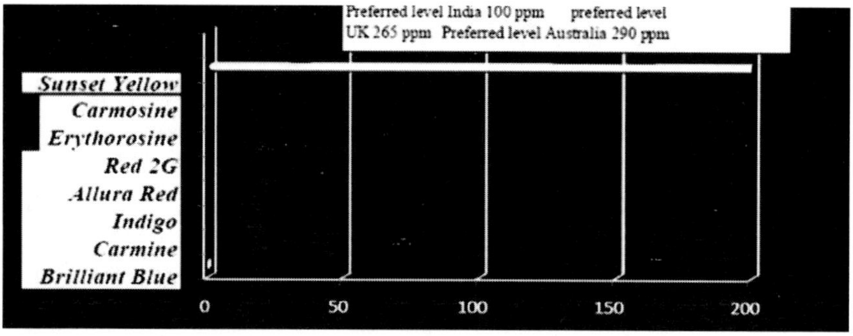

Graph 7. Comparison of Food color Additive Levels in Chocolate with GMP Limits across Different Countries (mg/kg).

Toxicological Impact of Synthetic Food Dyes

An Overview: Toxicology, as a diverse discipline and a wealth of data, has proven the usability of technology to study the mechanisms of adverse effects of exposure to exogenous substances, particularly those that affect human health.

An Overview

Toxicology, as a multidisciplinary field supported by extensive research, has demonstrated the value of technological advancements in exploring the mechanisms through which exposure to external substances—particularly synthetic compounds—can lead to adverse health effects in humans [2].

The toxicological impact of particulate matter on the cardiovascular system, particularly the role of magnetite nanoparticles in comparison to nonmagnetic nanoparticles, was noted to play a crucial role in the development of foam cells and in disrupting cholesterol metabolism as particulate matter-induced atherosclerosis advanced [3]. This paper was published and shown as stated above.

Kobylewski noted in his journal that food dyes, initially derived from coal tar and now from petroleum, have caused controversy due to safety concerns. Several dyes have been banned because either they would hurt lab animals or were not tested properly [4].

The study focused on the behavioral and neurochemical effects induced by tartrazine, a synthetic food color, in young rats. Depression and social anxiety measures were also used. Tartrazine-treated mice showed increased motivation as demonstrated by increased horizontal movement in the open-ended ELD test. From the study of Rotllan, it was observed that tartrazine produced an antigenic effect in the open field, elevated plus-maze, and dark-light transition tests.

Toxicological Evaluation of Some Synthetic Food Dyes

The greatest challenge during the evaluation process of foods with cancer-promoting potential and pathogenicity is the determination of the threshold value, or rather, the no side effect value. It was explained that the belief in the irreversibility of carcinogenesis was mainly based on chemical studies, experiments involving single or multiple doses, and the concept that somatic alterations serve as the initial step in cancer development following disruptions in cellular growth. It was also noted that a significant challenge remained in applying the results of animal studies, which often involved high doses, to rare and less frequent effects observed in humans. Additional concerns included issues such as service denial policies, low approval rates, and the use of safety margins. It was emphasized that acceptance thresholds should not remain fixed but should be revised as new information regarding safety measures becomes available. Preussmann, 1978 [5].

It was suggested that timely revisions should be carried out whenever new information related to safety evaluations becomes available.

It was indicated by the World Health Organization in 1991 that certified colorings agents might trigger adverse reactions, and it was also noted that some food items could carry a risk of causing cancer. It was reported that synthetic colorings agents could lead to significant alterations in neurotransmitter activity and changes in exploratory behavior. It was reported that synthetic colorings agents could lead to significant alterations in neurotransmitter activity and changes in exploratory behaviour. It was pointed out that the mechanisms through which artificial food colors, flavors, preservatives, and certain foods might lead to ADHD-like symptoms or other physical effects were still not well understood. A review published in Nutrition Reviews was said to explore possible contributing factors, including toxicological, anti-nutritional, and hypersensitivity-related elements. It was reported by the World Health Organization in 1991 that certified colouring agents could lead to adverse reactions, and research by Das and colleagues in 2001 indicated that certain food items might have carcinogenic potential. Further studies by Dalal and Poddar in 2009 suggested that synthetic food colorings could cause significant alterations in neurotransmitter levels.

Arnold in 2013 observed that individuals with conditions such as asthma, eczema, chronic urticarial, and migraines frequently exhibited negative responses to artificial food.

Colorings and specific food items. However, the precise mechanisms through which artificial food dyes, flavoring agents, preservatives, and certain foods contribute to symptoms resembling ADHD or other physical effects remain largely unknown. A review in Nutrition Reviews examined the potential roles of toxicological, anti-nutritional, and hypersensitivity-related factors in triggering these adverse reactions.

Health Effects of Artificial Food Colorings: A Toxicological Perspective

The World Health Organization reported in 1991 that certified food colourants may cause adverse reactions in specific individuals. In support of this concern, a study conducted by Das and colleagues in 2001 suggested that some food products might have carcinogenic potential. Further research by Dalal and Poddar in 2009 indicated that synthetic food colourings could lead to significant changes in neurotransmitter levels and alter exploratory behaviour.

According to Arnold and colleagues in 2013, gaining a clearer understanding of the relationship between dye allergies and ADHD-type symptoms would require further investigation into both the biochemical and behavioral effects of artificial food colourings in trained and untrained children. Over the past thirty-six years, numerous studies have examined the impact of synthetic food colourants on the behavior and physiology of children, laboratory animals, and the brain. However, there is still no comprehensive understanding of how these processes are interconnected.

In vitro evaluations were carried out on eight food coloring agents or commercially available color mixtures approved for use by the United States government, to determine their potential to induce transformation in Fischer rat embryo cells. (Kobylewski et al., 2012) Prior research has identified these cells as a sensitive indicator of chemicals that may cause cancer. (Shamberger, 1984) The commercial mixture of two dyes, Blue 1 and Yellow 5, led to the transformation of cells into a malignant state. Similar effects were also observed with Blue 2, Green 3 (in one of the two experiments), and Red 4. (Shamberger, 1984 It was observed that dyes which do not transform—such as Blue 1, Red 3, Yellow 5, Yellow 6, and Green 3—did not lead to any significant increase in tumor formation or mortality rates in the tested animals.

Das and colleagues in 2004 reported that three dyes—Blue 2, Green 2024, and Red 4 were found to increase tumor growth and/or mortality in certain strains of hamsters. Laboratory investigations suggested a possible link between some food dyes and cancer, and studies involving hamsters were said to support this observation. Studies have connected specific food colorings to problems with behavior, including hyperactivity. And attention deficit, particularly in children. Allergenic Reactions: Some individuals may experience allergic reactions to food dyes, including skin rashes and respiratory issues. Animal studies have demonstrated that certain dyes can lead to organ toxicity, particularly liver damage, when consumed in large quantities. However, ongoing research and periodic re-evaluations are necessary to ensure safety, particularly considering the potential for cumulative exposure from multiple food sources. Each food color authorized by the EFSA conducted a thorough safety assessment of products intended for use in Europe. It has been verified that tartrazine (TZ) is the root cause of biological damage to the kidneys and liver. It affects learning, working memory, and reproduction in Swiss albino mice. Toxicological studies on TZ are essential as it is used in many foods [6]. Most of the artificial food colour additives (AFCAs) are found mostly used in food products to increase visual attractiveness appeal, in particular to those marketed to children [7].

Discussion and Results

1. The above Table 1.1 representing all the colorings additives was tested in different foods and food products and results.
2. In Graph 1 it showed that Desserts like Jelly Red 2G(E128) and Brilliant Blue (E133) are High. In Graph 2 it is shown that Carmosine (Azorubine) (E122) is high. In Graph 3 it showed that Carmosine (Azorubine) (E122) is high. In Graph 4 it showed that Brilliant Blue is high. In Graph 5 it showed that Allura Red AC (E129) is High. In Graph 6 it showed that Tatrazine(E102) is high. in Graph 7 it showed that Sunset Yellow(E110) is High.

Conclusion

Addressing the concerns surrounding synthetic food dyes requires a multifaceted approach involving various stakeholders.

Firstly, numerous research studies are needed to explore these additives' health effects, their long-term impact on human health, and the underlying mechanisms of action. Rigorous scientific studies can provide valuable insights into the safety profile of synthetic dyes and inform regulatory decisions regarding their use in food products.

Public education serves as the second critical component, as it enables consumers to make informed decisions regarding the potential hazards associated with synthetic food dyes and raises awareness. Educational initiatives that inform individuals about the presence of these additives in processed foods, their potential health effects, and strategies for minimising exposure can help mitigate risks and promote healthier dietary habits.

Lastly, regulatory agencies should consider revisiting existing regulations and possibly implementing stricter guidelines to ensure the safety of synthetic food dyes. This might involve reassessing acceptable daily intake levels, enhancing labelling requirements to provide transparent information to consumers, and conducting regular monitoring and surveillance to detect any adverse effects associated with these additives.

Although synthetic food dyes have long been considered safe for consumption within regulatory limits, emerging evidence suggests potential health risks that warrant caution and action [8]. Addressing these concerns requires collaboration among researchers, policymakers, industry stakeholders, and consumers. By taking proactive steps to mitigate the risks associated with synthetic food dyes, we can safeguard public health and promote a safer food environment for all. Further research, public education, and possibly stricter regulation may be needed to address these concerns effectively.

References

[1] Bell et al., 20, the subject was "Just by Showing Up" *SAGE Journals* (clinical review).

[2] 'Anxiety-like symptoms, depression, and anti-social behavior in rats. *Journal of American Science*,' 7(6), 1211–1223. Kamel, M. M., & El-Lethey, H. S. (2011). The potential health hazard of tartrazine and levels of hyperactivity.

[3] Yu, H., Xu, L., Cui, T., Wang, Y., Wang, B., Zhang, Z., Su, R., Zhang, J., Zhang, R., Wei, Y., Li, D., Jin, X., Chen, W., & Zheng, Y. (2022). The foam cell formation is associated with imbalanced cholesterol homeostasis due to exposure to airborne magnetite nanoparticles. *Toxicological Sciences,* 189(2), 287–300. https://doi.org/10.1093/toxsci/kfac079.

[4] Kobylewski, S., & Jacobson, M. F. (2013). 'Toxicology of food dyes. *International Journal of Occupational and Environmental Health,*' 19(3), 220–246. https://doi.org/10.1179/1077352512Z.00000000034.

[5] Preussmann, R. (n.d.). *Toxicological Aspects of Food Safety — Carcinogenicity and Mutagenicity* (pp. 179–191). https://doi.org/10.1007/978-3-642-66896-8_7.

[6] Tartrazine Induced Oxidative Damage in Mice Liver and Kidneys.Latif Ali, Nabila Mehdi, Bonhado Youcef,Omar Kleroua & Djamel Saidi.South *Asian Journal of Experimental Biology*. DOI: 10.38150/Sajeb.7(6) p 271-278.

[7] Ahmed, M. A., Al-Khalifa, A. S., Al-Nouri, D. M., & El-Din, M. F. S. (2021). Dietary intake of artificial food colour additives in food products by school-going children. *Saudi Journal of Biological Sciences, 28(1), 27–34.* https://doi.org/10.1016/j.sjbs.2020.08.025.

[8] Mota, I. G. C., Das Neves, R. A. M., Nascimento, S. S. D. C., Maciel, B. L. L., De Araújo Morais, A. H., & Passos, T. S. (2021). Artificial dyes: Health risks and the need for revision of international regulations. *Journal of the American College of Nutrition,* 40(2), 1578–1593. https://doi.org/10.1080/87559129.2021.1934694.

Chapter 2

Simulation of Energy Loss in Proton Beam Therapy for Lung Cancer Treatment

Arkabrata Gupta[1,*]
and Priya Chakraborty[2]
[1]Department of Basic Science and Humanities, Institute of Engineering & Management, University of Engineering and Management, New Town, Kolkata, India
[2]Department of Physics, Acharya Prafulla Chandra College, New Barrackpore, West Bengal, India

Abstract

Proton beam therapy (PBT) is a form of external beam radiation therapy that utilises protons to treat cancer. This advanced radiological technique employs a proton beam, which interacts with various layers of the human body through nuclear and electronic processes, delivering its energy through its track. It is designed to precisely deliver the maximum dose to cancerous cells while leaving the surrounding healthy tissues unaffected. We use the Monte Carlo-based simulation software, Stopping Range of Ions in Matter, acronymed as SRIM, to estimate the energy loss through individual layers of the human body when treating lung cancer. This study supports the design of proton beam therapy for lung cancer.

Keywords: proton beam therapy, SRIM, TRIM

---

* Corresponding Author's Email id: arkabratagupta@gmail.com

In: Breakthrough Research and Technologies in Physics …
Editors: Koyel Ganguly, Soumyadipta Pal and Kamakhya Prasad Ghatak
ISBN: 979-8-90134-111-7

Introduction

Cancer is a group of diseases characterised by the uncontrolled growth and spread of abnormal cells. Various techniques, such as radiation therapy, chemotherapy, and drug introduction, are available to treat cancer. However, given their associated risks, there is a requirement for new methods to mitigate the drawbacks of these methods. Proton beam therapy (PBT) is an effective external beam radiation method that utilises proton beams to treat different types of cancer in the human body.

A charged particle beam, such as a proton beam, traversing the human body undergoes a series of interactions with atomic electrons and nuclei. Protons interact with matter via Coulomb forces due to their positive charges and the negative charges of orbital electrons of the matter, as well as nuclear forces with target atomic nuclei. Initially, protons interact with atomic electrons through frequent inelastic Coulomb interactions, losing their kinetic energy. Despite these interactions, protons maintain a nearly straight-line path because their rest mass energy is significantly greater than that of electrons. As the proton's energy decreases, it interacts with atomic nuclei through repulsive elastic Coulomb interactions. The much larger mass of the nucleus causes the proton to deviate from its original path [1, 2].

A proton's energy loss is inversely proportional to the square of its velocity. Thus, when the proton beam enters the human body with high velocity, the number of interactions is low, and only a small amount of energy is deposited. As the proton slows down at the targeted site, the number of interactions increases, depositing maximum energy with a rapid distal fall-off. The energy deposition profile of these interactions illustrates how protons transfer energy along their path [1]. Most of the proton's energy is lost near the end of its range, a phenomenon known as the Bragg peak. At this point, the proton beams effectively damage the DNA of cancer cells. Therefore, understanding the interaction characteristics of protons with the human body is crucial for treatment planning. These simulations require precise qualitative and quantitative data, such as the thickness and composition of the body parts through which the proton beam traverses, enabling accurate modelling of interactions and the selection of appropriate proton beam energy for treatment [2, 3].

In this study, we utilise the Monte Carlo simulation software SRIM-2013 (Stopping Range of Ions in Matter) [4] to analyse proton interactions and calculate the energy loss of proton beams through various tissues and organs. The paper is organised as follows: Section 2 details the methods of our simulation, and Section 3 presents the results of our findings. This section provides a detailed description of using PBT to treat lung cancer. Finally, Sec. 4 summarises our work.

Methods of Simulation

For our purpose, we have utilised the SRIM software. SRIM, a powerful software collection, facilitates the study of ion transport phenomena in various materials. Key applications include the calculation of ion stopping and range in multiple targets, determining stopping powers and modelling range and straggling distributions across a wide range of ions and energies. Furthermore, it supports applications like ion implantation, sputtering, ion transmission and ion beam therapy simulation. In case of a target having a multilayer composition, TRIM (integrated part of SRIM) is used, which provides several outputs such as ion ranges, backscattered/ transmitted ions, etc. We can easily calculate the energy loss of various layers, such as energy loss due to ionisation and energy loss due to recoils, and find out the position of the Bragg peak and the lateral distribution of protons in the target. For this scenario, the quick calculation damage option was chosen as it computes the final distribution of ions within the target, energy transferred to recoil atoms, ionisation energy loss by the ion, and the number of backscattered and transmitted ions, simulating the interaction between the beam and a target. The simulation of the interaction of the beam with any target, i.e., the energy loss profile of the beam, requires specific quantitative and qualitative information about the target and beam. The following subsection discusses in detail the requirements for the simulation inputs.

Specification of the Beam

The details of the beam, i.e., elementary properties of the projectile particles, such as mass, atomic weight, atomic number, and energy, are required as simulation inputs. The quantity of bombarding particles in the beam can also be selected based on the statistical precision necessary. In the case of PBT, a high-energy proton (~MeV) beam is needed. However, the selection of beam energy varies for different cancers, depending on the location and composition of the affected organ.

Specification of the Target

SRIM requires the detailed properties of the target material as well. The atomic number, atomic weight, and stoichiometry properties are fundamental requirements. Moreover, the composition of the target material (element-wise), its density, and the thickness of each of the layers are basic requirements for calculation in SRIM. In the case of PBT, the target details need to be chosen according to the cancer-affected organs and the path of propagation of the beam.

Results and Discussion

As previously noted, the beam and target details are highly case-specific, requiring a detailed analysis for each individual scenario. In the following section, we present the results of our findings in the treatment of cancer in human organs—lung cancer.

PBT Simulation to Treat Lung Cancer Using SRIM

The lung is well protected by the rib cage, except for the apex of the lung. It is covered by the parietal pleura and the visceral pleura. Visceral pleura divides the lung into various lobes known as bronchopulmonary segments. Any carcinoma or obstruction into the substance of the lung may metastasise to the brain.

Qualitative and Quantitative Specifications

To reach the lung cancer cells, the proton beam travels through the skin, adipose tissue, soft muscle, rib bones, and lung. The properties of each layer are provided in Table 1. From this table, it can be shown that to reach the lung, the proton beam needs to travel almost ~1.8 cm through various body parts. The composition of each layer varies, as detailed in Table 2. Carbon, nitrogen, oxygen, and hydrogen make a significant quantitative contribution to the structure of human body organs.

Table 1. Properties of layers through which the proton beam propagates to reach the lung cancer site [5, 6]

Layer	Layer Name	Layer width (mm)	Layer density(g/cm3)
1	Skin	4	1.09
2	Adipose tissue	3	0.92
3	Soft muscles	3.5	1.07
4	Rib bones	7	1.85
5	Lung	40	1.05

Table 2. Composition (in %) of various layers [4, 7]

Layer	H	C	N	O	Na	Mg	P	S	Cl	K	Density (g/cm^3)
Skin	10.00	20.40	4.20	64.50	0.20	-	0.10	0.20	0.30	0.10	1.09
Adipose tissue	63.47	28.40	0.30	7.77	0.01	-	0.02	-	0.01	-	0.92
Soft muscles	63.18	6.24	1.58	28.98	-	-	-	-	-	-	1.07
Ribbones	47.54	12.19	3.04	28.28	0.09	0.09	3.43	0.09	5.31	-	1.85
Lung	10.30	10.50	3.10	74.90	0.20	-	0.20	0.30	0.30	0.20	1.05

Selection of Beam Energy

The selection of beam energy is the first step in the simulation procedure. For the initial energy estimation, we consider layers of a water phantom equivalent to the thickness of the human body layer. The human body consists of 60% water, and the density of water is similar to that of human muscles [2]. Thus, by considering a single layer of water phantom and calculating the proton energy loss through this layer, we can estimate the proton's energy loss. Based on this estimated energy, we proceed with our simulation to obtain the energy loss profile.

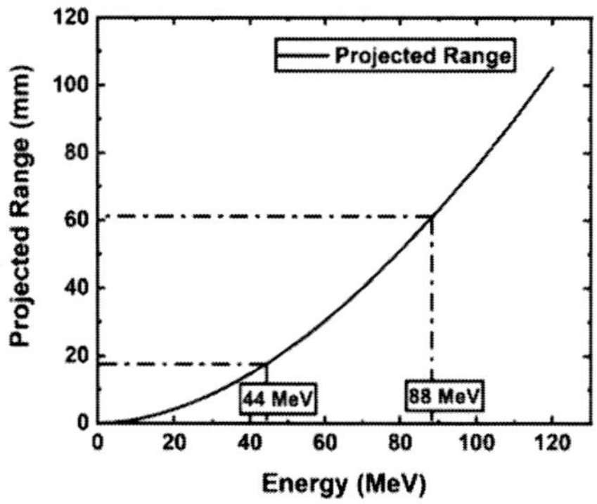

Figure 1. The projected range of the proton beam with beam energy.

Figure 1 illustrates the projected range of the proton beam through an equivalent water phantom, with a thickness corresponding to the total path length. It shows that, depending on the cancer's location in the lung, the required beam energy will range from 44 to 88 MeV. For this simulation, assuming the cancer is located 30 mm inside the lung, a beam energy of 85 MeV was selected to obtain the energy profile.

Energy Loss Due to Ionisation

When the proton beam interacts with atomic electrons through Coulomb interaction, it loses energy in various layers. Figure 2 illustrates the loss of energy per unit depth across different layers due to ionisation. In this plot, 'Ions' represents the energy transferred from protons to target electrons via direct collisions, and 'Recoils' represents the energy transferred to target electrons from recoiling target atoms [4]. It is clear from the plot that the proton beam loses energy primarily due to ions, while energy loss due to recoils is negligible.

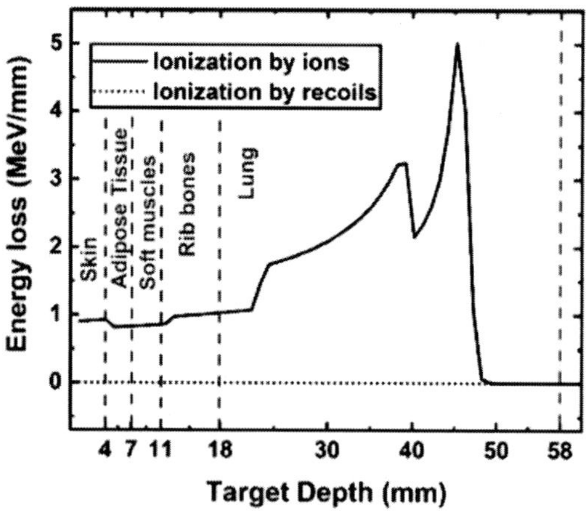

Figure 2. Energy loss of the proton beam in various layers due to ionisation.

Energy Loss Due to Phonon

Phonons are generated through nuclear interactions. When an energetic proton beam interacts with the target atom, phonons are produced due to atomic vibrations. The energy loss due to phonons occurs primarily when proton beams generate phonons, with additional energy loss arising from recoiling atoms transferring energy to phonons. The energy loss per unit depth through different layers is shown in Figure 3. It can be observed that the energy loss due to phonons is also dominated by ions over recoils.

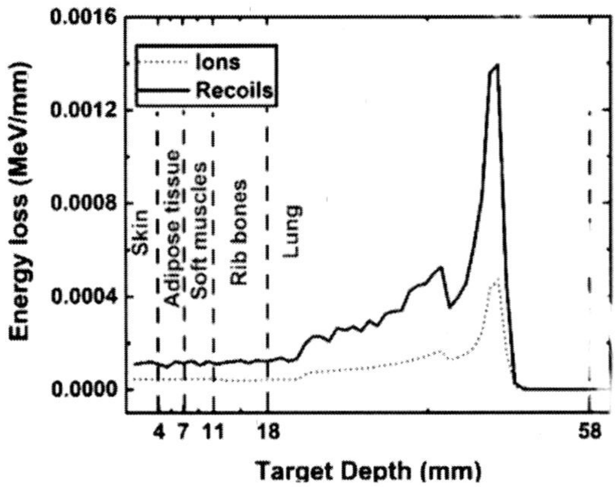

Figure 3. Energy loss due to phonons.

Table 3. Energy loss of the proton beam in various layers due to different interaction processes

Layer	Energy loss due to ions (MeV)	Energy loss due to recoils (MeV)	Energy loss due to Phonon (MeV)
Skin (0 mm to 4 mm)	2.47	1.26E-3	1.56E-2
Adipose tissue (4 mm to 7 mm)	2.39	3.44 E-4	1.12E-3
Soft muscles (7 mm to 10.5 mm)	3.29	9.22 E-4	5.05E-4
Ribbones (10.5 mm to 17.5 mm)	6.61	6.77 E-4	4.30E-4
Lung (17.5 mm to 57.5 mm)	69.42	1.09 E-2	5.25E-4

Table 3 compares the energy loss due to different interactions across various layers. It can be observed that 75 MeV of the 85 MeV is lost in the lung tissue alone. Additionally, it is essential to note that the energy loss due to ionisation is the highest, making it the dominant process. Proton beams lose most of their energy through ionisation, i.e., atomic interactions.

Ion Range

From Figure 4, it can be observed that the maximum energy is deposited at 45.7 mm, where the Bragg peak occurs. This plot indicates that the proton deposits its maximum energy at the cancer site in the lung. The 85 MeV proton beam slows down at this position, leading to increased interactions with surrounding atoms and depositing its maximum energy at the cancer site. By determining the cancer position using a CT image, we can design the treatment so that the Bragg peak occurs at the cancer site, ensuring the proton deposits its maximum energy at the cancerous tissue, thereby destroying the DNA of the affected cells and preventing metastasis. In this case, the proton average range is 45.7 mm and the proton average straggling is 0.780 mm.

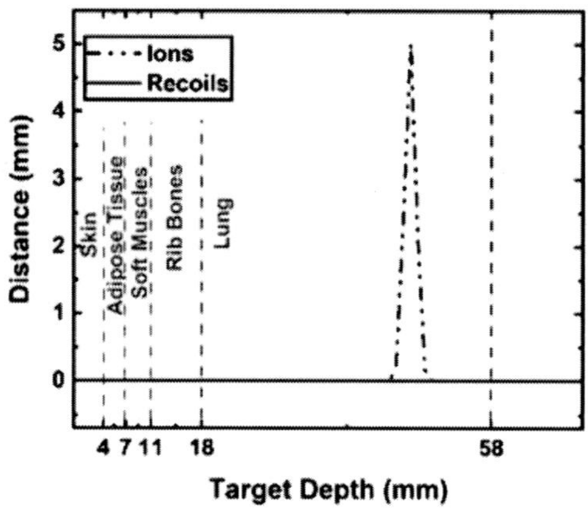

Figure 4. Ion range of 85 MeV proton beam (Bragg Peak).

Lateral Distribution

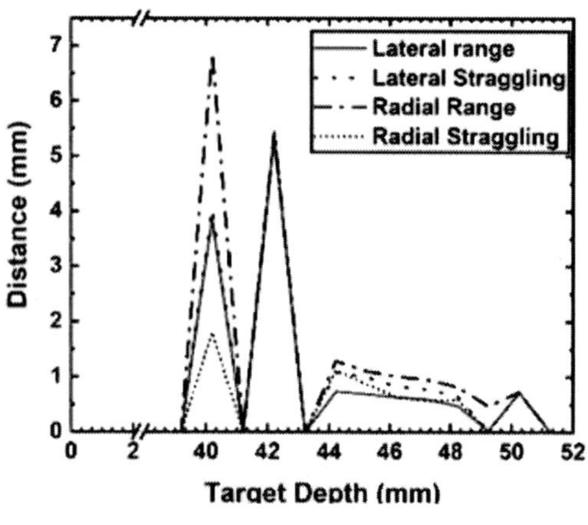

Figure 5. Lateral and radial distribution of 85 MeV proton beam.

Figure 5 depicts the ion spread within the target window, showcasing both lateral and radial distributions. The lateral projected range represents the average magnitude of displacements projected from the x-axis [3].

Conclusion

In this work, we estimate the required proton beam energy to treat lung cancer. Assuming the cancer cells are at a specific position in the lung, we selected a definite energy and performed simulations to analyse the energy loss profile. The results indicate that, although protons interact with matter through various types of interactions, 99.96% of the total energy loss occurs due to ionisation by ions. Our simulation also confirms the formation of the Bragg peak at the cancer site, where 82.45% of the proton beam's energy is deposited, while 2.93%, 2.83%, 3.91%, and 7.85% are lost through the skin, adipose tissue, soft muscles, and rib bones, respectively. Furthermore, the lateral and radial distribution of the proton beam demonstrates a minimal spread (~6 mm), enhancing the precision of the PBT process in targeting specific regions. This study highlights and emphasises the application of SRIM software in designing and planning ion beam radiation therapy.

Acknowledgments

The authors extend their gratitude to Dr Santanu Nandi, Associate Professor at Calcutta National Medical College and Hospital, for providing the essential information.

References

[1] Kenneth S. Krane, *Introductory Nuclear Physics* (John Wiley and Sons, Inc., 1998).

[2] Thomas J. FitzGerald, eds, *Interaction of Proton Beam with Human Tissues in Proton Therapy* (Intech Open, 2024).

[3] K. Giri, K. Khatiwada and P. Bista, *Journal of Nepal Physical Soc*iety **8(1)**, 48(2022).

[4] J. F. Ziegler, J. P. Biersack, *The stopping and range of ions in matter*, SRIM2013 (version 2013.00) [Online], www.srim.org.

[5] T. F. De Laney and H. Kooy, eds., *Proton and Charged Particle Radiotherapy* (Philadelphia, PA: Lippincott, 2007).

[6] Harald Paganetti, eds, *Proton Beam Therapy* (IOP Publishing Ltd, 2017).
[7] Metin Usta and Mustafa Cagatay Tufan, *Radiation Physics and Chemistry* **140,** 43 (2017).

Chapter 3

Unveiling the Novelty of Plasmon-Exciton Coupling in Hollow Gold Triangular Platelets and Cyanine Dye J-Aggregates Nanocomposites

Arpan Maiti*
Biswarup Gourd
and Spandan Aich
Centre of Excellence in Multifunctional Materials and Devices,
Department of Basic Science and Humanities, Institute of Engineering & Management,
University of Engineering and Management, New Town, Kolkata, India

Abstract

Hollow gold nanoparticles have recently garnered significant interest due to their unique optical properties. The capacity of these particles to maintain powerful amplified electromagnetic fields at the nanostructure's surface has drawn much interest in surface-enhanced spectroscopies, photothermal treatment, solar cells, and catalysis. The field generation depends on the physical properties of the nanoparticles and can thus be easily adjusted to suit specific needs. We examined the potential for detecting strong plasmon-exciton coupling using hollow gold nanotriangular (HGNT) platelets and cyanine dye J-aggregate nanocomposites. Several sets of HGNT platelets with different aspect ratios were synthesised by varying the seed concentration to observe the coupling behaviour. Hybrid nanosystems were created by adhering and assembling 1,1′-diethyl-2,2′-cyanine (also known as pseudocyanine, or

---

* Corresponding Author's Email: arpanmaiti25@gmail.com

In: Breakthrough Research and Technologies in Physics …
Editors: Koyel Ganguly, Soumyadipta Pal and Kamakhya Prasad Ghatak
ISBN: 979-8-90134-111-7

PIC) iodide onto the surface of HGN platelets. The UV-visible absorption spectra showed an apparent, transparent dip, indicating coherent coupling between HGN surface plasmons and PIC-J aggregate excitons. Creating such a strong, localised electromagnetic field through coherent coupling at quantum scales would have great potential in high-sensitivity sensing, quantum information processing, and nano-optics.

Keywords: plasmon-exciton coupling, localised surface plasmon resonance (LSPR), hollow gold nano-triangular platelets (HGNTs), cyanine dye J-aggregates, nanophotonics, plexcitons, hybrid nanostructures

Introduction

Due to their exceptional optical properties, noble metal nanoparticles (MNPs) have garnered immense attention in plasmonic research (Odom et al., 2017). One of the most fascinating properties of MNPs is the significant scattering and absorption of light in the visible to near-infrared range. These unique optical properties arise mainly from the ability of MNPs to sustain localised surface plasmons (LSPs). LSPs are coherent collective oscillations of conduction electrons at the nanoparticle surface when excited by an electromagnetic wave (for example, light). The resonance of these LSPs—known as localised surface plasmon resonance (LSPR)—can be tuned over a broad spectral range by manipulating their size, shape, and composition, as well as modifying the dielectric environment around them. Localised Surface Plasmons (LSPs), when resonantly excited by electromagnetic waves, produce intense electromagnetic field enhancement (Maiti et al., 2017; Roy et al., 2017; Maity et al., 2016) around nanoparticles as small as 10–50 nm, surpassing the diffraction limit. Such controlled field localisation is valuable for various applications, such as sensing, photovoltaics, optical trapping, waveguiding, and single-molecule detection. Among different nanostructures, hollow gold nanotriangular platelets (HGNTs) have particularly attracted remarkable attention in recent years owing to the presence of both the hollow and triangular geometry of these platelets, intensifying the electromagnetic field near the nanoparticle's surface and creating "hot spots" where field confinement is extreme. Various studies (Das et al. 2017; Hazra et al., 2016) have shown that HGNTs enable more efficient field confinement than solid nanoparticles, thanks to their unique geometry, which allows for larger surface areas and optimal tunability of plasmonic properties.

An exciton is a quasiparticle that arises when an electron is excited and travels to a higher energy state, creating a positively charged hole. The electron and hole, bound by Coulombic attraction, act as a single unit with energy transfer capabilities that can be harnessed in light-sensitive applications. J-aggregates of cyanine dyes are molecular assemblies that facilitate the formation of stable excitons (Das et al., 2017; Hazra et al.,2016). In these ordered structures, excitonic interactions are enhanced, producing narrow absorption bands that shift to longer wavelengths (a phenomenon called red-shifting) characteristic of these molecular assemblies. Moreover, the J-aggregates of cyanine form sharp structures that exhibit sharp, narrow-band absorption and can couple strongly with plasmonic fields in their vicinity. The stability and high oscillator strength of excitons within J-aggregates make them ideal for applications that require strong coupling with nearby plasmonic fields. The high degree of order within J-aggregates enables coherent dipole interactions among molecules, leading to sharper, more defined absorption bands. Various studies (Das et al., 2017; Hazra et al., 2016; Li et al., 2018; Pelton et al., 2019) have highlighted cyanine dye J-aggregates as exemplary systems for creating excitonic states that can effectively couple with LSPs, forming hybrid states that exhibit combined characteristics of both excitons and plasmons. This compatibility makes J-aggregates crucial for designing hybrid nanostructures, where effective coupling between excitons and plasmons can unlock new functionalities in nanophotonic and quantum applications. During plasmon-exciton coupling, an interaction between a nanoparticle's plasmonic field and a molecular assembly's excitonic field occurs under strong coupling conditions. Strong coupling between the two states is indicated by spectroscopic phenomena like Rabi splitting and transparency dips (Das et al., 2017; Hazra et al., 2016; Li et al.,2018; Pelton et al., 2019). When the electromagnetic field from an LSP interacts with the excitonic dipole of nearby J-aggregates, it can lead to energy sharing between the two states, creating a new hybrid system. This coupling alters the energy layout of the system, leading to the emergence of Rabi splitting—a clear indicator that the plasmon and exciton are in the strong coupling regime. The coupled system no longer behaves as independent plasmonic and excitonic states in this regime. Instead, it forms plexcitons, hybridised states that possess the characteristics of the plasmon and exciton. Rabi splitting is observed in the optical spectrum as two distinct peaks, representing the upper and lower energy states resulting from coherent energy exchange. The separation between these peaks, termed the Rabi splitting energy, is a quantitative measure of the coupling strength (Das et al., 2017; Hazra et al., 2016; Li et al., 2018; Pelton et al., 2019).

In this study, we have discussed the plasmonic properties of hollow gold nanotriangular platelets (HGNTs) and studied their coupling behaviour with PIC-J-aggregate. Our hybrid nanosystem displays a strong coupling among plasmons and excitons, manifesting as an apparent transparency dip in the UV-visible absorption spectra. These form hybrid energy states called plexitons, which are hybridised states that possess the characteristics of both plasmons and excitons. These hybrid systems display both plasmonic and excitonic states. They are, therefore, convenient in various fields like high-sensitivity sensing, quantum information processing and nano-optics, where manipulation of electromagnetic fields and light at the nanoscale is essential.

Experimental Section

Synthesis of Hollow Gold Nanotriangular (HGNT) Platelets

In the present work, HGNT were prepared using a sacrificial galvanic replacement approach following the procedure developed initially by Senapati et al. (Senapati et al., 2012) with slight modification. The HGNT platelets synthesis process was basically a two-step process. Silver nanoseeds were prepared as a precursor for the galvanic replacement reaction in the first stage. This was accomplished by adding 1 mL of 10^{-2} M $AgNO_3$ dropwise while stirring continuously in 40 mL of Milli-Q water. Subsequently, 200 μL of 2.5×10^{-2}M trisodium citrate was added to this $AgNO_3$ solution. Next, 60 μL of 10^{-1} M freshly prepared ice-cold $NaBH_4$ was added dropwise until the solution gradually turned translucent yellow. This solution was left undisturbed in the dark for two hours. Next, the CTAB solution was prepared by dissolving 0.49 g of CTAB into 45 mL of water with constant stirring. Subsequently, 2 mL of 10^{-2} M $AgNO_3$ was added dropwise while stirring the CTAB solution. Then, 300 μL of 10^{-2} M $HAuCl_4$ was added drop by drop. When the solution turned yellow-brown, 320 μL of 10^{-1} M ascorbic acid was added dropwise, turning the yellow-brown hue colourless. Another 300 μL of 10^{-2} M $HAuCl_4$ and 320 μL of 10^{-1} M ascorbic acid were rapidly injected into this colourless solution. Different seed quantities ranging from 250 μL to 4 mL were used to synthesise HGNTs with variable edge lengths and cavity sizes.

Preparation of HGNT-PIC Composite

First, we made a saturated aqueous solution of 1×10^{-4} M of PIC. Next, different volumes of the HGNT solutions were added to the saturated PIC solutions of the dye. This solution was left without interruption in the dark for 48 hours so that the PIC dye adsorbed on the surface of the HGNT. Extra cyanine dye molecules were removed using repeated centrifugation at 6000 rpm for 10 minutes, followed by washing, and the final precipitate was redistributed in water.

Characterisation

Morphology of the as-synthesised nanoparticles was characterised by field emission scanning electron microscope (FESEM) (Model No-Magellan 400) and high-resolution transmission electron microscopy (HRTEM) (Model No – FEI TECNAI G2 TF - 20). A few microliters of the diluted solution were drop-coated onto a pristine Si (100) substrate and a carbon-coated copper grid to gain deeper insights into the nanoparticles' morphology. Absorption Spectra of the as-synthesised hollow Au nano triangles for six different seed concentrations varying from 250 μL to 4 mL were recorded using a dual-channel UV–vis–NIR spectrophotometer (Jasco V670).

Results

Figure 1a shows an SEM image of the as-synthesised hollow gold nanotriangular (HGNT) platelets. The image clearly indicates the formation of almost uniform clusters of nanotriangular-shaped particles with a cavity at their centre of mass position.

Although the SEM image only indicates the central black spot, the cavity is not observable from the SEM image. In order to confirm the hollow structure, we have also performed transmission electron microscopy (TEM) on an isolated single nanostructure, as shown in Figure 1b. TEM image (Figure 1b) indicates the formation of a hole of 2-3 nm diameter at the centre of mass position of the equilateral gold nanotriangular plate. In Figure 2, we have depicted HGNT absorption spectra for six distinct seed concentrations ranging from 250 μL to 4 mL. When adding 4 ml of Ag seed, we found two peaks: around 600 nm and 640 nm. Now, as we gradually decrease the seed concentration, the 600 nm peaks shift a little bit. However, the 640 nm peak gradually red-shifted with decreasing seed concentration and ultimately appeared around 850 nm for the 250 μL concentration. It is well known that the LSPR peak position of HGNT can be altered by varying the cavity dimension and side length of the triangle. Increasing the aspect ratio results in a more red-shifted (Das et al., 2017) LSPR maximum, as indicated in UV- Vis absorption spectra (Figure 2). So, by altering the seed quantity, we tuned the LSPR peak position across an extensive spectral range, indicating the creation of various HGNTs with different side edge lengths and cavity diameters. (Das et al., 2017).

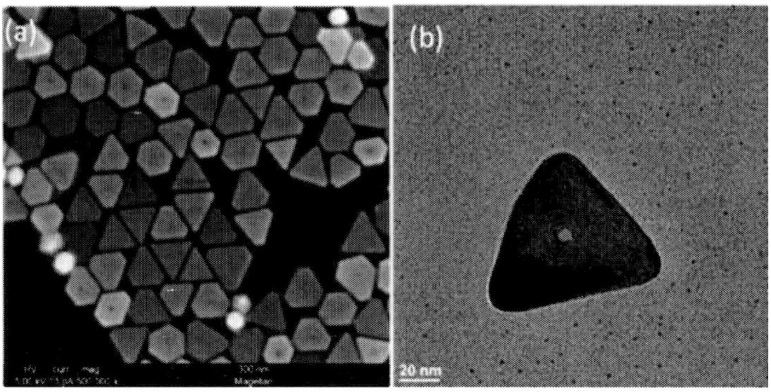

Figure 1. (a) SEM image of as-synthesised hollow Au nano triangles. (b) TEM image of an isolated HGNT plate indicating a hollow interior at the centre of the triangle.

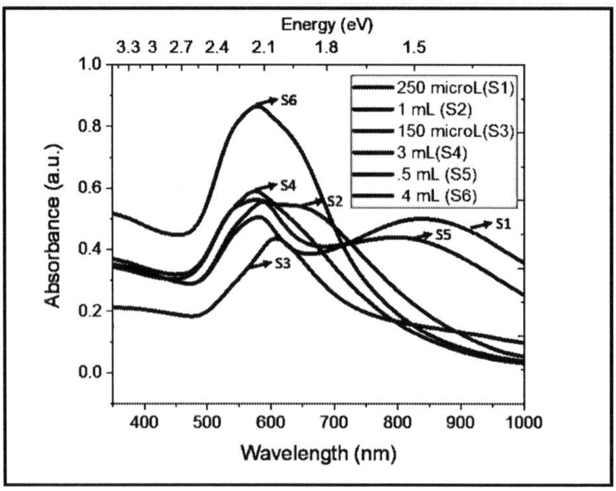

Figure 2. Absorption Spectra of the as-synthesised hollow Au nano triangles for six different seed concentrations varying from 250 μL to 4 mL.

Detection of Coupling

We have studied plasmon-exciton coupling in the hybrid nanostructure of HGNT and cyanide dye J-aggregate using UV-Vis absorption (Shown in Figure 3) spectroscopy in samples S1 and S2. The absorption spectrum of PIC dye in water (Figure 3) shows a peak around ~525 nm, which is a well-known peak position of the PIC monomer due to the formation of electronic π-π* transition in the cyanine dye's conjugated system (Das et al. 2017; Hazra et al., 2016; Li et al.,2018). When PIC monomer aggregates (e.g., forming dimers or higher-order aggregates), i.e., J-aggregates form, then a new red-shifted peak typically around ~570–600 nm (Das et al., 2017; Hazra et al., 2016).

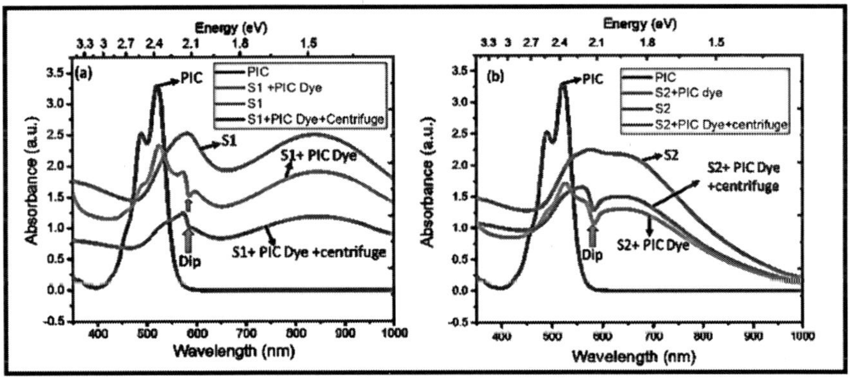

Figure 3. Absorption spectra of an aqueous solution of ~10^{-4} M PIC (black), as synthesised HGNT (S1, blue), S1 + PIC dye composite (red), S1 + PIC dye composite after centrifugation (green) for (a) Sample S1 and (b) Sample S2.

As in our experiment, we have used the concentration of PIC ~ 10^{-4} M, which is why no J-aggregate was formed. Now, the absorption spectra of the as-synthesised HGNT (S1, blue) sample show two peaks, one around 590 nm and another at 860 nm. Now, if we look at the absorption spectra (Fire 3a, red curve) of HGNT (S1) and PIC dye composite, then it primarily resembles the superposition spectra of HGNT (S1) and monomeric PIC dye, indicating minimal interaction between them under standard conditions. However, a slight dip was observed at ~ 590 nm (2.1 eV), suggesting a weak coupling effect or interference phenomenon. The dip could be related to plasmon-exciton coupling, which occurs when the HGNTs' LSPR and the PIC dye's excitonic transition exchange energy or hybridise. Such features often emerge in systems where the plasmon and exciton energies are near resonance but do not strongly overlap, leading to a minor perturbation in the absorption spectrum.

We also depicted the spectrum of the S1 + PIC dye composite monomer after centrifugation and washing in Figure 3a. This clearly shows that the 525 nm PIC monomer peak gradually disappears with washing. However, the transparency dip at ~ 2.1 eV (584 nm) is clearly visible. Now, if we look at the absorption spectra (Figure 3b) of HGNT (S2) and PIC dye composite, then we also observe the same phenomena. The HGNT-PIC-J-aggregate hybrid nanosystem exhibits coherent coupling between HGNT surface plasmons and PIC-J-aggregate excitons, as evidenced by the observed transparency at ~ 2.1 eV (590 nm). A similar type of observation has previously been recorded for various systems. (Das et al. 2017; Hazra et al, 2016; Li et al.,2018; Pelton et al., 2019; sun et al., 2019).

Discussion

The field of nanophotonics, which focuses on manipulating light at the nanoscale, greatly benefits from plasmon-exciton coupling. Hybrid systems like HGNTs coupled with J-aggregates provide controlled electromagnetic field localisation, which enhances the performance of devices such as nanoscale light emitters, lasers, and LEDs. The ability to create "hot spots" with high field intensity enables the precise control of light emission, which is crucial for developing more efficient photonic devices. Additionally, the unique optical properties of plexcitons offer novel approaches to light manipulation, potentially enabling faster and more compact photonic circuits. In sensing applications, the high sensitivity of plasmon-exciton coupled systems allows for enhanced detection of minute environmental changes. Since the optical response of plexcitons is highly sensitive to their surroundings, even minor variations in the local dielectric environment, such as the presence of biomolecules, can cause measurable shifts in the optical spectrum. This sensitivity makes plasmon-exciton systems highly effective for biosensing and chemical detection applications, where detecting low concentrations of target molecules is often essential. Quantum information processing also stands to benefit from plasmon-exciton coupling due to its ability to support coherent energy transfer at quantum scales. The hybrid plexcitonic states produced by strong plasmon-exciton coupling may create quantum materials that enable the manipulation, storage, and transfer of information at the nanoscale. By tuning the properties of these hybrid states, researchers can achieve control over quantum information processes, helping advancements in quantum computing and communication. This is a great launching pad for future research based on specific applications of the hybrid plexitons.

Acknowledgment

We thank Professor Dr. Jon Camden of the University of Notre Dame, USA, for many helpful discussions.

References

Das, K., Hazra, B., & Chandra, M. Exploring the coherent interaction in a hybrid system of hollow gold nanoprisms and cyanine dye J-aggregates: role of plasmon-hybridisation mediated local electric-field enhancement. *Physical Chemistry Chemical Physics*, 2017, *19*(41), 27997-28005.

Hazra, B., Das, K., Das Chakraborty, S., Verma, M. S., Devi, M. M., Katiyar, N. K. & Chandra, M. Hollow gold nanoprism as highly efficient "single" nanotransducer for surface-enhanced Raman scattering applications. *The Journal of Physical Chemistry C*, 2016, *120*(44) 2016, 25548-25556.

Li, X., Zhou, L., Hao, Z. and Wang, Q. Q. Plasmon–exciton coupling in complex systems. *Advanced Optical Materials*, 2018, *6*(18), p. 1800275.

Maiti, A., Maity, A., Satpati, B., Large, N., & Chini, T. K. Efficient excitation of higher order modes in the plasmonic response of individual concave gold nanocubes. *The Journal of Physical Chemistry C*, 2017, *121*(1), 731-740.

Maity, A., Maiti, A., Satpati, B., Patsha, A., Dhara, S., & Chini, T. K. Probing localised surface plasmons of trisoctahedral gold nanocrystals for surface enhanced raman scattering. *The Journal of Physical Chemistry C*, 2016, *120*(47), 27003-27012.

Odom, T. W., & Schatz, G. C. Introduction to plasmonics. *Chemical Reviews*, 2011, *111*(6), 3667-3668.

Pelton, M., Sheldon, M., & Khurgin, J. Plasmon-exciton coupling. *Nanophotonics*, 2019, *8*(4), 513-516.

Roy, A., Maiti, A., Chini, T. K., & Satpati, B. Annealing-induced morphology of silver nanoparticles on pyramidal silicon surface and their application to surface-enhanced Raman scattering. *ACS applied materials & interfaces*, 2017, *9*(39), 34405-34415.

Senapati, D., Senapati, T., Wate, P. S., Kanchanapally, R., Fan, Z., Singh, A. K., Ray, P. C. Length-dependent NLO Properties of 2D Hollow Gold Nanoprisms Formed by Guided Assembly. *Chemical Communications* 2012, 48, 6034–6036.

Sun, L., Li, Z., He, J., & Wang, P. Strong coupling with directional absorption features of Ag@ Au hollow nanoshell/J-aggregate heterostructures. *Nanophotonics*, 2019, *8*(10), 1835-1845.

Chapter 4

Lung Pneumonia Detection: A U-Net Mobile Net-Based Sustainable Approach

Arunima Patra*
Krishnangshu Paul
and Prithwineel Paul, PhD
Department of Computer Science and Engineering,
Centre of Excellence for Quantum Computing, Institute of Engineering & Management,
University of Engineering and Management, Kolkata, India

Abstract

This paper presents a novel image segmentation approach utilising a modified version of the U-Net architecture, where the MobileNet architecture is employed as the encoder to reduce computational complexity. This paper involves transfer learning by including user fine-tuning and pre-trained ImageNet weights. This decoder block utilises precise spatial information in conjunction with more abstract semantic information through skip connections, thereby improving segmentation results. The training process of the model is highly flexible, as it utilises multiple callbacks, resulting in improved accuracy. We have also used generators for data augmentation and customised loss functions and evaluation metrics. This paper has also demonstrated the trade-off between accuracy and computational sustainability. Using an encoder, the model significantly reduces computational overhead, making it highly applicable for deployment on energy-conserving edge devices and mobile platforms. It reduces resource consumption and is beneficial in the development of sustainable AI, as it lowers energy consumption

* Corresponding Author's Email: patraarunima07@gmail.com

In: Breakthrough Research and Technologies in Physics …
Editors: Koyel Ganguly, Soumyadipta Pal and Kamakhya Prasad Ghatak
ISBN: 979-8-90134-111-7

without compromising accuracy. Moreover, the study also examines the impact of hyperparameter tuning on the model's performance.

Keywords: image segmentation, U-Net, MobileNet, transfer learning, ImageNet, skip connections, callbacks, data augmentation, loss function, evaluation metric

Introduction

Lung pneumonia identification through the use of U-Net and MobileNet combines the strong segmentation functions of U-Net with the efficient function extraction capabilities of MobileNet. U-Net was developed as an architecture for image segmentation; its encoder-decoder structure is used to identify specific areas in chest X-ray images that have pneumonia. In this setup, MobileNet serves as the encoder, which utilises depth-wise separable convolutions to enable lightweight and practical feature extraction. During the training process, the combined model learns to segment pneumonia-affected areas from labelled chest X-ray images, facilitating automated analysis for enhanced diagnostic accuracy.

This approach combines medical accuracy with computational sustainability. MobileNet offers a lightweight architecture that significantly reduces processing power requirements, making the model well-suited for deployment on energy-efficient devices—such as mobile health platforms and edge computing systems. This approach reduces the environmental impact of AI-powered medical imaging while making it more accessible to underprivileged regions by supercharging deep learning for resource-constrained environments. The practical application of responsible AI solutions must therefore balance the trade-offs in accuracy, sustainability, and environmental footprint, ultimately leading to a greener AI future.

Literature Review

Lung Pneumonia is a dangerous lung infection that has caused a considerable number of deaths in the past, which marks the need for a less time-consuming diagnosis for better treatment of the patients. Deep learning has contributed to the factor by introducing several detection mechanisms using chest X-ray [18]. Several segmentation techniques, including U-Net and MobileNet, have demonstrated excellent results in the early detection of pneumonia. The hybrid model, comprising these two structures, has been shown remarkable improvements in accuracy and efficiency [1, 2]. The U-Net is used to segment the image, and MobileNet performs its part in lightweight feature extraction. The U-Net is a convolutional neural network that comprises an encoder block and a decoder block. The encoder captures the contextual information from the data, and the decoder block is responsible for the precise localisation of the lung pneumonia. Chest X-rays are used to train the model to segment the areas affected by pneumonia in the X-ray images [3, 4]. MobileNet is a compact convolutional Neural network used in solving visual tasks in embedded and mobile devices. By using depth-wise separable convolutions, MobileNet reduces the complexity and parameter count, making it well-suited for environments with resource constraints, thus resulting in a sustainable solution [5, 6].

In the hybrid model mentioned above, Mobilenet replaces the encoder in the U-Net, which is conventionally used to extract high-level features from the input of chest X-rays, before passing it to the decoder block, where it undergoes segmentation [7, 8]. The model has been trained on annotated chest X-ray datasets. They learn to generate masks to point out the area affected by pneumonia. The masks help radiologists make diagnoses much faster than usual, thereby speeding up the overall process [9, 10]. The use of MobileNet in U-Net ensures proper feature extraction without compromising the accuracy of the segmentation performed. Thus, making it particularly more advantageous for deployment in devices with limited resources [11, 12]. This approach not only improves the accuracy but also gives a sustainable solution in the field of healthcare using AI. The hybrid model has shown considerably less energy consumption, making this sustainable solution for pneumonia detection much more accessible in areas where resources are limited. The traditional deep learning models that have been used until recently have shown to consume much more computational power, thus showing an increased amount of carbon footprint, but the architecture of this model, which is lightweight, aims to solve the challenge by providing high performance in devices where there is a constraint on the resources.

There have been multiple studies that have shown the application of MobileNet U-Net in detecting lung Pneumonia. For example, Rajpurkar et al. [13] described a modified U-Net architecture with a pre-trained MobileNet encoder for pneumonia detection. Meanwhile, Yadav et al. [14] introduced a multi-task U-Net-MobileNet model that simultaneously detects pneumonia and localises it. There have been other studies as well that have explored different loss functions to highlight the improvement in performance [15], the use of techniques such as data augmentation [16], and methods like transfer learning strategies [17], which are used to refine the model's performance further. This model not only enhances the complexity but also improves the performance of the U-Net-MobileNet model in medical diagnosis, providing sustainable AI applications in healthcare. This aligns with the vision of discovering breakthrough technologies in physics and computing, which will enable the creation of a more sustainable future in healthcare.

A detailed study evaluating a few pre-trained models, including ResNet50, VGG16, Xception, ResNet152V2, DenseNet121, DenseNet201, EfficientNet, and MobileNet, achieved the highest accuracy of 94.23% and 93.75% on two different datasets using the MobileNet model [18]. This research evaluated essential hyperparameters, including various optimisers, batch sizes, and the number of epochs, to identify the most appropriate model through comparative analysis. A paper in which MobileNet-UNet for lesion segmentation has designed a system that can detect both COVID-19 and pneumonia from Computed Tomography scans. It has shown excellent results [21]. This method has shown an accuracy of 98.05%, an F1-score of 98.70%, a precision of 98.70%, a recall of 98.70%, and a specificity of 96.05% in lung and lesion segmentation.

Advanced pneumonia detection techniques employed in MobileNet-UNet involve evaluating segmentation performance across various transfer learning models. A study comparing U-Net, SegNet, and DeepLabV3 revealed that DeepLabV3 outperforms the others [19]. It achieved an accuracy of 84.4% and an IoU of 81%. The advantage of Transfer Learning Models in automating the pneumonia-affected region segmentation process with less human intervention is also confirmed through this research.

Several other architectures have been introduced over time to detect affected areas of the lungs. PneumNet, a dual-version deep learning model, has shown exceptional performance with 99.84% accuracy, 99.87% F1-score, 99.74% sensitivity, 100% specificity, 100% positive predictive value (PPV), and 99.58% negative predictive value (NPV) [20]. These results highlight the transformative role of deep convolutional neural networks in advancing the diagnosis of pneumonia.

Ensemble approaches are typically employed to enhance accuracy. Using UNET with MobileNet as a model to enhance pneumonia detection precision [22]. This approach highlights the benefits of combining MobileNet-UNet with other architectures to achieve improved performance. Additionally, MobileNet can be used effectively and utilised as a method for feature extraction in pneumonia classification, achieving an accuracy rate of 95.67% in the binary classification of normal and pneumonia lungs [23].

Multi-scale approaches offer insights into probable improvements for U-Net-based architectures. The (DEMSU-NET) Differential Evaluation Multi-Scale U-NET Architecture includes multi-scale feature maps, which are extracted from every convolutional layer of the U-NET encoder [24]. This approach handles issues related to noisy annotations, thus improving accuracy.

The development of lightweight models, employing a similar approach to MobileNet-UNet, emphasises the need for efficiency in resource-constrained environments. The LWSNet (Lightweight Stacking Network), a novel deep learning architecture, has demonstrated high performance in distinguishing between Covid-19, pneumonia, and normal chest x-ray images [25, 26]. This study attained classification accuracies of up to 98.54% using a quadruple stack mechanism.

This review highlights the results of MobileNet-UNet and similar lightweight architectures in pneumonia detection from chest X-ray images. Future research could focus on further optimising these models for environments where resources are limited, exploring multi-scale approaches, and exploring ensemble methods to enhance the accuracy and robustness in pneumonia detection and classification.

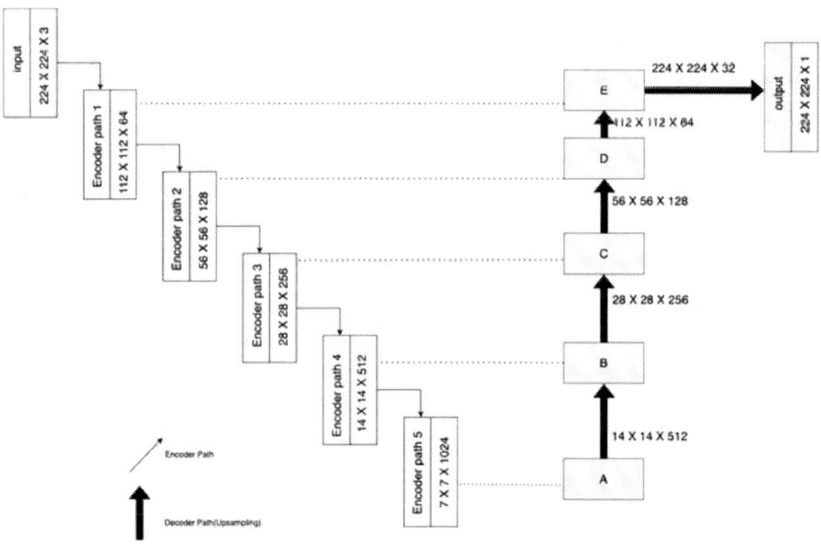

Figure 1. U-Net MobileNet Architecture.

Methodology

Functions and Class Definitions

Define IoU and Jaccard Loss Function

The previous code snippet describes a loss function based on IoU. It takes two input tensors, the actual mask and the predicted mask. It reshapes tensors into 1D vectors. Next, it calculates the intersection of the two vectors, which is the sum of the product of the two vectors element-wise. The union is calculated by summing two vectors, with the intersection subtracted from the result.

The final loss value is calculated as the complement of the ratio between the intersection and union, with a small constant added to both the numerator and denominator to prevent division by zero.

The second code is a function to calculate the mean IoU across multiple classes. These are two input tensors: the first is the actual mask, and the second is the predicted mask. First, the mask tensor predictions are rounded to 0s and 1s. Next, this function calculates the intersection and union for each class along the specified axis. Intersection: This is obtained by multiplying the two input tensors element-wise, followed by a sum of the resulting values over the axis. To get the union, we add the two input tensors along the axis and subtract the intersection from the result. We introduce a small constant to both the numerator and denominator, which prevents the same problem of division by zero that was encountered above.

Finally, the function computes the mean of Intersection over Union values for all the classes and returns it. These are commonly used to measure the performance of computer vision tasks, such as pixel segmentation and object detection, where the goal is to predict pixel-level labels or accurately localise objects with bounding boxes. In medical imaging, IoU is frequently used as a metric for computing similarity between a predicted mask and a ground truth mask, making it a realistic loss function for refining the object mask to represent the actual value.

Two Input Arguments: Labeled Bounding Box, Predicted Bounding Box A bounding box is represented as either a tuple or a list of four values:(x min, y min, x max, y max), where (x min, y min) denotes the top-left corner coordinates, and (x max, y max) denotes the bottom-right corner coordinates.

By looking at the largest, the function gets coordinates; thus, one can get the dimensions of the intersection rectangle from the top-left coordinates and the smallest from the bottom-right coordinates.

Then, it computes the area of the intersection rectangle by multiplying the dimensions of that rectangle. It returns 0 for a high IoU value if the intersection area is zero.

After that, the function calculates the areas of the labelled bounding box and the estimated bounding box separately by multiplying their respective dimensions.

Finally, the IoU value is the ratio of the intersection area to the total area covered by both the labelled and predicted bounding boxes, which is the sum of their areas minus the intersection area. This ratio also represents the degree of overlap between the two bounding boxes; a value of 1 shows complete overlap, and a value of 0 means no overlap. The function returns the computed IoU value as the output.

This function is used in object detection tasks, aiming to predict the bounding boxes in an image. The IoU metric is used here to enhance the overlap between bounding boxes, and it can be utilised for calculating the effectiveness of object detection models or as part of the loss functions used in model training.

Define Masked Image, Unet-Train, Unet-Test

The first code snippet defines a function that takes three inputs: a set of images, a set of masks, and an index. You are taking the mask at that index from the set and laying it on top of the image from the set at the same index. The result of this product is a new image where the original pixel values are multiplied by the mask. And the image object is plotted using a plotting library.

The other code snippet declares a class that inherits from another class. This class allows us to create batches of data which we can feed to a deep learning model for training. It initialises with two data structures: the first contains identifiers and paths to images, and the second contains identifiers, labels, target vectors, and class values.

The class provides methods to fetch batches of data. One of the methods loads images from the specified paths, resizes them to a fixed size, and preprocesses them for input to the model. It also generates ground-truth masks based on the bounding box coordinates provided in the input data structure. The method returns the batch of preprocessed images and corresponding ground truth masks as arrays.

The third code snippet defines another class similar to the second one, but it is designed for generating batches of data for testing or making predictions with a trained model. It takes the same inputs during initialisation as the previous class.

This class also provides a method to fetch batches of data. The method loads images, preprocesses them, and generates ground truth masks based on bounding box coordinates. Additionally, it fetches patient identifiers, resized coordinates, class targets, and ground truth coordinates/labels from the input data structure. The method returns the batch of preprocessed images, ground truth masks, patient identifiers, resized coordinates, class targets, and ground truth coordinates/labels as arrays.

Create a U-Net Model with MobileNet

The function begins by obtaining the MobileNet model, which includes its weights pre-trained on the ImageNet dataset. The model is loaded without the top classification layer, and the input shape is set to the desired image size (e.g., 224 x 224 x 3 for RGB images). The alpha parameter of the model is fixed to a specific value, which controls the width multiplier of the MobileNet Architecture. Then, it loops through all the layers of the Mobilenet model and sets their trainability depending on the argument 'trainable.'

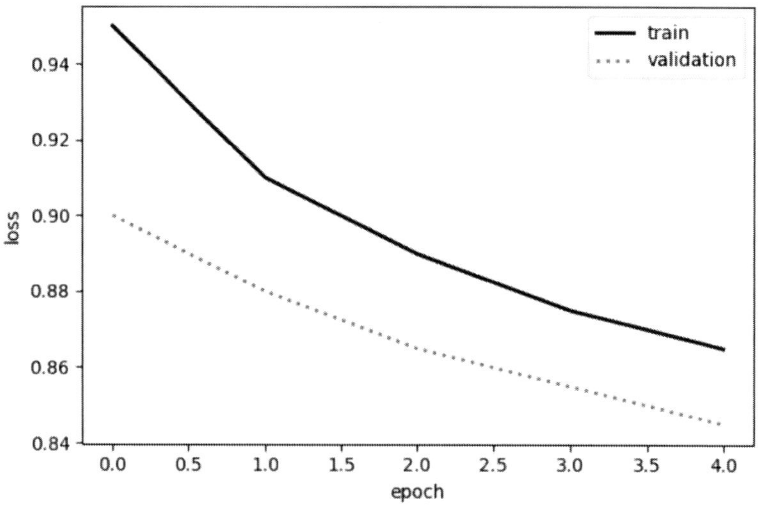

Figure 2. Model IOU and Learning Rate.

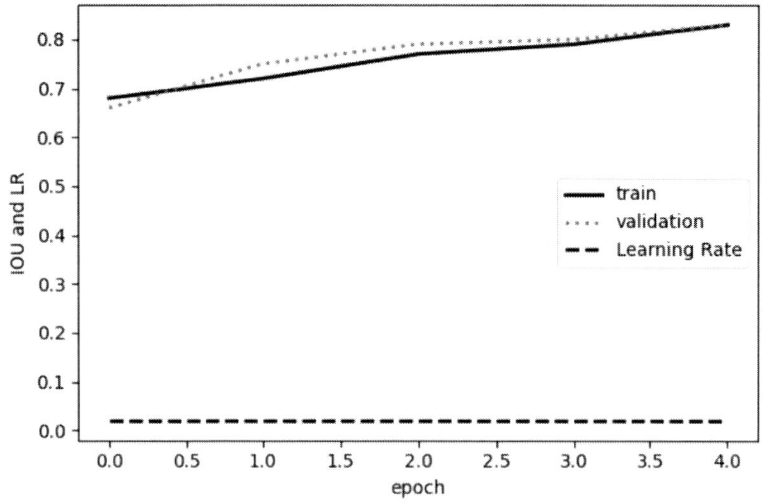

Figure 3. Model Accuracy Graph.

The previously trained weights can, therefore, be adjusted or left frozen. During training, the model must either incorporate these changes or find a model that already does, depending on the desired behaviour. The following line of the function retrieves the output tensors for specific layers from MobileNet. The new model architecture enables us to obtain five levels of feature maps for later use in the U-Net decoder part, corresponding to spatial resolutions of (112 x 112, 56 x 56, 28 x 28, 14 x 14, and 7 x 7).

The decoder takes the output tensor from the lowest resolution (7 x 7) and up-samples it while wrapping it in branches of the exact resolution (14 x 14) from the MobileNet model. This is done iteratively, upsampling the combined tensor and appending it to the following resolution's higher-resolution feature map from the MobileNet model. Finally, the last upsampling step reshapes the tensor into the desired output resolution (i.e., 224 × 224). The heatmap is generated by a 1x1 conv layer, followed by a sigmoid activation function and is used to produce the output mask. Finally, the output tensor is reshaped to the specified dimensions (e.g., 224 × 224). Finally, the function returns a Keras Model object that accepts an input image tensor and outputs the predicted mask tensor, which is used to generate the final result.

Model Training

In this code, a boolean flag is set to determine whether a particular process should be executed or not. It then specifies the number of iterations for a specific operation. Next, it defines a file path location and creates an object with particular hyperparameters. Several callback objects are designed to handle different scenarios during the training process. One callback is responsible for saving the current state of the model to a file whenever a specific metric improves. Another callback stops if the monitored metric does not show improvement over a defined number of iterations; the training process may be halted. The third callback adjusts a hyperparameter (learning rate) of the optimisation algorithm if the monitored metric does not improve for a certain number of iterations. A function is called to create a specific model architecture. The model is later built using a defined loss function, an optimisation algorithm, and a set of evaluation metrics. The code loads pre-trained values (weights) from a file into the model, A method utilised to convey knowledge from an earlier trained model to the present one. The code checks the value of the boolean flag set earlier. If the flag is set to a particular value, the following steps are executed: All components (layers) of the model are set to be adjustable during the training process. The model is trained using both a training data generator and a validation data generator, with the training process guided by the callbacks confirmed earlier. The information about the model's performance during the training process (e.g., loss, evaluation metrics) is stored in a data structure and saved to a file for future analysis and visualisation.

Results

Model Evaluation

The evaluation metrics are as follows:

Precision: 0.91
Recall: 1.0
Accuracy:0.95
F1-Score: 0.96

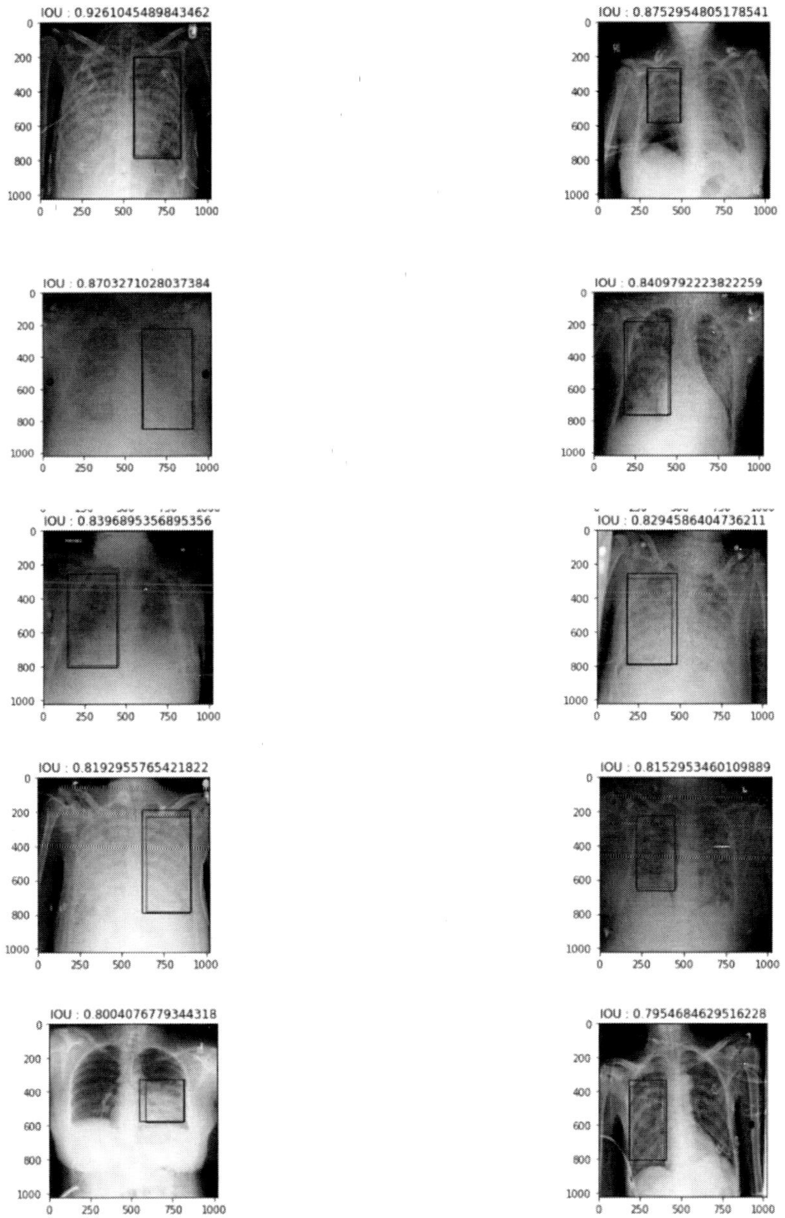

Figure 4. Predicted Result.

Discussion

This study demonstrates the effectiveness of combining MobileNet and U-Net models for image segmentation tasks. The proposed model utilises transfer learning from pre-trained ImageNet weights, enabling efficient knowledge transfer and adaptability to various segmentation challenges. The implementation of skip connections in the U-Net decoder effectively integrates conceptual insights with detailed spatial information, thereby enhancing the accuracy of segmentation. The flexible training options and robust callback system provide a framework for optimising model performance across different scenarios. The use of data augmentation through generators further enhances the model's generalisation capabilities. The customisable loss function and evaluation metric allow for task-specific fine-tuning, making the model versatile for diverse applications.

While the results are promising, future work should focus on extensive comparative studies across different datasets and segmentation tasks. Additionally, exploring the impact of various hyperparameter combinations could further optimise the model's performance and efficiency in real-world applications.

Disclaimer

None.

References

[1] Ronneberger, Olaf, Philipp Fischer, and Thomas Brox. "U-net: Convolutional networks for biomedical image segmentation." Medical image computing and computer-assisted intervention–MICCAI 2015: *18th international conference*, Munich, Germany, October 5-9, 2015, proceedings, part III 18. Springer International Publishing, 2015.

[2] Howard, Andrew G., et al. "Mobilenets: Efficient convolutional neural networks for mobile vision applications." *arXiv preprint arXiv*:1704.04861 (2017).

[3] Civic¸ek, ¨Ozg¨un, et al. "3D U-Net: learning dense volumetric segmentation from sparse annotation." Medical Image Computing and Computer-Assisted Intervention–MICCAI 2016: *19th International Conference,* Athens, Greece, October 17-21, 2016, Proceedings, Part II 19. Springer International Publishing, 2016.

[4] Milletari, Fausto, Nassir Navab, and Seyed-Ahmad Ahmadi. "V-net: Fully convolutional neural networks for volumetric medical image segmentation." 2016 fourth.

[5] Sandler, Mark, et al. "Mobilenetv2: Inverted residuals and linear bottlenecks." *Proceedings of the IEEE conference on computer vision and pattern recognition.* 2018.

[6] Howard, Andrew, et al. "Searching for mobilenetv3." *Proceedings of the IEEE/CVF international conference on computer vision.* 2019.

[7] Li, Xiaomeng, et al. "H-DenseUNet: hybrid densely connected UNet for liver and tumour segmentation from CT volumes." *IEEE Transactions on Medical Imaging*, 37(12), 2663-2674 (2018).

[8] Huo, Fushuo, et al. "Real-time one-stream semantic-guided refinement network for RGB-thermal salient object detection." *IEEE Transactions on Instrumentation and Measurement* 71 (2022): 1-12.

[9] Rajpurkar, Pranav, et al. "Chexnet: Radiologist-level pneumonia detection on chest x-rays with deep learning." *arXiv preprint arXiv*:1711.05225 (2017).

[10] Deng, Li. "The MNIST database of handwritten digit images for machine learning research [best of the web]." *IEEE signal processing magazine* 29.6 (2012): 141-142.

[11] Krizhevsky, Alex, Ilya Sutskever, and Geoffrey E. Hinton. "ImageNet classification with deep convolutional neural networks." *Communications of the ACM* 60.6 (2017): 84-90.

[12] Huang, Gao, et al. "Densely connected convolutional networks." Proceedings of the *IEEE conference on computer vision and pattern recognition.* 2017.

[13] Rajpurkar, Pranav, et al. "Mura: Large dataset for abnormality detection in musculoskeletal radiographs." *arXiv preprint arXiv:*1712.06957 (2017).

[14] Yadav, Samir S., and Shivajirao M. Jadhav. "Deep convolutional neural network-based medical image classification for disease diagnosis." *Journal of Big Data 6.1* (2019): 1-18.

[15] Sudre, Carole H., et al. "Generalised dice overlap as a deep learning loss function for highly unbalanced segmentations." Deep Learning in Medical Image Analysis and Multimodal Learning for Clinical Decision Support: Third International Workshop, DLMIA 2017, and 7th International Workshop, ML-CDS 2017, Held in Conjunction with MICCAI 2017, Qu'ebec City, QC, Canada, September 14, Proceedings 3. Springer International Publishing, 2017.

[16] Castro, Daniel C., Ian Walker, and Ben Glocker. "Causality matters in medical imaging." *Nature Communications* 11.1 (2020): 3673.

[17] Shin, H. C., Roth, H. R., Gao, M., Le, L., Eno, J., Woodbridge, G., ... & Xu, D. (2016). Deep convolutional neural networks for computer-aided detection: CNN architectures, dataset characteristics and transfer learning. *IEEE transactions on medical Imaging*, 35(5), 1285–1298.

[18] Reshan, M. S. A., Gill, K. S., Anand, V., Gupta, S., Alshahrani, H., Sulaiman, A., & Shaikh, A. (2023, May). Detection of pneumonia from chest X-ray images utilising the Mobilenet model. In *Healthcare* (Vol. 11, No. 11, p. 1561). MDPI.

[19] Pyar, K. (2024). Segmentation Performance Analysis of Transfer Learning Models on X-Ray Pneumonia Images. *Journal of Future Artificial Intelligence and Technologies,* 1(1).

[20] Aurangzeb, K., Jamil, S., & Alhussein, M. (2024). Introducing PneumNet—A Groundbreaking Dual-Version Deep Learning Model for Pneumonia Disease Detection. *International Journal of Imaging Systems and Technology*, 34(4), e23116.

[21] Motta, P. C., Cortez, P. C., Silva, B. R., Yang, G., & Albuquerque, V. H. C. D. (2023). Automatic COVID-19 and common-acquired pneumonia diagnosis using chest CT scans. *Bioengineering,* 10(5), 529.

[22] Bakır, H., Oktay, S., & Tabaru, E. (2023). Detection Of Pneumonia From X-Ray Images Using Deep Learning Techniques. *Journal of Scientific Reports-A,* (052), 419-440.

[23] Goyal, S., & Singh, R. (2023). Detection and classification of lung diseases for pneumonia and COVID-19 using machine learning and deep learning techniques. *Journal of Ambient Intelligence and Humanised Computing*, 14(4), 3239-3259.

[24] Balaji, A., & Rajakumari, S. B. (2023, December). Differential Evaluation Multi-scale U-NET (DEMSU-NET) Architecture for Classification of Lung Diseases from X-Ray Images. In *International Conference on Advancements in Smart Computing and Information Security* (pp. 340-355). Cham: Springer Nature Switzerland.

[25] LWSNet - a novel deep-learning architecture to segregate COVID-19 and pneumonia from X-ray imagery

[26] Lasker, A., Ghosh, M., Obaidullah, S. M., Chakraborty, C., & Roy, K. (2023). LWSNet: A novel deep-learning architecture to segregate COVID-19 and pneumonia from X-ray imagery. *Multimedia Tools and Applications*, 82(14), 21801-21823.

Chapter 5

Solar Auto Drive: A Renewable Energy-Powered Electric Vehicle with Automated Mode Switching

Asmita Dey
Anushka Pandit
Antara Mondal
Swapnonil Samui
Sulagna Jana
and Arunava Mukhopadhyay*
Department of Electronics and Communication Engineering, Institute of Engineering & Management, University of Engineering and Management, Kolkata, West Bengal, India

Abstract

Social responsibility is one of the integral tasks of any citizen. One of the most important of them is environmental awareness. Environmental awareness calls for eco-friendly ideas, innovation and products that utilise renewable energy sources. With sunlight being one of the most vibrant, widely available renewable energy sources, our work highlights the same. This work is a solar-powered electric vehicle that ingeniously employs an Arduino Uno and a Bluetooth module. These technical components allow the vehicle to seamlessly switch between solar power and battery power based on the availability of sunlight, significantly reducing reliance on electricity sources, and thereby reducing electricity

* Corresponding Author's Email: arunava.mukhopadhyay@iem.edu.in.

In: Breakthrough Research and Technologies in Physics …
Editors: Koyel Ganguly, Soumyadipta Pal and Kamakhya Prasad Ghatak
ISBN: 979-8-90134-111-7

consumption and wastage. It underlines the importance of renewable energy and the usage of sustainable resources.

Keywords: solar-powered electric vehicle, renewable energy, sustainable transportation, greenhouse gas reduction, climate change mitigation

Introduction

With the increasing pollution levels, the need of the hour is environmental awareness. Steps that lead to the usage of eco-friendly products and alternatives have increased demand in the market. Manufacturing of eco-friendly products, specifically for vehicles that use renewable energy sources. This work here focuses on the same goal. A solar-powered vehicle, designed with Arduino Uno and Bluetooth, utilises both solar and battery power, making it convenient to use in both sunlight and darkness. The work also reduces the dependence of electric vehicles on electricity and uses renewable energy as an alternative. This change can contribute one step towards a greener world and sustainable methods.

Literature Review

S. Sykiotis, C. M. Aikateriniadis, A. Doulamis, N. Doulamis, and P. S. Georgilakis proposed using Deep Reinforcement Learning (DRL) to optimise solar-powered electric vehicle (EV) charging and investigated real-time tariffs for improved efficiency in 2022. Previously, in the year 2020, P. Arras, G. Tabunshchyk, S. Korotunov, and V. Okhmak focused on a cost optimisation simulation for EV charging infrastructure to minimise delivery costs. Later, in 2022, Mats Gezelius and Reza Mortazavi examined the correlation between owning solar panels and Battery Electric Vehicles (BEVs) using a recursive bivariate probit model. Simultaneously, M. Yamaguchi, T. Masuda, K. Araki, Y. Ota, and K. Nishioka evaluated high-efficiency solar cell modules for photovoltaic (PV)-powered vehicles in 2022. Lately, N. M. Manousakis, P. S. Karagiannopoulos, G. J. Tsekouras, and Fotios D. Kanellos reviewed the integration of Renewable Energy Sources (RES) and EVs into power systems, summarising key research contributions in 2023. In the same year, M. A. Bandora, Visarion Catalin Ifrim, and M. Artiom proposed a solar power-assisted battery balancing system for EVs, simulating a pack of four 13.8 V batteries. S. Khan, M. S. Alam, MS Jamil Asghar, Mohd Aiman Khan, and Ahmad Abbas reviewed level 2 EV charging systems, discussing standards, characteristics, and safety measures in 2018. Later, M. H. Mobarak, Rafael N. Kleiman, and Jennifer Bauman studied how panel tilt and shading affect solar energy capture for 150 drivers in 2020. A. N. Oishi et al. conducted an exegesis on the development of SPEC, stating an account of its prototype design and evaluated the performance in 2022. M. U. Nawaz, Shayan Umar, and Muhammad Salik Qureshi conducted an elaborate study on the life cycle analysis (LCA) to understand the effect on the environment and the economy by solar-powered electric vehicles (SEVs) as compared to conventional EVs in 2023. Afterwards, G. Alkawsi, Yahia Baashar, Dallatu Abbas U, Ammar Ahmed Alkahtani, and Sieh Kiong Tiong conducted extensive studies in the field of alternative energy charging infrastructure providing detailed research covering essential features including resources, potentiality, planning, control, and pricing. A. K. Singhal et al. researched in 2015 a similar work to ours, in which he used the battery to operate the vehicle after charging it with a solar panel. The multi-charging hybrid vehicle can use both solar and electric power, depending on availability. Z. Taha et al. in 2009, built a solar car that utilises solar energy to charge its battery in the Centre for Product Design and Manufacture (CPDM), the University of Malaya, which had a combination of mechanical and electrical components at a sustainable cost margin. A. Sinuraya, Denny Haryanto Sinaga, Yoakim Simamora, and Ridho Wahyudi used photovoltaic systems in 2022, as a strategic approach for supplying energy to electric vehicles, as an alternative for BLDC electric motors for wheel propulsion. It aims to sketch out a battery charging system that uses photovoltaic technology to drive BLDC electric motors. Y. Ota, Kenji Araki, Akira Nagaoka, and Kensuke Nishioka in 2022 researched the coverage ratio, roof shapes, and sizes needed to fix a solar panel on it. it was seen that the potential coverage ratio was 96% for a hemispherical roof with ROC 1m. F. Mohammadi, in 2018, electrified an electric vehicle that was used in traditional plug-in technology. After the operation performed the vehicle charged with both solar power and battery power. The motor can run on both AC and DC power, as well as be powered by stored batteries and solar energy. F. A. In 2020, Tiano, Gianfranco Rizzo, Matteo Marino, and Antonio Monetti developed a model to predict the temperature effects on photovoltaic panels on a car under real weather conditions. The results were that parking phases are the most critical, where an increase in photovoltaic temperatures leads to a decrease in efficiency. M. F. Ezzat and I. Dincer created a photovoltaic fuel cell Li-ion battery integrated system in 2016. He compared it with a base system consisting of a PEM fuel cell and a Li-ion battery, investigating the effects of photovoltaic arrays, overall energy, energy efficiencies, and hydrogen consumption.

Methodology, Experimental Setup and Implementation

The following components are used to implement the hardware version of the work. Inclusion of proper methodology along with it results in actual, real-time implementation.

A. ***Solar Panel -*** A solar panel contains silicon-based photovoltaic cells that absorb sunlight to generate electricity. Encased in tempered glass and a durable backing, these cells energise electrons in the semiconductor. They can directly power devices in sunlight and charge batteries for low-light use. Figure 1 is a picture of a solar cell that was used in the process.

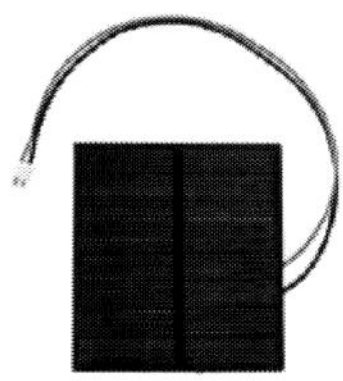

Figure 1. Solar Panel.

B. ***Arduino Uno-*** This acts as the central controller, managing voltage from the solar panel and facilitating smooth transitions between solar and battery power based on set thresholds. This ensures efficient operation in varying sunlight conditions. The picture of an Arduino is attached in Figure 2 below.

Figure 2. Arduino Uno.

C. ***Relay Switch-*** The relay switch is crucial for switching between solar and battery power based on sunlight availability. Controlled by an Arduino Uno, it optimises energy use and enhances the system's efficiency, safety, and adaptability. An image of the relay switch is provided below in Fig. 3 to showcase its versatility.

Figure 3. Relay Switch.

D. ***Bluetooth Module-*** The Bluetooth module is essential for wireless control and monitoring. It enables remote access, eliminating the need for manual switches, and allows users to adjust and monitor the system in real time via a mobile device or computer. This promotes efficient management of the Solar EV-based research. The respective component is shown in Figure 4 below.

Figure 4. Bluetooth Module.

E. ***DC Motor***- The DC motor converts electrical energy into motion, enabling tasks with flexibility in power sources. It rotates using direct solar energy or batteries charged by the panel, all managed through Arduino and relay switches for efficient automation that adapts to environmental changes. Figure 5 is a picture of a DC motor that was used at work.

Figure 5. DC Motor.

F. ***Rechargeable Battery-*** The rechargeable battery stores excess vitality from the solar panel, ensuring the motor operates even without sunlight. This improves efficiency, promotes energy independence, and supports the Solar Electric Vehicle (EV) research during low or no sunlight. Figure 6 is a picture of a rechargeable battery taken for the work.

Figure 6. Rechargeable Battery.

G. ***LED-*** LEDs indicate the system's operating status, showing whether it is solar- or battery-powered. This feedback allows for easy monitoring and troubleshooting, enhancing user understanding, efficiency, and the overall reliability of the Solar EV-based work. Figure 7 is the respective component diagram.

Figure 7. LED.

H. ***PN Junction Diode***- The PN junction diode is crucial for blocking reverse current flow and protecting the involved solar panel and batteries. It allows electricity to flow correctly—from the solar panel to the motor and batteries—preventing damage from backflow. This improves the system's efficiency and reliability, ensuring optimal operation of the Solar EV work. Figure 8 here is an image of a PN junction diode highlighting its versatility and innovative design.

Figure 8. PN Junction Diode.

I. ***Resistor-*** Resistors are crucial for controlling current and protecting sensitive components. They ensure that LEDs and motors receive the correct voltage and current, preventing damage. By stabilising electrical flow, resistors improve the reliability and efficiency of the Solar EV's work. Fig. 9 represents the pictorial representation of a resistor.

Figure 9. Resistor.

The Flow Chart in Figure 10 depicts a workflow for a system that can be powered by either a battery or solar panels. The working mechanism of the model is summarised through the flow chart here.

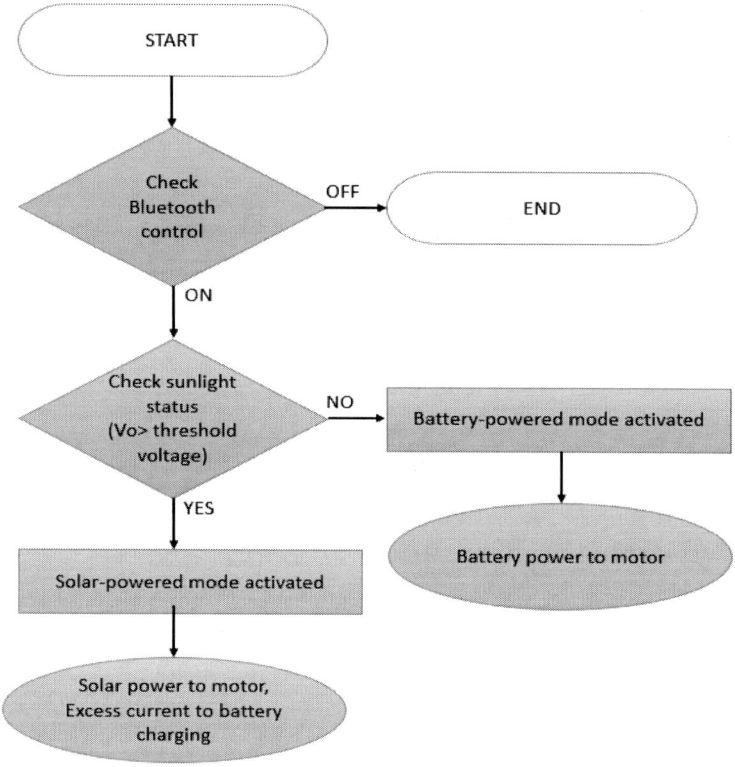

Figure 10. Flow chart depicting the working principle of the model.

The system first checks the status of the classic Bluetooth control. If the classic Bluetooth control is off, the system will not proceed further. When the classic Bluetooth control is activated, the system checks the sunlight conditions. If the sunlight voltage is above a certain threshold, the solar-powered mode activates, using solar energy to run the motor and charging the battery with any excess current.

If there is insufficient sunlight and the classic Bluetooth control is activated, the battery-powered mode will be engaged, and the system will utilise battery power to operate the motor. This particular model is,

$$I \ \eta \times A \times I_{Illuminance} \quad (1)$$

where,

- I is known to be the output current generated by the respective solar cell in Amperes (A).
- η is taken to be the conversion efficiency of the solar cell, representing the percentage of incident light energy which is converted into corresponding electrical energy.
- A is the effective area of the solar cell exposed to sunlight, typically measured in square meters (m^2).
- $_{Illuminance}$ is the intensity of light which is coming and falling on the solar cell, usually measured in Lux (lx) or Watts per square meter (W/m^2).

This mathematical equation that describes the conversion of solar photons to electricity in a solar panel is typically represented using the photovoltaic (PV) equation, which relates the current (I) produced by the solar cell to the incident light intensity (Illuminance), the conversion efficiency (η), and the cell area (A).

Now,

$$\eta = \frac{P_{out}}{P_{in}} \tag{2}$$

$$L = A \times W \tag{3}$$

$$I_{lluminance} \frac{\phi \times K}{A_{flux}} \tag{4}$$

Now, the generated current,

$$I = \frac{P_{out}}{P_{in}} \times L \times W \times \frac{\phi \times K}{A_{flux}} \tag{5}$$

In this equation,

- Pout is the output power of the solar panel in Watts (W). This is the electrical power produced by the solar panel.
- *P*in is the input power from sunlight in Watts (W). This is the solar irradiance or sunlight intensity falling on the solar panel.
- L is the measured length of the solar panel taken in meters (m).
- W is the measured width of the solar panel taken in meters (m).
- Φ is the radiant flux in watts (W), the total power emitted by the light source across all wavelengths.

- K is the luminous efficacy, measured in lumens per watt (lm/W), which represents the efficiency of the light source in producing visible light relative to the total power consumed. It is the surface area in square meters (m^2) over which the luminous flux is distributed.

Result and Discussion

The model is tested under different physical environmental conditions, and the results are discussed below.

A. Sunny Conditions: During sunlight, the solar panel generates electricity. Current flows from the solar panel to the motor, causing it to rotate. The relay connects to the battery to light up, indicating the mode of power source as solar power. In this case, sunny condition is considered and the corresponding model under experimentation is shown in Figure 11.

Figure 11. Model under sunny conditions.

B. No Sunlight: When there's no sunlight, the batteries power the motor. The motor's RPM will depend on the battery voltage and the factors mentioned above for sunny conditions. As batteries discharge, their voltage drops, potentially leading to a decrease in RPM. In this case, the overcast condition is chosen and the respective model under consideration is depicted in Figure 12.

Figure 12. Model under overcast conditions.

The motor's rotation speed (RPM) depends on several factors:

A. *Solar Power Intensity:* Higher sunlight intensity translates to higher voltage output from the solar panel. This, in turn, supplies more current to the motor, resulting in a faster RPM.
B. *Motor Selection:* The chosen motor's specifications, particularly its torque and speed-current curve, determine its RPM under varying current conditions.
C. *Load on the Wheel*: The weight and friction of the wheel attached to the motor affect the required torque to keep it moving.

In the graph depicted in Figure 13, the x-axis represents some readings taken at increasing points in time, and the y-axis represents the voltage with which the motor operates. Here, it is seen that after a specific period when the threshold voltage is crossed, the voltage does not drop any further and remains approximately 350 mV as the record of readings goes on. The initial value indicates the motor is moving through the battery as the voltage generated is less than the threshold voltage. The initial rise in voltage (moving from left to right) indicates increasing sunlight intensity in the morning. This would cause the solar panel to generate more voltage. The highest voltage reading might represent the point of peak sunlight intensity. A subsequent drop in voltage reading could indicate a switch to battery power due to decreased sunlight intensity, such as in the afternoon.

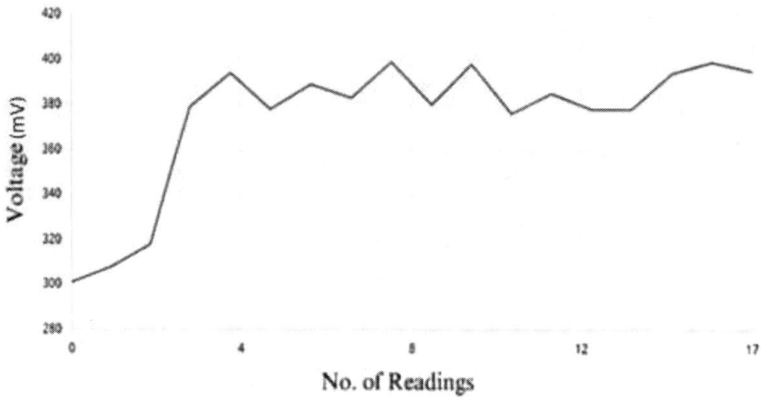

Figure 13. Variation in Voltage Reading.

Conclusion

In conclusion, this study effectively demonstrates the autonomous capabilities of Arduino's solar-powered electric vehicle (EV) circuit. The solar panel harnesses sunlight during the day to operate the motor and charge the batteries for backup power at night. The Arduino manages the relays to ensure smooth transitions between solar and battery power, depending on the available sunlight. This research showcases the potential of solar energy as an environmentally friendly solution. However, there is room for further investigation. Future iterations could incorporate a larger solar panel and a more robust battery pack to extend operational duration.

Acknowledgments

The authors are highly thankful to the University of Engineering and Management and also to the Institute of Engineering and Management for providing support at every stage of the research work.

References

Alkawsi G, Baashar Y, Dallatu Abbas U D, et al. "Review of renewable energy-based charging infrastructure for electric vehicles." *Applied Sciences* 11, no. 9 (2021): 3847.

Arras P, Tabunshchyk G, Korotunov S, et al. "Cost optimisation simulation for electric vehicle charging infrastructure." In *2020 IEEE European Technology and Engineering Management Summit (E-TEMS),* pp. 1-4. IEEE, 2020.

Bandora MA, Ifrim VC, Artiom M. "Battery Balancing System for Electric Vehicles Solar Power Assisted." In 2023, the *10th International Conference on Modern Power Systems (MPS),* pp. 1-6. IEEE, 2023.

Ezzat MF and Dincer I. "Development, analysis and assessment of a fuel cell and solar photovoltaic system powered vehicle." *Energy conversion and management* 129 (2016): 284-292.

Gezelius, M, and Mortazavi, R. *"Effect of having solar panels on the probability of owning a battery electric vehicle." World Electric Vehicle Journal* 13, no. 7 (2022): 125.

Khan S, Alam MS, Asghar MSJ, Khan MA, et al. "Recent development in level 2 charging system for xEV: a review." In *2018 International Conference on Computational and Characterisation Techniques in Engineering & Sciences (CCTES),* pp. 83-88. IEEE, 2018.

Manousakis NM, Karagiannopoulos PS, Tsekouras GJ, et al. "Integration of renewable energy and electric vehicles in power systems: a review." *Processes 11*, no. 5 (2023): 1544.

Mobarak MH, Kleiman RN, Bauman J. "Solar-charged electric vehicles: A comprehensive analysis of grid, driver, and environmental benefits." *IEEE Transactions on Transportation Electrification* 7, no. 2 (2020): 579-603.

Mohammadi, F. "Design, analysis, and electrification of a solar-powered electric vehicle." *Journal of Solar Energy Research* 3, no. 4 (2018): 293-299.

Oishi AN, MSS Tanjim, Ali H, et al "Analysis for Solar-Powered Electric Car Development with Wireless Control Features." In *2021 IEEE International Conference on Robotics, Automation, Artificial-Intelligence and Internet-of-Things (RAAICON)*, pp. 98-101. IEEE, 2021.

Ota Y, Araki K, Nagaoka A, et al. "Facilitating vehicle-integrated photovoltaics by considering the radius of curvature of the roof surface for solar cell coverage." *Cleaner Engineering and Technology* 7 (2022): 00446.

Singhal AK, Shukla L, Gupta A, et al. "Solar electric powered hybrid vehicle." *Journal of Electronic Design Technology* 6, no. 3 (2015): 1-12.

Sinuraya A, Sinaga DH, Simamora Y, et al. *"Solar photovoltaic application for electric vehicle battery charging."* In *Journal of Physics: Conference Series*, vol. 2193, no. 1, p. 012075. IOP Publishing, 2022.

Sykiotis S, Menos-Aikateriniadis C, Doulamis A, Doulamis N. et al. "Solar power driven EV charging optimisation with deep reinforcement learning." In *2022, the 2nd International Conference on Energy Transition in the Mediterranean Area (SyNERGY MED)*, pp. 1-6. IEEE, 2022.

Taha Z, Sah JM, Passarella R, et al. "A solar vehicle based on sustainable design concept." In *Proceedings of. The IASTED International Conference on Solar Energy*, pp. 16-18. 2009.

Tiano FA, Rizzo G, et al. "Evaluation of the potential of solar photovoltaic panels installed on vehicle body, including temperature effect on efficiency." *ETransportation* 5 (2020): 100067.

Yamaguchi M, Masuda T, et al. "Impact and recent approaches of high-efficiency solar cell modules for PV-powered vehicles." *Japanese Journal of Applied Physics* 61, no. SC (2022): SC0802.

Chapter 6

Renewable Energy Solutions: Advances in Solar Cell Technology Using Graphene-Based Nanostructures

Chandan Adhikari*, **PhD**

Department of Basic Science and Humanities, Institute of Engineering & Management, University of Engineering and Management, Kolkata, India

Abstract

Solar energy is one of the most appealing options because it is prolific and has great potential in reducing carbon emissions. The efficiency of converting solar energy into usable electricity remains a barrier. Graphene has become a central focus of photovoltaic research due to its exceptional electrical, thermal, and mechanical properties. Due to its high electron mobility and flexibility, it can be utilised in various applications, including transparent conductive electrodes, electron transport layers, and as an active layer in organic and perovskite solar cells. This section draws readers' attention to the physical characterisation of graphene and its derivatives, including graphene oxide and reduced graphene oxide, as well as their impact on overall photovoltaic performance through enhanced charge transport, increased light absorption, and reduced recombination losses. Furthermore, the function of graphene-based nanostructures is investigated, along with their use in creating new hybrids, such as graphene/quantum dot and graphene/nanoparticle composites, which aim to enhance the efficiency of solar cells further. The critical fabrication methods and integration techniques for incorporating graphene into actual photovoltaic systems are examined, addressing their real-world significance and their potential to scale to

* Corresponding Author's Email: chandan.adhikari@iem.edu.in

In: Breakthrough Research and Technologies in Physics …
Editors: Koyel Ganguly, Soumyadipta Pal and Kamakhya Prasad Ghatak
ISBN: 979-8-90134-111-7

commercially relevant dimensions. We also analyse contemporary experimental advancements and compare the efficiency increases obtained for various types of graphene-based solar elements. Focusing on the interplay between graphene and other advanced materials, this chapter highlights the potential of using graphene-based structures in the design of next-generation solar devices. Such knowledge and developments are presented to guide further research and development efforts aimed at enhancing the efficiency and durability of low-cost solar energy devices.

Keywords: graphene nanostructures, solar cell efficiency, renewable energy, photovoltaics, advanced solar materials, energy solutions

Introduction

Due to climate change and the growing global energy demands, the transition to renewable energy is becoming increasingly necessary. Solar, wind, and hydroelectric power are renewable energy sources that can replace fossil fuels, leading to reduced greenhouse gas emissions and a more positive overall impact on the environment. Solar energy, in particular, is also touted as one of the most promising and widely available renewable resources. The sun alone emits more energy in an hour than the world consumes in a year, showcasing its capability to provide energy to the entire globe. However, there have been significant difficulties in efficiently capturing this energy, particularly in terms of the efficiency and cost of solar cells. Yet, recent developments in nanotechnology, particularly those related to graphene, suggest alternative approaches to the problems of cost and efficiency, allowing us to envision other prospects for solar energy [1].

In the move from fossil fuels to sustainable energy systems, renewable energy strategies are key. These strategies not only reduce reliance on finite resources but also help tackle issues such as pollution and habitat destruction. Over the past few decades, the technology of renewable energy sources has also undergone significant improvements, with extensive investment in areas such as efficiency, scalability, and cost reduction. Due to its versatility and scalability, solar power is particularly well-suited for a wide range of applications, from powering a single house to supplying energy to sizable industrial operations. With the development of technology and the further integration of solar energy into the energy grid, its share in the global energy portfolio is increasing [2].

As an energy source, solar energy is also crucial in achieving sustainable development goals, as it provides clean, renewable, and abundant energy. Given that solar energy does not generate harmful emissions or use up exhaustible resources, it is an effective way to address future energy demands. Remote and rural communities lacking conventional energy services can also benefit from solar energy, which facilitates economic development and improves quality of life. By increasing energy supply and reducing the cost of energy, solar energy supports people's social and economic progress while helping to meet the international temperature goal of 1.5°C [3].

Solar installations can range from small systems serving individual users to giant solar farms that provide electricity to thousands. The most critical limitation on solar energy development as a renewable energy source remains the effectiveness of solar cells, despite the promising prospects of solar energy. This occurs because most commercial solar panels, made from silicon-based technology, have efficiency rates between 15% and 22%, which limits the amount of captured solar energy that can be converted into usable electricity. Additionally, highly efficient solar cells are also costly, which contributes to slow market penetration rates. Solar cells experience low performance due to various factors, including ageing, adverse weather events, heat-related energy loss, and electron recombination within the solar cell. Consequently, there is a growing demand for new types of materials and technologies to enhance efficiency and reduce costs [4].

These efficiency issues might be addressed with the help of nanotechnology, especially graphene-containing nanostructures. Scientists believe that this material exhibits qualities that enhance charge movement, increase transparency, and reduce recombination losses in solar cells, as graphene is a hexagonal lattice formed from a single layer of carbon atoms. Additionally, graphene's excellent conductivity, combined with its flexibility, enables the incorporation of the material in various regions of solar cell devices, ranging from transparent electrodes to electron transport layers. The integration and design development of graphene-based materials are expected to yield substantial benefits in terms of efficiency and lifetime performance, harnessing solar energy to its maximum potential [5].

This chapter explores the advancements in solar cell technology facilitated by graphene-based nanostructures. It provides an overview of graphene's unique properties, including its electrical conductivity, optical transparency, and flexibility, which make it a valuable asset in photovoltaic applications. The chapter examines various types of graphene-based materials, including graphene oxide and quantum dots, and discusses their roles in enhancing the performance of solar cells. Key fabrication techniques and integration strategies are also outlined, emphasising scalability and cost considerations for practical applications.

Through a review of recent experimental breakthroughs, this chapter aims to highlight the potential of graphene-based nanostructures in overcoming the current efficiency limitations of solar cells. The objective is to provide insights into how graphene can transform solar technology, making it more efficient, affordable, and adaptable for large-scale deployment. By addressing the technical and practical aspects of integrating graphene into solar cells, this chapter seeks to contribute to ongoing research and development efforts in the renewable energy sector, with a focus on achieving sustainable and impactful solutions.

Graphene and Its Unique Properties

Graphene, which is essentially a single layer of carbon atoms arranged in a two-dimensional, hexagonally structured lattice, has emerged as one of the most advanced materials in nanotechnology. First isolated from graphite in 2004, graphene has since become the focus of intense examination due to its unique combination of electrical, thermal, mechanical, and optical properties, which no other material possesses. In this section, we examine the structure and characteristics of graphene, its extraordinary properties, and the benefits it offers for enhancing solar cell performance [6].

Graphene is the structural building block from which most carbon allotropes are constructed, including graphite, carbon nanotubes, and fullerenes. This single layer of carbon atoms, confined to a two-dimensional honeycomb lattice, gives graphene its extraordinary properties, which are due to its atomic and electronic structure. Each carbon atom is bonded covalently to three other carbon atoms to make a stable, flexible, and thin sheet, just one atom thick. One of the aspects that will always define graphene is its two-dimensionality, thus making it the thinnest material known, yet it still possesses incredible strength. Its mechanical strength is over two hundred times that of steel, making it highly resilient and durable. Furthermore, because graphene is lightweight and flexible, it can be incorporated effortlessly into a range of applications, such as photovoltaics and flexible electronics. Graphene is widely known as a zero-bandgap semiconductor. This indicates that its conduction and valence bands merge at a single point, allowing electrons to move freely without the typical bandgap limitations found in other semiconductor materials. As a consequence of this peculiar band structure, the electrical conductivity of graphene is very high, with high electron mobility achievable even at room temperature. Moreover, the high surface area of graphene (up to approximately 2630 square meters per gram) enhances its efficacy in various disciplines, as it increases the number of interaction sites per mass unit, which is essential for rapid charge transfer and other physicochemical applications [7].

The unique structure of graphene possesses a string of remarkable properties that enable it to be widely used in various areas, particularly in electronics, energy storage, and photovoltaic systems.

The unique structure of graphene possesses a string of remarkable properties that enable it to be widely used in various areas, particularly in electronics, energy storage, and photovoltaic systems. The outstanding electrical conductivity of graphene is one of its key features. Graphene's unique band structure allows its electrons to be quasi-particles (Dirac fermions) that move with very high velocity. The electron mobility can reach up to 200,000 cm^2/Vs in an optimal environment, which is significantly higher than that of common materials, such as silicon. This high mobility enables graphene to conduct electric current with minimal energy loss through heating and resistance, which is essential in applications where electrons need to move quickly, such as in photovoltaic cells. Its high conductivity enables graphene to be used readily as a transparent conducting electrode, a crucial part of solar cells, as it meets the need for electrical conduction combined with transparency. Graphene's thermal conductivity, with an approximate value of around 5000 W/mK, is also an exceptional feature. This thermal property enables fast heat dissipation, preventing overheating in devices where it is used. In solar cells, integrating graphene enhances thermal management, improving the efficiency and performance of the solar cells while safeguarding sensitive device structures. Excessive heat is known not only to degrade performance but also to damage components. Integrating graphene into solar cells enhances thermal management, enabling devices to operate more efficiently under various environmental conditions [8].

Graphene's thickness corresponds to one single atom; however, it is optically very transparent, absorbing only about 2.3% of incoming light across the visible spectrum. This distinctive feature, which combines transparency and conductivity, complements each other and is highly advantageous in photovoltaic applications. This combination of high transparency, allowing light to reach the active layer, and high electrical conductance is proper in solar cells, where graphene can act as an electrode, enabling better energy conversion efficiency. The tunable bandgap and wavelength dispersion in graphene make it a promising candidate, especially in multijunction or tandem solar cells, where various wavelengths need to be absorbed efficiently [9].

The incorporation of graphene into solar cells is driven by its complementary electrical, optical, and thermal properties, enabling new possibilities in the design of future photovoltaic devices. One of the biggest concerns in solar cell design is charge transport, due to the propensity for charge recombination losses. The generally high mobility and conductivity of electrons in graphene create a practical pathway for charge carriers, reducing recombination and increasing the overall efficiency of the cell. This is most useful in perovskite and organic solar cells, where charge carrier mobility can be a limiting factor. The use of graphene in transport layers allows electrons to be moved faster and more efficiently, thereby increasing the power conversion efficiency. Graphene's remarkable thermal conductivity provides a promising solution to the thermal management challenges of solar cells. In photovoltaic systems, particularly those operating under high sunlight intensities, heat dissipation is crucial for long-term performance and stability. Graphene helps conduct heat and facilitates the dispersion of excess thermal energy, preventing damage to the cell's active material and allowing it to perform normally under various environmental conditions. Such a thermal advantage renders graphene-based materials very attractive for large-scale solar applications where temperature management is critical [10].

Another critical aspect of graphene is its optical transparency with high electrical conductivity; therefore, it can be used as a material for transparent electrodes in solar cells. Indium tin oxide (ITO) is a common transparent conductor; however, it is brittle and prone to deterioration over time. Unlike indium tin oxide, graphene is flexible while also offering decent conductance, which means solar cells may be more durable. Thin, light, and flexible solar cells made of graphene electrodes also enable new types of applications, including wearable solar devices that can be integrated into various surfaces, potentially expanding the possibilities for solar cells. Graphene's mechanical strength and elasticity make it suitable for use in flexible solar cells, which have potential applications in building-integrated photovoltaics (BIPV), flexible electronic devices, and other unconventional photovoltaic (PV) applications. Apart from being cost-effective, graphene-enhanced solar cells are highly adaptable, allowing for new applications that require bending or conforming to non-rigid surfaces, unlike conventional solar cell panels, which are rigid.[11] Such flexibility also increases the operational life span of solar cells since physical damage and degradation risks are reduced due to graphene's superior longevity [11].

Although still under development, favourable scalability factors for graphene production point to a future reduction in advanced solar cell prices. CVD and chemical exfoliation are just a few of the scalable methods that can be used to synthesise graphene. Producing graphene at a lower cost and applying it to existing solar technologies represents an affordable way to increase the efficiency of solar energy. Graphene's extraordinary properties can bring substantial benefits for the sustainable development of solar cell technology. These properties will enhance both charge transport and thermal management, in addition to transparency and flexibility, positioning graphene as a crucial component for future solar cells. As research progresses, the use of graphene in solar cells may lead to significant enhancements in efficiency, durability, and functional diversity, thereby advancing solar energy solutions [12].

Graphene-Based Nanostructures in Solar Cells

Nanostructures based on graphene represent a novel direction in the development of solar cell technology, offering enhanced efficiency, flexibility, and stability. Various components of solar cells are being modified with these materials because of their remarkable characteristics. To date, several graphene-based nanomaterials, including graphene oxide (GO), reduced graphene oxide (rGO), and graphene quantum dots (GQDs), have been incorporated into various solar cell architectures to enhance their performance [13].

Graphene oxide can be described as a graphite derivative, characterised by numerous oxygenated functional groups, including hydroxyl, epoxide, and carboxyl groups. These groups enhance the dispersion of GO into water and organic solvents, thereby facilitating its processing into various materials. GO is hydrophilic and can make composites with other materials, which enhances its use in charge transport layers in solar cells. Its large surface area also helps in better interactions with the active layer of solar cells, resulting in improved efficiency. Reduced graphene oxide (rGO) is produced by chemically reducing GO, a process that removes some oxygen-containing functional groups, thereby restoring the conductive properties of graphene. rGO has many favourable characteristics of graphene, including high electrical conductivity and mechanical strength, but is less expensive to produce than pristine graphene. In solar devices, rGO can be incorporated into transparent conductive electrodes and electron transport layers, thereby enhancing charge mobility and improving overall device performance. Graphene quantum dots (GQDs), which are nanometer-sized fragments of graphene, exhibit unique optical properties due to quantum confinement effects. GQDs can be modified to enhance their energy levels and band gaps; therefore, these materials are suitable for use in photovoltaic systems. They can increase light absorption and improve charge separation, thereby increasing the efficiency of solar energy conversion [14-16].

Graphene-based nanostructures are also functional in various elements of a photovoltaic system. They support the mobility of charge carriers and therefore reduce recombination losses, one of the main drawbacks of conventional solar cells. These nanomaterials can be utilized to form electrodes, charge transport layers, or even part of the active layer, thereby enhancing light absorption efficiency, improving charge transport, and facilitating energy conversion. Additionally, their high tensile strength and mechanical properties enable the production of lightweight and flexible solar cells, opening up new possibilities in portable and wearable energy devices. Solar cells containing graphene nanostructures may be more efficient, have better scalability potential, and exhibit more extended durability, making them an integral part of future solar energy technologies [17].

Applications in Solar Cell Technology

Graphene, with many special properties such as high conductivity, structural flexibility, and excellent transparency, can also benefit passive components or layers of solar cells. With advancements in research, graphene is being progressively integrated into multiple layers of photovoltaic devices to enhance efficiency, reduce costs, and provide improved durability. This section discusses several key applications of graphene in solar cell technology, including its use in transparent conductive electrodes, electron transport layers, graphene/quantum dot composites, and perovskite solar cells.

Solar cells rely heavily on transparent conductive electrodes for efficient electron flow and for transmitting light. Materials such as indium tin oxide (ITO) have served the purpose effectively. However, such materials have limitations in terms of cost, mechanical strength, and flexibility. Graphene has emerged as a potential substitute due to its high electrical conductivity and transparency. A distinguishing feature of graphene is its two-dimensional structure, which gives rise to a very high surface area, consequently providing excellent conductivity while retaining high optical transparency, absorbing only about 2.3% of light in the entire visible spectrum. For these reasons, graphene is an excellent fit for use as transparent electrodes in organic, perovskite, and dye-sensitised solar cells. The lightweight and flexible nature of graphene enables the creation of flexible solar panels that can be embedded on various surfaces, such as windows or clothing, thereby facilitating a wider range of applications for solar energy. Such electrodes made from graphene provide the advantages of increased scalability and cost-effectiveness. Graphene is also very suitable for cost-effective mass production, as it can be synthesised by low-cost techniques such as CVD and solution-based methods. Furthermore, the mechanical characteristics of graphene provide solar cells with enhanced durability and longevity compared to cells with ITO electrodes, which are more susceptible to fracturing and degradation over time [18].

As the name suggests, the electron transport layer is a crucial component of a solar cell's overall design, enabling electrons to migrate from the active layer to the electrode. Efficient transport of charge on the framework is crucial for reducing recombination losses and enhancing the operational efficiency of the device. Due to its high electron mobility, graphene is well-suited for applications in ETLs, as it facilitates charge extraction and reduces energy losses. Graphene-based materials, such as reduced graphene oxide (rGO), have found applications in ETLs for enhancing efficiency in various types of solar cells, including organic and perovskite solar cells. This is made possible because graphene's large surface area and high graphitic content enable efficient electron transport. At the same time, its nanoscale structure facilitates good interaction with the active layer, thereby enhancing charge extraction. Additionally, the flexibility and ease of processing of graphene make it a suitable candidate for use as an ETL in lightweight and thin-film solar energy devices. It has been established that graphene-based ETLs also enhance the stability of solar cells. For instance, it has been reported that endothermally functionalised graphene-based materials were able to improve device stability by performing a protective role in perovskite solar cells undergoing continuous illumination. This aspect is essential for the economic viability of solar technologies as it guarantees that the operational efficiency of solar cells is retained over their useful lifetime [19].

Graphene and quantum dots (QDs) are two materials that have garnered significant attention in the field of photovoltaics due to their complementary features. Quantum dots are tiny semiconductor particles, measuring a few nanometers in diameter, that exhibit distinct optical properties due to quantum confinement. In contrast, graphene possesses high electrical and mechanical properties. The combination of graphene and quantum dots forms composites that can improve solar cell performance. They have the potential to absorb light in a broader range and generate charge carriers, for which graphene creates an efficient transportation path. When utilised fully, this synergy improves solar cell performance, as both light absorption and charge transport are enhanced. Additionally, these composites have the potential to elevate the effectiveness of next-generation solar technologies, such as quantum dot solar cells, which utilise quantum dots to generate excitons for power generation. The inclusion of graphene in these composites enables the more efficient separation of photo-generated excitons, thereby improving charge transfer and reducing recombination loss. Graphene also allows the development of flexible quantum dot solar cells, making them suitable for use in portable and wearable solar devices [20].

Perovskite solar cells (PSCs) have gained popularity due to their high efficiency and low production cost. The light-absorbing material in these cells consists of a perovskite-structured compound, and a gradual improvement in their performance has been observed over the last decade. However, several issues, including charge recombination, instability, and scalability, must be addressed to enable the commercial exploitation of perovskite solar cells. The integration of graphene oxide (GO) and reduced graphene oxide (rGO) with perovskite solar cells has addressed some of these challenges. These structures can achieve effective charge transport and minimise recombination losses in graphene layers integrated into the electron transport layer (ETL) and hole transport layer (HTL) of the solar cells. Additionally, the use of graphene in PSCs can enhance device stability by reducing the permeability of moisture and oxygen through the perovskite layer. Furthermore, graphene can improve the mechanical properties of perovskite solar cells, enabling the manufacture of lightweight and flexible devices that can be applied to a wide range of surfaces, including textiles, roofs, and windows. It has also been established in recent studies that very high-power conversion efficiencies (PCEs) alongside the long-term stability of PSCs are achievable through the inclusion of different graphene-based materials within the cells. Through the improvement of material stability and charge transport, graphene is unlocking the potential for the commercial uptake of perovskite photovoltaics [21].

Graphene is unique in its properties, has wide applications, and is therefore an excellent material that can accelerate the development of solar cells. It can be seamlessly integrated into various components of solar cells, including transparent conductive electrodes, electron transport layers, or graphene/quantum dot composites, as well as perovskite solar cells, thereby significantly increasing their efficiency, stability, and scalability. As research continues to explore other methods for integrating graphene with photovoltaic devices, it is expected that such materials will become the building blocks for future technologies, enabling these PV devices to be used for more practical and affordable energy solutions [22].

Fabrication Techniques and Integration Strategies

To integrate graphene-based nanostructures into solar cells, advanced fabrication techniques are needed that can ensure both high performance and cost-effectiveness. These techniques include the synthesis of graphene materials with specific characteristics and the embedding of these materials in tailored designs of solar cells. This section outlines various fabrication methods for graphene-based nanostructures, discusses particular issues related to integrating graphene within solar cells, and addresses their mass production and scalability. Several effective strategies have been developed for synthesising graphene-based nanostructures, each with its pros and cons regarding ease of scaling, material quality, and applications. Such techniques can be broadly classified into chemical, physical, and biological methods [23].

One of the advantages of chemical vapour deposition (CVD) is that it enables the synthesis of high-quality graphene on a mass scale, as it is easy to control the growth parameters. In the CVD process, carbon-bearing molecular gases (such as acetylene or ethylene) are decomposed at high temperatures in the presence of a metal substrate (like copper or nickel), forming a graphene film on the substrate. This controlled, gas-phase growth makes CVD technology robust and suitable for the electronics industry, and it is an excellent method for synthesising large-area graphene sheets. Graphene oxide (GO) can be synthesised by oxidising graphite powder, resulting in the material's surface bearing oxygenated functional groups. Removing the oxygen-containing groups by chemically reducing GO restores electrical conductivity, forming what is known as reduced graphene oxide (rGO). This process is straightforward and economical; however, the quality of the produced rGO depends on the reduction process. This material helps create electron transport layers and serves as an alternative conductive material for flexible solar cells.

A solvothermal method, or hydrothermal method in general, involves the application of high pressure and high temperature to synthesise graphene or graphene-based nanomaterials. Other synthesis processes, such as flame preparation, are also capable of accomplishing this, making these techniques essential for graphene synthesis, as well as for producing graphene oxide, graphene quantum dots, and composites of graphene and other materials. These techniques are preferred as they can be conducted in a one-pot reaction, which is beneficial in terms of scalability and control over the dimensions and shape of the nanostructures being synthesised. In this methodology, graphite is broken down into graphene layers using a solvent and mechanical shear force, such as ultrasonication. Graphene oxide, along with reduced graphene oxide, has been effortlessly incorporated into active and charge transport layers. Liquid-phase exfoliation is a straightforward and efficient method for the mass production of graphene nanomaterials [24].

Although the positive aspects of graphene are well-known, particularly its applications in solar cells, integrating the material into devices presents several challenges. The most challenging task, however, is the uniform and controlled integration of graphene-containing materials into the solar cell's structure. To avoid aggregation and enable effective charge transport, it is crucial to achieve a uniform distribution of graphene in both the active layer and the transport layers. A further problem is the compatibility of graphene with other materials commonly used in solar cells. Due to its hydrophobic nature and strong van der Waals forces, graphene struggles to form stable interfaces with various materials, including polymers, perovskites, and metal electrodes. This, in turn, can adversely affect the solar cell's stability and performance. In addition to these technological and scientific barriers, the weak mechanical adhesion of graphene-based nanostructures to the substrate or other materials also prevents achieving device stability and long-term performance. To address this, either the functionalisation of graphene with specific functional groups or the application of graphene composites can be used to improve adherence and increase the overall stability of the solar cell [25].

There is a need to develop scalable and cost-effective fabrication methods for commercial and industrial utilisation of graphene-based solar cells. One possible method is to apply substrate-coated materials using solution-based processing techniques, such as inkjet printing, spraying, or spin coating. These are relatively inexpensive, can be up-scaled for large-area manufacturing, and can be used in the fabrication of lightweight, flexible solar cells. Roll-to-roll (R2R) processing, an efficient and rapid production method, provides a practical approach for mass-producing graphene-based solar cells on an industrial scale. In this method, flexible substrates are printed with graphene-based inks or films, which results in the efficient production of flexible solar panels. R2R processing is highly effective for producing thin-film solar cells that can be widely used for mounting on surfaces such as windows, clothing, building facades, and others. Another cost-effective and scalable alternative is the adoption of graphene-based inks for printing on flexible substrates. Inks comprising graphene oxide or its reduced form, combined with solvents, allow for easy application through screen printing or inkjet printing. This method permits the manufacture of high-performing solar cells at low material and production costs. Additionally, printing technologies provide an improvement in design. It is also possible to manufacture irregularly shaped or large-area solar panels. CVD is still the current best practice for producing high-efficiency graphene-based solar cells. The development of large-scale CVD systems can make the process more cost-effective and scalable. New types of CVD reactors and coated catalyst supports are expanding the possibilities for growing large-area graphene films, making these techniques suitable for mass production. This is important because high-quality, large-area graphene films are crucial for integrating graphene into transparent electrodes or other components of solar cells [26, 27].

Promising strategies include the use of graphene-based nanostructures and their incorporation into solar cells to enhance both the efficiency and cost-effectiveness of photovoltaic devices. Though challenges exist in terms of the suitability and morphology of the materials, solution-based processing and roll-to-roll technology offer promising strategies to address these problems. Future scaling up of graphene synthesis and integration technologies will be crucial for expanding the scope and applications of graphene-based solar cells, as well as accelerating their market penetration in the renewable energy sector.

Future Prospects and Challenges

The adoption of graphene-based nanostructures into solar cells has yielded very encouraging results, particularly in enhancing the efficiency of solar energy utilization. With a transition to a future powered primarily by renewables rapidly approaching, the special characteristics of graphene make it an attractive material for advanced solar technology. However, despite the potential, the barriers to entry in terms of material development, scale-up, and commercialisation are substantial. In this section, the authors discuss the potential ways graphene can facilitate the development of advanced solar technologies, identify gaps worth studying, and outline the barriers to commercialisation and scaling [28].

Graphene has opened several exciting new avenues that can be utilised in enhancing solar cells and their efficiency. One of the most intriguing opportunities is that graphene can improve the efficiency of more established solar technologies, including silicon solar power cells, by enhancing specific features such as the transparent conductive electrode, charge transport layer, and the solar cell's active layer. The high electrical conductivity, mechanical strength, and flexibility of graphene make it a suitable component for these applications, enhancing power conversion efficiency, reducing production costs, and enabling more efficient and simplified solar devices. The application of graphene in polymer-based solar cells provides yet another opportunity. Solar cells, based on perovskite materials, have already begun to transform the solar world, thanks to their high efficiency and low-cost, scalable production. The improved charge collection, stability, and durability of perovskite solar cells could result from integrating graphene into these solar cells. Additionally, graphene's capacity to create composites with perovskite materials will assist in resolving some of the stability challenges, optimising the potential of PSCs for enhanced commercial viability and application. The role of GQDs in solar cells is also another focus of future work, as GQDs can be used to leverage their unique photonic properties. GQDs are instrumental in modifying the optical and electronic properties of solar cells, thereby enhancing photovoltaic performance, due to their unique optical and electronic characteristics. Incorporating GQDs into hybrid solar cells or multi-junction cells may increase efficiency by expanding the spectrum of sunlight available for absorption. Another avenue to increase innovation is through the use of graphene in flexible and transparent solar cells. It will be essential to be able to manufacture lightweight, flexible, and transparent solar cells, as the need for portable electronics and building-integrated photovoltaics (BIPV) increases. Graphene is a flexible and transparent material that can be used to manufacture devices and open up new opportunities in fields such as wearable electronics and solar windows [29, 30].

Despite the highlighted promise of graphene in solar cell applications, several issues must be addressed before its full potential can be unlocked. One such challenge is the issue of scale, or scalability, as well as process consistency. Currently available graphene synthesis techniques (chemical vapour deposition, CVD, or liquid-phase exfoliation) that can produce high-quality but low volumes of graphene materials struggle to be commercialised for mass production. The ability to fabricate large-area, uniform graphene films or nanomaterials with exceptionally high electrical conductivity remains a significant obstacle to the large-scale adoption of graphene in solar cells. Another challenge is the problem of interfacial incompatibility. Graphene tends to have weak bonding with active materials, such as silicon or perovskites, which are often used in solar cells. Such interfacial incompatibility results in reduced charge transport efficiency, which, in turn, affects the overall device performance. It is necessary to improve existing methods of graphene functionalisation and to integrate graphene with other materials within solar cell architectures more effectively. Moreover, although graphene has high electrical conductivity and is therefore well-suited for several applications, its low light absorption capacity in some instances may be a significant drawback for use in the active layer of a solar cell. There is still room for further work in modifying the optical properties of graphene or its combination with other materials, such as quantum dots or organic semiconductors, to improve sunlight absorption. The longevity of graphene-based solar cells also requires focus. Such devices may be susceptible to temperature, humidity, and exposure to UV light. There will be a need to investigate the encapsulation and protection of graphene in solar cell devices so that they can withstand real-world conditions [31].

Due to the mesmerizing prospects of graphene-based solar cells, there are still few implementations of these technologies in real-world applications. The key hurdle is the production of large-scale graphene. Methods of production in laboratory conditions have shown encouraging prospects; the only problem now is how to approach large-volume production economically without compromising material quality, a challenge that remains the focus of many studies. While methods for the industrial-scale production of graphene that guarantee specific standards of quality, purity, and performance are being developed, they currently limit the transition of graphene-based solar technologies from research labs to commercial products. Also, it is about the costs. Graphene itself is cheaper than precious metals or exotic semiconductors; however, the cost of synthesis and integration when incorporating graphene into solar cells remains relatively high. For these technologies to be commercially viable and competitive with traditional solar cells, methods for synthesising graphene over large areas at low cost and reducing the overall cost of producing graphene-enhanced solar cells will be critical. Another challenge concerning the commercialisation of solar technologies is the regulatory and market assimilation of graphene-based solar technologies. While there is an increasing interest in the potential of graphene from both academic and industrial perspectives, guidelines and standards for certifying graphene-based products are still being formulated. To facilitate the successful launch of these technologies into the market, manufacturers must demonstrate product robustness and offer them at prices sufficiently attractive to target potential consumers and investors. Additionally, integrating graphene into existing solar cell production lines will necessitate a substantial investment and modifications to the infrastructure. Many present processes for solar cell manufacturing are optimized for materials such as silicon and cadmium telluride, which are inherently dissimilar from graphene-based materials. Transitioning to production lines capable of using graphene will entail significant adaptations to existing structures, thereby increasing the cost of implementing graphene technologies in solar cell systems [32].

With a clear emphasis on cost-effective, flexible, and efficient solar devices, the outlook for graphene in solar cell technology is auspicious. Nevertheless, a remaining gap in the field remains, which includes challenges in material fabrication, material integration, and scalable commercialisation. To address this, researchers, engineers, and industry leaders must unite to develop efficient, scalable, and low-cost pathways for integrating graphene with solar cells. With the advancement of innovation and increased research, graphene-based solar technology has the potential to become a game-changer for renewable energy, helping to meet global energy needs while contributing to the achievement of sustainable development objectives [33].

Conclusion

This chapter examines the promising prospects for utilising graphene-based nanostructures to improve solar cell technology. With its outstanding electrical, thermal, and optical characteristics, graphene has quickly become a frontrunner in endeavours aimed at improving the efficiency and performance of solar cells. After an in-depth examination of its structure, properties, and various morphologies, including graphene oxide, reduced graphene oxide, and graphene quantum dots, it has been demonstrated that graphene can enhance the performance of different types of solar cells, such as silicon, perovskite, and their hybrids. Among the most significant conclusions from the chapter is that graphene enhances the transparency and conductivity of electrodes, while maintaining the mechanical flexibility of the solar cells. Its applications as transparent conductive electrodes, electron transport layers, and composites with quantum dots could enhance charge transport and light absorption, leading to higher solar energy conversion efficiencies. Additionally, graphene's exceptional characteristics make it a suitable candidate for flexible and portable solar technologies, which would extend its applications to consumer devices, wearables, and building-integrated photovoltaics, among other technologies.

Graphene holds outstanding promise for designing new, advanced solar cells with unprecedented efficiency, flexibility, and scalability. Its exceptional properties, together with its versatile composition, aim to address many of the shortcomings that contemporary photovoltaic devices face, such as high cost, rigidity, and inefficiency. As research progresses in graphene production coupled with integration methodologies and material formulation, this unique material has the potential to disrupt the solar energy sector by making solar cells more efficient, cheaper, and versatile for various end-use applications. Beyond its functional aspects in solar technology, graphene has the potential to support the broader objectives of renewable energy development, as it would help meet the increasing global demand for clean and sustainable energy. The potential of graphene in reducing energy costs and enhancing the efficiency of solar cells could facilitate the adoption of solar energy systems, contributing to the achievement of global targets for transitioning to a low-carbon economy [34].

As promising as it is, a few barriers remain regarding graphene synthesis techniques, their scalability, and ensuring long-term performance. Overcoming such challenges would require significant developments in both research and industry. However, as the understanding of graphene's properties and its incorporation into solar cells advances, the prospects of utilising graphene for future solar technologies seem promising. In conclusion, graphene is a highly effective material that can significantly contribute to the advancement of solar energy systems. Further research on this topic will enhance the efficiency of solar cells and make solar energy systems more accessible, affordable, and sustainable in the long run.

Disclaimer

The content of this chapter is intended for academic and informational purposes only. The authors declare that there are no conflicts of interest associated with the research, analysis, or publication of this work. The information provided here is based on current scientific understanding and is subject to change as new research emerges in the field of renewable energy and graphene-based technologies.

References

[1] Panwar, N. L., S. C. Kaushik, and S. Kothari, Role of renewable energy sources in environmental protection: A review. *Renewable and sustainable energy reviews*, 2011. **15**(3): p. 1513-1524.

[2] Nwokediegwu, Z. Q. S., et al., Renewable energy technologies in engineering: A review of current developments and future prospects. *Engineering science & technology journal,* 2024. **5**(2): p. 367-384.

[3] Obaideen, K., et al., Solar energy: Applications, trends analysis, bibliometric analysis and research contribution to sustainable development goals (SDGs). *Sustainability*, 2023. **15**(2): p. 1418.

[4] Oni, A. M., A. S. Mohsin, M. M. Rahman, and M. B. H. Bhuian, A comprehensive evaluation of solar cell technologies, associated loss mechanisms, and efficiency enhancement strategies for photovoltaic cells. *Energy Reports*, 2024. **11**: p. 3345-3366.

[5] Niazi, Z., A. Hagfeldt, and E. K. Goharshadi, Recent progress on the use of graphene-based nanomaterials in perovskite solar cells. *Journal of Materials Chemistry A*, 2023. **11**(13): p. 6659-6687.

[6] Sharma, N., et al., Graphene: An overview of its characteristics and applications. *Materials Today: Proceedings,* 2021. **47**: p. 2752-2755.

[7] Urade, A. R., I. Lahiri, and K. Suresh, Graphene properties, synthesis and applications: a review. *Jom,* 2023. **75**(3): p. 614-630.

[8] Sang, M., J. Shin, K. Kim, and K. J. Yu, Electronic and thermal properties of graphene and recent advances in graphene based electronics applications. *Nanomaterials*, 2019. **9**(3): p. 374.

[9] Kumar, V., Linear and Nonlinear Optical Properties of Graphene: A Review. *Journal of Electronic Materials*, 2021. **50**(7).

[10] Huang, P., et al., Graphene film for thermal management: A review. *Nano Materials Science,* 2021. **3**(1): p. 1-16.

[11] Muchuweni, E., B. S. Martincigh, and V. O. Nyamori, Perovskite solar cells: current trends in graphene-based materials for transparent conductive electrodes, active layers, charge transport layers, and encapsulation layers. *Advanced Energy and Sustainability Research*, 2021. **2**(9): p. 2100050.

[12] Bagade, S. S., S. Patel, M. Malik, and P. K. Patel, *Recent advancements in applications of graphene to attain next-level solar cells*. C, 2023. **9**(3): p. 70.

[13] Wang, J., J. Song, X. Mu, and M. Sun, Optoelectronic and photoelectric properties and applications of graphene-based nanostructures. *Materials Today Physics*, 2020. **13**: p. 100196.

[14] Velasco Davoise, L., A. M. Díez-Pascual, and R. Peña Capilla, Application of graphene-related materials in organic solar cells. *Materials*, 2022. **15**(3): p. 1171.

[15] Mahmoudi, T., Y. Wang, and Y.-B. Hahn, Graphene and its derivatives for solar cells application. *Nano Energy*, 2018. **47**: p. 51-65.

[16] Panachikkool, M. and T. Pandiyarajan, Graphene quantum dots as game-changers in solar cell technology: a review of synthetic processes and performance enhancement. *Carbon Letters*, 2024. **34**(1): p. 445-475.

[17] Znidi, F., M. Morsy, and M. N. Uddin, Recent advances of graphene-based materials in planar perovskite solar cells. *Next Nanotechnology*, 2024. **5**: p. 100061.

[18] Jain, P., et al., Recent advances in graphene-enabled materials for photovoltaic applications: a comprehensive review. *ACS omega*, 2024. **9**(11): p. 12403-12425.

[19] Ndlovu, S., M. A. Ollengo, E. Muchuweni, and V. O. Nyamori, Current advances in perovskite oxides supported on graphene-based materials as interfacial layers of perovskite solar cells. *Critical Reviews in Solid State and Materials Sciences*, 2023. **48**(1): p. 112-131.

[20] Shilpa, G., et al., Recent advances in the development of high efficiency quantum dot sensitized solar cells (QDSSCs): A review. *Materials Science for Energy Technologies,* 2023. **6**: p. 533-546.

[21] Lim, E. L., et al., A mini review: can graphene be a novel material for perovskite solar cell applications? *Nano-micro letters*, 2018. **10**(2): p. 27.

[22] Su, H., et al., The application of graphene derivatives in perovskite solar cells. *Small Methods,* 2020. **4**(10): p. 2000507.

[23] Zhang, J., et al., Graphene-based materials in planar perovskite solar cells. *SOLAR* RRL, 2020. **4**(11): p. 2000502.

[24] Alasmari, A., et al., Enhancing perovskite solar cells with graphene-based nanocomposites for sustainable energy: A comprehensive review. *Diamond and Related Materials*, 2024. **148**: p. 111517.

[25] Quesnel, E., et al., Graphene-based technologies for energy applications, challenges and perspectives. *2D Materials*, 2015. **2**(3): p. 030204.

[26] Qin, J., et al., Towards operation-stabilizing perovskite solar cells: fundamental materials, device designs, and commercial applications. *InfoMat,* 2024. **6**(4): p. e12522.

[27] Phogat, P., R. Jha, and S. Singh, Emerging advances and future prospects of two dimensional nanomaterials based solar cells. *Journal of Alloys and Compounds*, 2024. **1001**: p. 175063.

[28] Belkhode, P., et al., Nanomaterials applications in solar energy: Exploring future prospects and challenges. *Materials Today: Proceedings*, 2024.

[29] Hossain, M. T., et al., Progress, prospects and challenges of MXene integrated optoelectronics devices. *Chem Electro Chem*, 2024. **11**(8): p. e202400008.

[30] Ferry, D. B., T. Rasheed, M. T. Anwar, and M. Imran, Graphene and Graphene Derived Nanomaterials as Versatile Candidates for Organic Solar Cells and Smart Windows Applications–A Review. *Chemistry Select*, 2024. **9**(1): p. e202301442.

[31] Xie, B., et al., Advances in graphene-based electrode for triboelectric nanogenerator. *Nano-Micro Letters*, 2025. **17**(1): p. 17.

[32] Chu, Q. Q., et al., Encapsulation: The path to commercialization of stable perovskite solar cells. *Matter,* 2023. **6**(11): p. 3838-3863.

[33] Scalia, T., L. Bonventre, and M. L. Terranova, From protosolar space to space exploration: the role of graphene in space technology and economy. *Nanomaterials,* 2023. **13**(4): p. 680.

[34] Barkan, T., et al., Mapping the landscape for graphene commercialization. *Nature Reviews Physics,* 2024. **6**(11): p. 646-647.

Chapter 7

DFT-Based Analysis of Toxic VOCs to Evaluate Their Structural Stability and Sustainability

Indranil Maity*, PhD
Aman Kumar Jha, BTech
and Yash Shaw, BTech
Department of Electronics and Communication Engineering (ECE),
Institute of Engineering & Management (IEM),
University of Engineering and Management (IEM), Kolkata, West Bengal, India

Abstract

This study describes the electronic properties of two industrially important VOCs, categorised as toxic pollutants, namely ethane and toluene. Density Functional Theory (DFT) calculations were performed on both compounds using Gaussian 09W and GaussView 6.0, with an emphasis on key electronic parameters. It covered the molecular orbitals (MOs), molecular electrostatic potential (MEP), band structure, infrared (IR) spectra and electron localisation function (ELF). From IR spectra analysis, it was found that the symmetry was much higher in ethane than in toluene. The energy gap for molecular orbitals was obtained as +0.452 Hartree for ethane and +0.241 Hartree for toluene. Ethane was chosen here due to its widespread presence in industrial processes and natural gas, making it a relevant baseline for comparative analysis. Toluene, on the other hand, is a more toxic aromatic compound commonly found in solvents and fuel emissions, necessitating closer investigation of its electronic properties. It presents an essential study on the identification and detailed characterisation of these toxic VOCs, which is crucial for

* Corresponding Author's Email: indrasanu026@gmail.com

In: Breakthrough Research and Technologies in Physics …
Editors: Koyel Ganguly, Soumyadipta Pal and Kamakhya Prasad Ghatak
ISBN: 979-8-90134-111-7

the development of selective gas sensors and for understanding their environmental and health impacts in detail.

Keywords: DFT, HOMO-LUMO, MEP, PDOS, band structure, IR spectra

Introduction

The increasing presence of toxic VOCs in industrial effluents has become one of the most pressing issues, highlighting the need to conduct detailed studies to develop selective gas sensors and subsequently assess their implications for health and environmental parameters. In this scenario, ethane (C_2H_6), with its numerous industrial applications, takes precedence over other compounds. It is widely used in the manufacture of ethylene for plastics production, as a clean-burning fuel, and as a refrigerant in cryogenic systems, where its reactivity is essential for these applications [1]. Ethane has been known to form tropospheric ozone, according to a study conducted by Friedreich et al., which can lead to respiratory problems. In addition, ethane indirectly increases the atmospheric lifespan of methane by consuming hydroxyl radicals (OH), thereby contributing to increased global warming [2]. Ethane is also much less toxic than many hydrocarbon species; however, this particular type of hydrocarbon still actively participates in air pollution and contributes to the formation of smog, as observed during Karczewski et al.'s investigation [3]. Similarly, toluene is another VOC that is widely used in various industrial processes. Toluene is known for its diverse applications. It serves as a solvent in paints, coatings, and adhesives, as well as in the manufacture of chemicals such as benzene and xylene. Toluene toxicity poses severe risks because it can cause neurological damage, respiratory issues, and long-term health effects in humans [4]. Toluene exposure in gasoline station workers can lead to several health issues, including fatigue, headaches, dizziness, muscle cramps, and irritability. The weight loss and overall weakness can also be related to prolonged exposure. It primarily affects the nervous system, causing irreversible damage and central nervous system depression at high concentrations, with potential liver, kidney, and respiratory complications contributing to its toxicity [5]. It is also found to cause cognitive impairment, anxiety, and depression, while also affecting the cardiovascular system (arrhythmias and heart issues), immune and reproductive systems [6].

Primarily, the research goal of this chapter is to offer a thorough analysis of the electronic properties of two toxic industrial VOCs, ethane and toluene.

Although these molecules have been the subject of several studies, a detailed examination of each molecule's electronic characteristics has not yet been conducted. Therefore, filling the lacuna (knowledge gap) is the focus of our investigation. This chapter aims to enhance our understanding of these molecules by presenting a DFT study of these fundamental chemicals through a thorough examination of their molecular orbitals, electronic properties, and reactivity analysis. The precise control of electronic properties in nanoscale devices is of great interest, as it enables innovative applications in computing, sensing, and medical diagnostics.

Computational Methods

Using Gaussian 09W and GaussView 6.0, the computational analysis was performed on the molecular structures of ethane and toluene. The ground-state optimisation was performed using the B3LYP method with the 6-31G(d,p) basis set in DFT calculations to obtain molecular orbitals and perform MEP analysis. The band structures for both ethane and toluene were subsequently obtained using Material Studio and CASTEP software, based on the optimised orbital data. GaussView 6.0 software was used to calculate the infrared (IR) spectra of both molecules, capturing key vibrational modes and intensities.

Results and Discussion

Optimised Geometry and Molecular Orbital Analysis

The results of the considered structures (obtained after optimisation) are presented in Fig. 1(a–b). After the optimisation process, the bond lengths were observed as follows: for ethane, the C-C bond length was 1.535 Å, and the C-H bond length was 1.096 Å. For toluene, the C-C bond length was 1.399 Å, the C-H (ring) bond length was 1.085 Å, and the C-H (methyl) bond length was 1.095 Å.

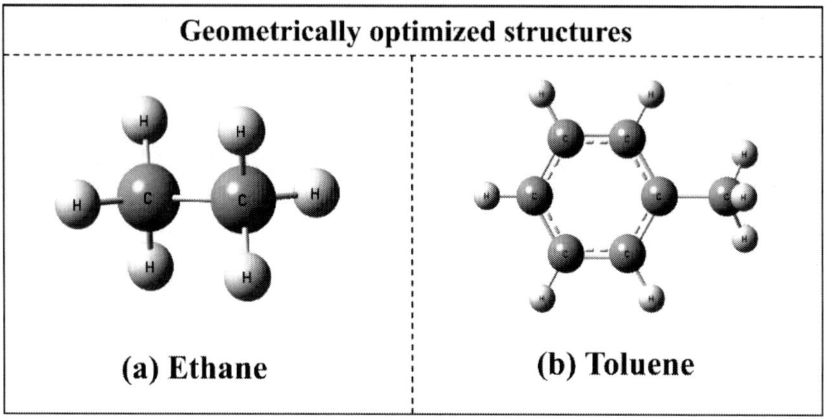

Figure 1. Geometrically optimised structures of (a) Ethane, and (b) Toluene.

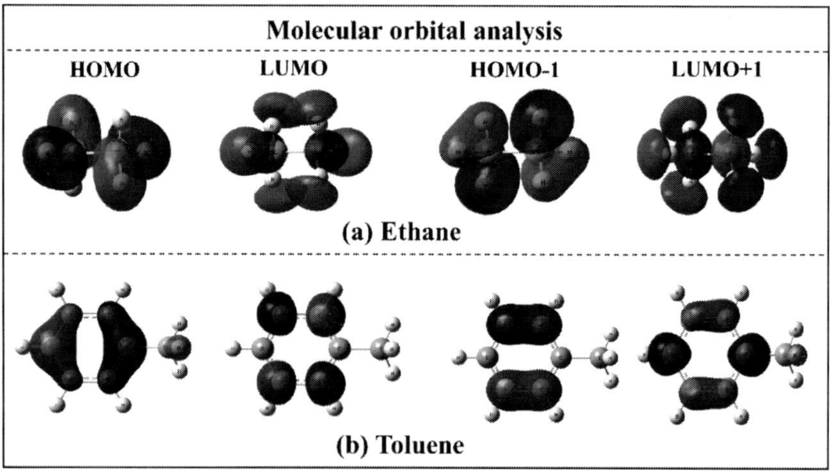

Figure 2. Visual representation of molecular orbitals (HOMO, LUMO, HOMO-1, LUMO+1) of (a) Ethane, and (b) Toluene.

The vital parameters that provide essential information about the reactivity and stability of a molecule are its lowest unoccupied molecular orbital (LUMO) and highest occupied molecular orbital (HOMO). These orbitals are crucial for gaining insight into a molecule's ability to accept or donate electrons. The HOMO indicates the molecule's ability to donate electrons, while the LUMO indicates its ability to accept electrons [7-9]. In the representation of these orbitals, regions with higher electron density are shown in dark grey, while areas of lower electron density are shown in light grey. The molecule's reactivity is influenced by it, because a larger energy gap value signifies reduced reactivity and increased chemical stability [10-11]. Conversely, a lower energy gap implies higher reactivity.

Table 1. Molecular orbital energies and energy gap of ethane and toluene

Molecular orbitals energies (in Hartree)	For Ethane	For Toluene
HOMO	-0.346	-0.236
LUMO	+0.106	+0.005
HOMO-1	-0.346	-0.245
LUMO+1	+0.165	+0.007
HOMO-LUMO gap	+0.452	+0.241
(HOMO-1) - (LUMO+1) gap	+0.511	+0.252

In the case of ethane, it was observed that HOMO was dispersed throughout the molecule, with an energy calculated to be -0.346 Hartree. In contrast, +0.106 Hartree was found to be the value of the LUMO energy. Thus, the reactivity and stability determining factor, i.e., the HOMO-LUMO energy difference ($\Delta E_{(HOMO\text{-}LUMO)}$), was calculated to be +0.452 Hartree. The HOMO-LUMO energy gap provided a comprehensive understanding of the reactivity of ethane. This gap indicated that ethane was susceptible to reactions, and that electrons can transfer among these orbitals [7-9]. Moreover, examining the adjacent molecular orbitals, it was found that the magnitude of the HOMO-1 energy was similar to that of HOMO, i.e., -0.346 Hartree, as shown in Table 1. However, the LUMO+1 energy was determined to be 0.165 Hartree; therefore, the corresponding energy gap was +0.511 Hartree. In the case of toluene, it was observed that the HOMO was spread over carbon with an energy magnitude of -0.236 Hartree. At the same time, the LUMO was distributed over notable portions of the molecular structure, except for the methyl group. For LUMO, the evaluated energy was +0.005 Hartree. Thus, the HOMO-LUMO energy gap was +0.241 Hartree. Corresponding energy of HOMO-1 was found to be -0.245 Hartree, while the energy magnitude of LUMO+1 was found to be +0.007 Hartree. Thus, +0.252 Hartree was the corresponding energy gap.

A strong correlation exists between the calculated electronic properties and the environmental behaviour of ethane and toluene. The lower HOMO-LUMO gap of toluene (+0.241 Hartree) compared to ethane (+0.452 Hartree) suggests a higher chemical reactivity, which aligns with its increased toxicity and potential to interact with biological systems [4].

Molecular Electrostatic Potential (MEP) Analysis

Another valuable tool in the study of chemical reactivity is molecular electrostatic potential (MEP) plots [10]. It enables the projection of sites of reactivity by providing a three-dimensional (3D) plot of the molecular charge distribution. This tool has proven helpful in establishing relationships between physicochemical properties and molecular structure, particularly when mapping is done in conjunction with electron density overlay for in-depth results [10-11]. Colour changes in the molecular electrostatic potential are based on the colour scheme used for MEP visualisation, with dark grey representing higher electron density and light grey denoting regions of neutral or slightly charged states [12].

The MEP plot of the ethane molecule is shown in Fig. 3(a), which allows for the visual analysis of the reaction-prone sites of the molecule. To facilitate analysis of the MEP orientations of ethane, illustrated in Fig. 3(a), two specific orientations were chosen (front view and side view). The MEP plot was generated using a range of -4.553 × 10^ (- 3) eV to 4.753 × 10^ (-3) eV. The negative potential surrounded the carbon atoms, whereas the positive potential dominated over the hydrogen atoms. It indicated the susceptibility of the hydrogen atoms to nucleophilic attack, while the carbon atoms would be susceptible to electrophilic attack [11-13]. The considered isovalue for this plot was 0.0004. The significant positive electrostatic potential surrounding the hydrogen atoms dominated over the carbon atoms in that region.

The MEP plot of toluene, shown in Fig. 3(b) with two orientations (front view and side view), can be interpreted like that of ethane. The considered range was between -2.478 × 10^ (-2) eV and 1.300 × 10^ (-2) eV during the computation and plotting. The MEP surface of toluene exhibited a continuous carbon ring of negative electrostatic potential, as depicted in dark grey, encircling the electron-rich region around the aromatic central ring. The darker colour along the edges of the molecular structure, surrounding the hydrogen atoms, indicated a positive electrostatic potential. In toluene, the presence of the methyl (CH_3) group made the net dipole moment a non-zero quantity. The net dipole moment was approximately +0.43 D towards the direction of the associated carbon atom, present in the central aromatic carbon ring (i.e., the red-colored zone of Fig. 3(b)) from the methyl group, indicating a high probability of an electrophilic attack at that higher electron density zone (red area).

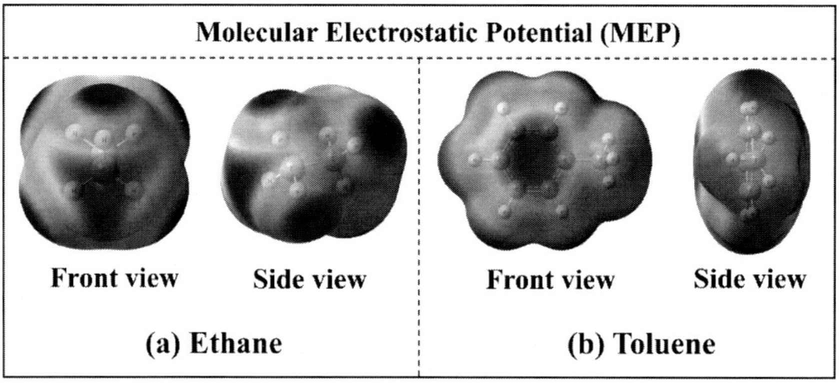

Figure 3. Molecular electrostatic potential (MEP) plots of (a) Ethane, and (b) Toluene.

Toluene exhibits electron-rich regions over the aromatic ring, making it more susceptible to electrophilic attacks and radical reactions in the atmosphere, thereby contributing to the formation of secondary organic aerosols and photochemical smog [5-6]. Conversely, ethane exhibited relative stability, resulting in a longer atmospheric lifetime and an indirect contribution to global warming by depleting hydroxyl radicals [2-3].

Band Structure Analysis

The Projected DOS spectra for ethane and toluene decomposed the total density of states (DOS) into orbital-specific contributions (s, p, d and f orbitals). The s and p orbitals exhibited significant contributions around the Fermi level, underscoring their key roles in bonding and electronic interactions [14]. Peaks in the negative energy range correspond to occupied states, primarily governed by s and p orbitals. At the same time, unoccupied states above the Fermi level (positive energy range) also reflected contributions from these orbitals [15-16]. The pronounced s-p orbital hybridisation observed in ethane further emphasised its influence on its electronic structure. Additionally, the band structures of ethane and toluene (Fig. 4(a-b)) were computed, showcasing the electronic states across the high-symmetry points (G, M, K) in the Brillouin zone, where the Fermi level is at 0 eV. These plots illustrate the distribution of valence and conduction bands relative to the Fermi level [17-19]. The band gaps for ethane and toluene were determined to be 7.699 eV and 4.784 eV, respectively.

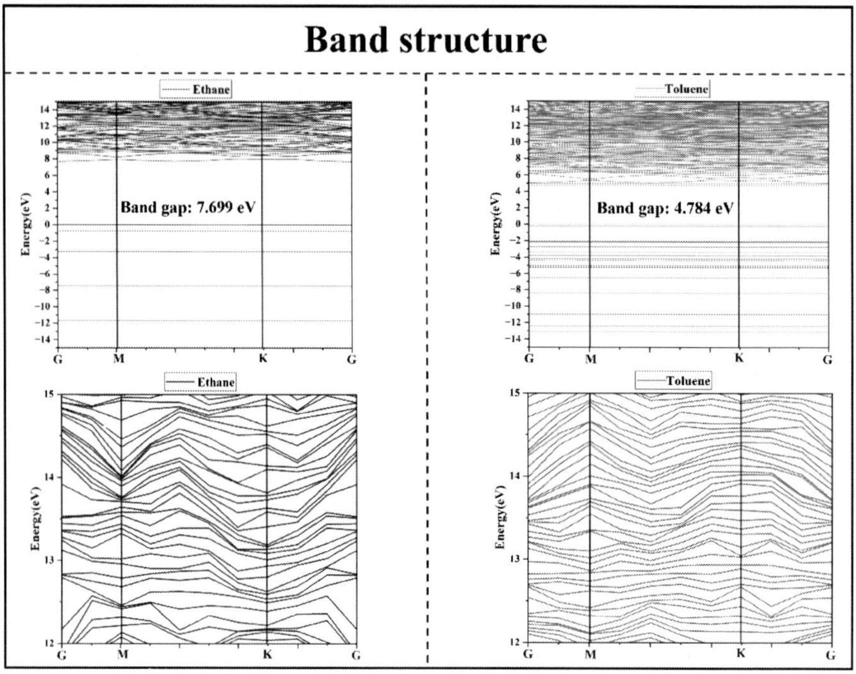

Figure 4. Band structure plots of (a) Ethane, and (b) Toluene

IR Spectra Analysis

In the IR spectra depicted, the frequency in wavenumbers (cm^{-1}) was shown on the horizontal axis. In contrast, the vertical axis displays intensity as both molar extinction coefficient (ε) (in M^{-1} cm^{-1}) on the left and dipole strength (D) (in 10^{-40} esu^2 cm^2) on the right. The peaks corresponded to specific vibrational transitions of ethane and toluene, with each peak's position and height indicating the vibrational frequency and intensity, respectively [20]. Ethane, consisting of eight atoms (two carbon and six hydrogen), exhibits 18 vibrational modes. In contrast, toluene consisting of 15 atoms (7 carbon and eight hydrogen), exhibited 39 vibrational modes in their respective infrared (IR) spectra spanning from 700 to 3500 cm^{-1} as depicted in Fig. 5. Notably, for ethane, bands were absent within the 750-1350 cm^{-1} and 1600-3200 cm^{-1} range, whereas for toluene it was in the range of 1700-3000 cm^{-1}. Five primary vibrational modes dominated the IR spectra of ethane, with strong peaks observed at 3384.38 cm^{-1}, 3267.18 cm^{-1}, 1518.04 cm^{-1}, 1394.06 cm^{-1}, and

739.30 cm^{-1}. For toluene, four major peaks were observed at 3369.66 cm^{-1}, 3081.22 cm^{-1}, 3072.26 cm^{-1}, and 815.63 cm^{-1}, while minor peaks were observed at 1575.92 cm^{-1}, 1539.03 cm^{-1}, 1503.72 cm^{-1}, and 1053.54 cm^{-1}. The limited number of IR-active vibrational modes in the case of ethane, as compared to toluene, suggested that most bond vibrations occur at nearly identical or very closely spaced frequencies, a consequence of the high symmetry within the system's structure [21-23].

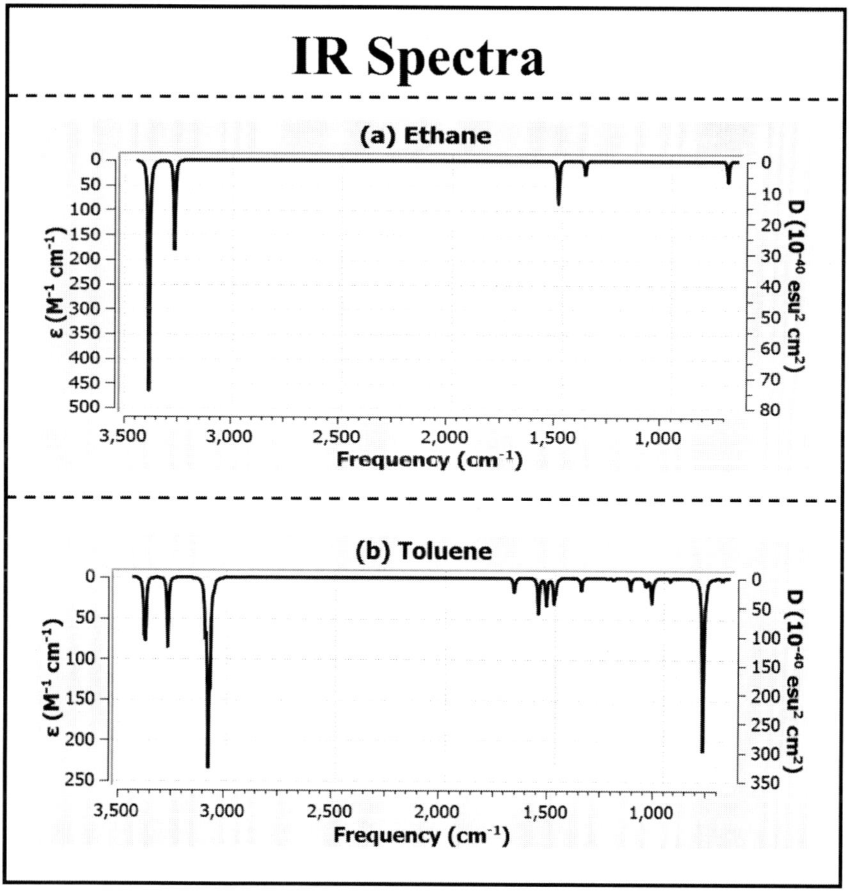

Figure 5. IR spectra of (a) Ethane, and (b) Toluene.

Comparative Analysis with Previously Reported Literature

To the best of our knowledge, no previous studies have directly analysed the molecular properties of ethane and toluene in the same context as our study. Due to the absence of direct comparisons in the literature, we have incorporated a few earlier reported/referenced studies, where similar volatile organic compounds (VOCs) or their substitutes were examined by varying computational platforms and basis sets. In particular, we focus on comparing the HOMO-LUMO energy gap of our simulated results with those reported in prior research, as tabulated in Table 2. This comparison assesses the effectiveness and reliability of our computational approach. By addressing this gap in the literature, we aim to contribute meaningful insights into the molecular properties of these compounds, particularly in the context of their toxicological and environmental relevance.

Table 2. Comparison with previously reported literature

Parameter	Molecule	Our Value	Reported molecule	Previously reported values	Reference
HOMO-LUMO energy gap (in Hartree)	Ethane	+0.452	Ethane (DFT: B3LYP)	+0.477	[24]
	Toluene	+0.241	4-chloro-3-nitro toluene	+0.220	[25]
			α,α,α-trichlorotoluene	+0.158	[26]

Conclusion

In this chapter, we made a detailed study of the two toxic industrial VOCs, ethane and toluene, focusing on their molecular orbitals, molecular electrostatic potential (MEP), band structure, and IR spectra. The molecular structures, electron density distributions, and kinetic stabilities of these two toxic molecules were examined. The energy gap between their molecular orbitals was found to be +0.452 Hartree for ethane, and +0.241 Hartree for toluene. The calculated bandgap values were 7.699 eV for ethane and 4.784 eV for toluene. These results were further supported by detailed assessments of their electronic properties, including changes in the band structure, MEP, and IR spectra. Such analyses are crucial for understanding the molecular mechanisms underlying the toxicity of these compounds and their potential environmental and health implications. These electronic properties not only define the reactivity and kinetic stability of the studied VOCs but also correlate with their observed ecological behaviours, such as persistence, degradation potential, and health impacts. Such correlations are crucial for selecting suitable materials for environmental monitoring and sensor development.

Acknowledgments

We acknowledge the support we received from Saurabhi Biswas and Souvik Bhanja from the Electronics and Communication Engineering (ECE) department at the Institute of Engineering & Management (IEM), Kolkata.

References

[1] Xiang Y, Wang H, Cheng J, Matsubu J. Progress and prospects in catalytic ethane aromatisation. *Catalysis Science & Technology* (2018), 8(6):1500-1516.

[2] Friedrich M, Beutner E, Reuvers H, Smeekes S, Urbain JP, Bader W, Franco B, Lejeune B, Mahieu E. A statistical analysis of time trends in atmospheric ethane. *Climatic Change* (2020) 162:105-125.

[3] Karczewski M, Szamrej G, Chojnowski J. Experimental assessment of the impact of replacing diesel fuel with CNG on the concentration of harmful substances in exhaust gases in a dual fuel diesel engine. *Energies* (2022), 15(13):4563.

[4] Geraldino BR, Nunes RF, Gomes JB, da Poça KS, Giardini I, Silva PV, Souza HP, Otero UB, Sarpa M. Evaluation of exposure to toluene and xylene in gasoline station workers. *Advances in Preventive Medicine* (2021) 1:5553633.

[5] Sarkingobir Y, Bello M, Yabo HM. Harmful effects of plastics on air quality. *Academia Letters* (2021) 2967.

[6] Saeedi M, Malekmohammadi B, Tajalli S. Interaction of benzene, toluene, ethylbenzene, and xylene with human's body: Insights into characteristics, sources and health risks. *Journal of Hazardous Materials Advances* (2024) 16:100459.

[7] Choudhary V, Bhatt A, Dash D, Sharma N. DFT calculations on molecular structures, HOMO–LUMO study, reactivity descriptors and spectral analyses of newly synthesised diorganotin (IV) 2-chloridophenylacetohydroxamate complexes. *Journal of Computational Chemistry* (2019) 40(27):2354-2363.

[8] Maity I, Nagasawa H, Tsuru T, Bhattacharyya P. Correlation between ammonia selectivity and temperature dependent functional group tuning of GO. *IEEE Transactions on Nanotechnology* (2020) 20:129-136.

[9] Isravel AD, Jeyaraj JK, Thangasamy S, John WJ. DFT, NBO, HOMO-LUMO, NCI, stability, Fukui function and hole-electron analyses of tolcapone. *Computational and Theoretical Chemistry* (2021) 1202:113296.

[10] Rajalakshmi K, Nayak D. HOMO–LUMO analysis of dasatinib. *International Journal of Material Science* (2017) 12:32-37.

[11] Ekramipooya A, Valadi FM, Farisabadi A, Gholami MR. Effect of the heteroatom presence in different positions of the model asphaltene structure on the self-aggregation: MD and DFT study. *Journal of Molecular Liquids* (2021) 334:116109.
[12] Shan S, Junhong W, Wu L. The pristine BN nanocone as a toluene gas sensor: a computational study. *Sustainable Chemistry and Pharmacy* (2020) 17:100277.
[13] Maity I, Chatterjee S, Bhanja S. A first-principles-based comparative study between pristine and Au-modified graphene nanosheet towards acetaldehyde sensing performance. *38th International Conference on VLSI Design and 23rd International Conference on Embedded Systems (VLSID)* (2025) 499–504.
[14] Sundaram S, Jayaprakasam R, Dhandapani M, Senthil TS, Vijayakumar VN. Theoretical (DFT) and experimental studies on multiple hydrogen bonded liquid crystals comprising between aliphatic and aromatic acids. *Journal of Molecular Liquids* (2017) 243:14-21.
[15] Gber TE, Louis H, Owen AE, Etinwa BE, Benjamin I, Asogwa FC, Orosun MM, Eno EA. Heteroatoms (Si, B, N, and P) doped 2D monolayer MoS_2 for NH_3 gas detection. *RSC Advances* (2022) 12(40):25992-26010.
[16] Maity I, Chatterjee S, Bhanja S. Acetone sensing performance of pristine and gold doped graphene sheet: A comparative analysis. *28th International Symposium on VLSI Design and Test (VDAT)* (2024) 1–6.
[17] Cao J, Zhao M, Xiao T, Xu J, Chen J, Xu W, Ma C, Ma P. Regulation of external electric field on the high-energy polynitrogen compound 1,5-diaminotetrazole-4 N-oxide. *Journal of Molecular Modeling* (2023) 29(1):28.
[18] Mazumder JT, Jha RK. Theoretical insights into gas sensing properties of MXene. *Sensors and Actuators Reports* (2023) 100174.
[19] Maity I, Bhanja S, Dutta R. Investigation on electrochemical sensing properties of Rh-decorated graphene nanosheet in the exposure of 1-propanol vapor: A DFT study. *IEEE Calcutta Conference (CALCON)* (2024) 1–6.
[20] Sambandam S, Sarangapani B, Paramasivam S, Chinnaiyan R. Molecular structure, vibrational spectral investigations (FT-IR and FT-Raman), NLO, NBO, HOMO-LUMO, MEP analysis of (E)-2-(3-pentyl-2,6-diphenylpiperidin-4-ylidene)-N-phenylhydrazinecarbothioamide based on DFT and molecular docking studies. *Biointerface Research in Applied Chemistry* (2021) 11(4):11833-11855.
[21] Zhou L, Zhu H, Zeng W. Density functional theory study on the adsorption mechanism of sulphide gas molecules on α-Fe_2O_3 (001) surface. *Inorganics* (2021) 9(11):80.
[22] Njapba AS, Galadanci GS. Electronic structure, optical and thermodynamic properties of germanium-phenyl nanocluster in the gas-phase: A DFT study. *International Journal of Engineering and Applied Science* (2021) 2(1).
[23] Maity I, Das S, Gangopadhyay M, Maity I. Understanding 2-propanol sensing mechanism of Pd modified graphene based gas sensor devices using DFT study. *37th International Conference on VLSI Design and 23rd International Conference on Embedded Systems (VLSID)* (2024) 37-42.
[24] Burmeister D. Exploring graphitic carbon nitrides for (opto)electronic applications. Humboldt-Universität zu Berlin (2023).

[25] Govindarajan M, Karabacak M, Periandy S, Xavier S. Vibrational spectroscopic studies, NLO, HOMO–LUMO and electronic structure calculations of α,α,α-trichlorotoluene using HF and DFT. *Spectrochimica Acta Part A: Molecular and Biomolecular Spectroscopy* (2012) 94:53-64.

[26] Govindarajan M, Karabacak M, Suvitha A, Periandy S. FT-IR, FT-Raman, ab initio, HF and DFT studies, NBO, HOMO–LUMO and electronic structure calculations on 4-chloro-3-nitrotoluene. *Spectrochimica Acta Part A: Molecular and Biomolecular Spectroscopy* (2012) 89:137-148.

Chapter 8

Data Storage and Energy Harvesting with Low Energy Consumption: Utilising Nonvolatile Spintronic Devices

Madhurima Sen
Koyel Ganguly, PhD
and Saswati Barman*, PhD

Department of Basic Science and Humanities, Institute of Engineering & Management, Salt Lake, Kolkata, India
University of Engineering and Management, University Area, Newtown, Kolkata, India

Abstract

Spintronic devices and solid-state batteries with improved energy conversion efficiencies and longer lifespans have been developed as a result of substantial advancements in energy harvesting that utilise spin information and energy-efficient storage devices over the past decade. These innovations have the potential to revolutionise portable electronic gadgets. Walker breakdown and low energy density are two significant obstacles that must be addressed. Optimising the stability and controllability of the magnetic domain requires the use of advanced materials engineering and nanofabrication techniques. Tunnelling magnetoresistance (TMR) in magnetic tunnel junctions (MTJs) has been used to great advantage in magnetoresistive random access memory (MRAM) and other areas of magnetic informatics. For speed, low energy consumption, and high density, Spin Hall spintronic memory systems utilise the Spin Hall Effect, which leverages the electrons' spin-orbit

* Corresponding Author's Email: saswati.barman@iem.edu.in

In: Breakthrough Research and Technologies in Physics …
Editors: Koyel Ganguly, Soumyadipta Pal and Kamakhya Prasad Ghatak
ISBN: 979-8-90134-111-7

interaction. Perpendicular magnetic anisotropy is necessary for spintronics devices to store data correctly, and domain wall memory offers low energy consumption and scalability potential. Magnetisation is controlled by spin-orbit torque devices to enhance storage inside the devices.
An antiferromagnet in domain wall memory stabilises chiral magnetic textures that enhance the efficiency and speed of spintronic devices. The magnetic anisotropy energy barrier can be controlled by voltage; hence, spintronic devices work faster. Antiferromagnets exhibit exchange interactions that can protect domain walls and increase the Walker-breakdown field outside the disturbance region. Due to the ability of nanofabrication techniques to achieve accurate fabrication, these methods are promising candidates for next-generation spintronic devices. This results in magnetic memory devices with higher storage capacities and faster speeds. In this article, we have discussed in detail the current state-of-the-art research, developments, existing bottlenecks, and future trends of using spintronics devices for data storage and memory devices.

Keywords: energy harvesting, data storage, low energy consumption, spin information, spintronic devices

Introduction

Large volumes of data are generated in our day-to-day lives, and these require huge data storage facilities. High-capacity hard disk drives and cloud storage techniques can solve this issue. Compression algorithms and data deduplication techniques can also significantly reduce storage requirements. As a result, there is a demand for research on highly efficient data storage technology. Memory devices have reduced in size considerably in the last 20 years. Nowadays, it is possible to store more data in smaller spaces on mobile devices.

These can also be exchanged online due to the rise in storage capacities. Additionally, cloud storage technologies are used to sync data across multiple devices. By utilising spintronics, researchers have developed smaller and more efficient storage devices, including hard drives and memory chips, by leveraging the spin properties of electrons.

It became possible to change the principle of operation of the memory devices as soon as giant magnetoresistance was discovered (Fert and Van Dau, 2019). In these devices, conduction of electricity, depending on the orientation of the spins, is achieved through the collective magnetisation of the localised spins in ferromagnetic layers (Fert and Van Dau, 2019; Tsymbal and Pettifor, 2001; Parkin, 1995). Binary states 0 and 1 correspond to the two distinct spin configurations, up and down spins. Hence, information in digital form can be stored. This technological breakthrough in memory technology has opened up a new approach to data storage and information processing devices. Moreover, the ability to control electronic conduction through the spin of electrons is essential for security and improving the size of memory devices. Many fields, including computing, telecom, and data storage, are all gaining profits thanks to this improvement.

It is also notable that there is a proportional increase in energy usage with this increase in storage capacity. Environmentally sustainable and economical data storage methods are put to the test with this upsurge in energy consumption. Everyone agrees that our world in digital terms is developing at a rapid pace. In all terms, this is why there must be green, efficient energy-enabling technologies to help address this widening phenomenon. Thus, big data centres, which facilitate cloud-based data storage and data exchange over the internet, account for a significant portion of the world's energy consumption (Puebla et al., 2020). However, the increase in power consumption of the major players in this cloud computing narrative, who store perpetual data, is on the rise.

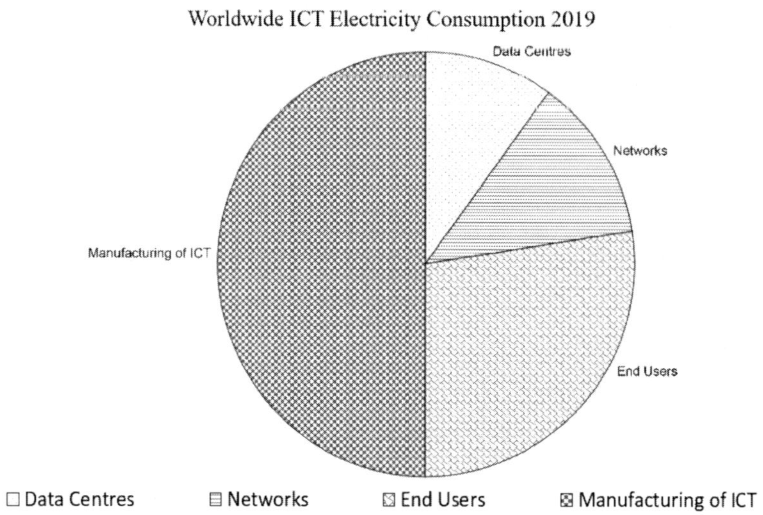

Figure 1. The information and communication technology (ICT) industry's electricity use in 2019.

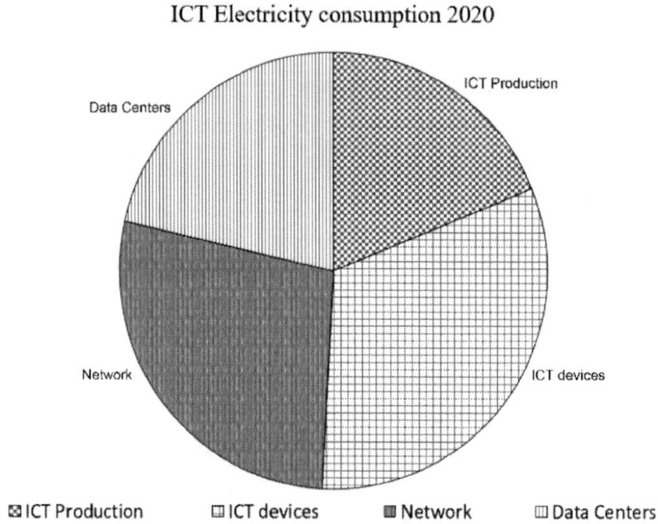

Figure 2. The information and communication technology (ICT) industry's electricity use in 2020.

Figure 1 and 2 represents the electricity consumptions by the information and technology sector. This calls for immediate intervention in finding and designing strategies that promote the energy efficiency of data storage systems and minimise their impact on the environment. To bring about environmental sustainability with these facilities, wind or solar energy generation systems may also be implemented. Moreover, the introduction of newer cooling facilities and proper management of maximum server workloads can also improve energy efficiency and reduce the anticipated increase in power usage.

Systems that provide additional energy to the information and communication technology (ICT) sector, as well as those that enhance energy-saving industrial practices, are also beneficial in efforts to reduce global energy consumption further. This increase in demand for ICTs can be addressed by encouraging energy efficiency and conservation measures, such as powering servers with renewable energy and reducing the number of underutilised servers through server virtualisation. Moreover, creating awareness and educating people about the responsible use of ICT and its peripherals may help both individuals and businesses mitigate their contribution to environmental degradation. Around half of the energy consumed by a typical data centre is devoted to powering the cooling infrastructure that supports the centre's servers, which account for 26% of the total energy consumption. This situation calls for a design available in most major data centres, which utilises modern cooling methods because active cooling accounts for a large percentage of energy consumption. In addition, the efficiency of energy consumption in these centres can be further improved by applying energy-efficient computing devices, including their servers and solid-state disks. Not only are there energy supply problems facing the information and communication technology industry, but there are also growing concerns nowadays regarding public health and ecological impacts, particularly global warming and greenhouse gas emissions.

Spintronics Logic Devices and Memory for Data Storage

Currently, digital information storage technologies include hard disk drives (HDDs), flash memory, and dynamic random-access memory (DRAM), which have weaknesses in terms of speed, power consumption, and data density. These limitations, however, can be mitigated by spintronics, which stores and processes information through the spin of electrons (Puebla et al., 2020). In other words, spintronics can revolutionise the field of digital information storage by uncoupling spins, paving the way for future storage devices. The primary issue with HDDs is the mechanical movement of the read/write head and the storage disk (Puebla et al., 2020). Reliability is compromised by mechanical movement, as the operation rate is slowed and power consumption increases. In contrast, DRAM, which is devoid of moving parts, is highly efficient in terms of handling I/O operations per second due to its speed in random access. However, DRAM does not serve as a secondary storage system as its primary functionality is employed in temporary storage or trafficking while processing, known as volatile memory. Raw IOPS data for DRAM, in terms of read/write operations per second, are very high, often ranging in the millions or even billions. Therefore, DRAM is capable of multitasking at lower power consumption than that observed for SRAM.

Nevertheless, engineers are still searching for an optimal substitute for DRAM due to its volatility, which means that the capacitor releases electrical charge information when the device is turned off. This instability of DRAM results in the loss of data most of the time; in each execution of functions and instructions, one is forced to save and back up the data constantly. Also, the constant power required to keep data in DRAM may pose a disadvantage in terms of energy consumption compared to memory using spintronics (Apalkov et al., 2013). However, one disadvantage of flash memory is that it has lower data retention compared to other types of memory. Additionally, it is nonvolatile, as it does not use capacitors. That may be why flash or Nonvolatile memory is used for storage in portable devices like smart mobile phones and USB drives, as it is not affected when power is switched off. This makes it more convenient for long-term storage and eliminates the hassle of saving and backing up. Additionally, capacitors are not used in flash memory, unlike in DRAM, which helps save energy during use, making it more energy-efficient than DRAM.

For this reason, we consider spintronic ideas and technologies as promising substitutes, as they demonstrate nonvolatile properties with unlimited endurance.

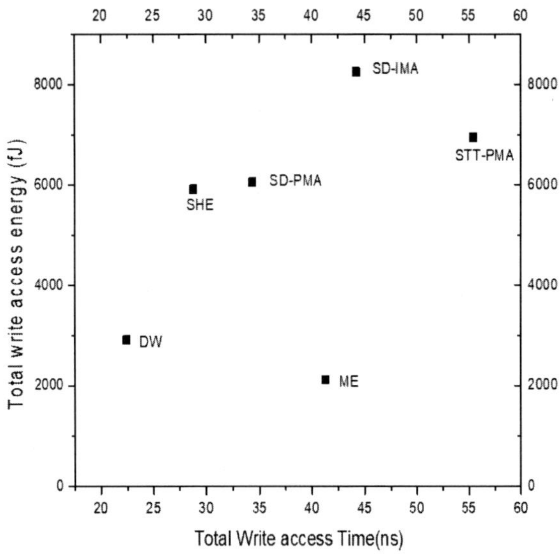

Figure 3. Performance estimates for different spintronic memory cells, expressed as energy against time. SD: spin diffusion, STT: spin-transfer torque, SHE: spin Hall effect, DW: domain wall motion, ME: magnetoelectric switching, IMA: in-plane magnetic anisotropy, PMA: perpendicular magnetic anisotropy (Puebla et al., 2020).

Figure 3 illustrates the performance estimates of various spintronic memory cells. Note that the nonvolatile nature of the storage implies that the data can be preserved even if the power supply is cut off; this makes it even more efficient and easy to use when storing digital data. In contrast to flashes, used in most modern spintronic devices, data is not lost over time, and the storage device's performance remains stable. This makes them highly desirable for applications that require long-term storage and large amounts of data, such as archival systems or cloud servers. Moreover, spintronic devices make a good fit for use as energy-efficient replacements for traditional memory solutions. Currently, spintronics-based storage devices, such as spin-transfer torque magnetic random-access memory (STT-MRAM) (Puebla et al., 2020; Sbiaa et al., 2011), do not require mechanical elements. This not only improves reliability and speed but also minimises power consumption. Additionally, spintronics has the added benefit of being less susceptible to external magnetic fields, thereby reducing the likelihood of data corruption.

In addition, due to the nonvolatile characteristic of spintronics-based memory, the start-up time will be short, and frequently used applications or files can be accessed easily.

Main trends in writing time and energy aspects associated with the basic phenomena in nonvolatile, spin-based magnetic memory with high endurance are described in terms of performance standards. These guiding principles contain important information concerning possible optimisation of energy consumption and speed for spintronic devices. Also, the fact that the spin-based memories can store information even when there is no power supply makes them quite reliable, especially in applications requiring stringent data accuracy. In magnetic devices, the most critical consideration is between perpendicular magnetic anisotropy (PMA) (Honolka et al., 2012) and in-plane magnetic anisotropy (IMA) (Zhang, 2000). PMA is superior in some aspects compared to IMA, including faster performance and lower energy consumption. On the other hand, the IMA device is likely to be slower and consume more energy. Compared with the in-plane magnetic anisotropy devices, such as spin-diffusion writing and SHE memory (Liu et al., 2022), the first one has the benefit of demonstrating low power consumption. Furthermore, spin-diffusion writing and SHE memory have been identified as providing faster data processing than those supplied in MAG anisotropy devices. This makes them less energy-intensive, which is beneficial for applications that require extended periods of use or operate on battery power. These are suitable in cases where vast amounts of information need to be processed within the shortest time possible, for example, in computational time. Additionally, their speed is still attributed to the fact that attaining the overdrive condition is relatively easy compared to the critical current. Here, the term “SHE” encompasses a wide range of spin-orbit torque and not just bulk spin-orbit. Moreover, the high variability of SHE means that it is easier to incorporate other materials and structures into the system, which extends the possibilities of adjusting devices for specific application needs.

Domain wall (DW) memories (Kumar et al., 2022) are expected to operate with comparatively low energy, as they are designed to toggle at considerably lower driving currents. This is because DW memories utilise the movement of magnetic domain walls to store and retrieve data, resulting in reduced power consumption. Furthermore, DW memories can maintain data even when power is off, making them energy-efficient and suitable for portable structures or devices, as well as systems that are sometimes powered down. Remember that the domain wall devices under consideration herein include a domain wall motion that spans only one RAM cell. Thus, incorporating domain wall motion into a single cell of RAM results in faster data access than with other conventional RAM technologies. Moreover, due to the scalability of domain wall devices, their future use in new-generation memory applications with enhanced capacity and functionality is credible. Unlike the RAM, racetrack memory (Gu et al., 2022) is a serial memory device in which a domain wall passes over different positions containing bits (Fert and Van Dau, 2019; Puebla et al., 2020). Compared to RAM, racetrack memory lacks random-access ability. However, it holds promise for a promising future memory technology, offering higher storage density and superior read-write operation capabilities. Despite being different from other types of memory, it has several advantages, including the flexibility of data access and the potential for further utilisation. It can become the most interesting candidate for the role of the next-generation memory.

This is true because magnetoelectric (ME) memories depend on capacitor charging rather than torque created by current, and therefore cost the least amount of energy out of all the types considered here. In comparison to other data storage techniques, ME memories are also nonvolatile, meaning they can store data even in the event of a momentary power failure. They are therefore ideal for use in scenarios that demand quick access to data and operating environments, such as in mobile devices and data centres. Switching speed may be slightly worse, depending on the value of the ME field. However, it is worth mentioning that only spin-transfer torque (STT) memories are currently available on the market (Fert and Van Dau, 2019; Puebla et al., 2020; Wong and Salahuddin et al., 2015). The ME memories include STT memories, which have now been liked due to their low power consumption and nonvolatile nature. They are packaged to provide dependable and effective solutions for various uses, such as mobile devices and data centres. Moreover, recent research is being conducted to enhance the switching speed of MRAMs, thereby improving their performance. The remaining ones are at different levels of investigation and development, with ME memories being at the lowest level. Switching speed can be slightly slower, depending on the value of the ME field. It is also worth noting that only spin-transfer torque (STT) memories are currently available commercially. The switching speed may be slower than that and depends on the value set in the ME field. The remaining ones are at different stages of research and development, where ME memories have yet to undergo a considerable level of research.

Notably, MTJ (Samanta et al., 2024) structures can be considered the origin of most studies and new developments in memory technology. These structures offer prospects for higher switch rates and a desirable power consumption profile as compared to STT memories. Additionally, further exploration is conducted to determine the optimal value of the ME field for achieving higher density ME memories. MTJ structure consists of a free layer, which is a ferromagnetic material, and a fixed layer, which is also a ferromagnetic material separated by an oxide barrier layer. While the orientation of the magnetisation in the free layer can be altered by applying an external magnetic field, the orientation of the magnetisation in the pinned layer remains constant. The tunnelling barrier allows electrons to tunnel from one layer to another, and this change in resistance is used for information storage. The distinctive style of design enables the introduction of nonvolatile memory, thereby mitigating all the other biases that point to its high stability and scalability prospects. The operation of reading is based on the property known as tunnelling magnetoresistance (TMR), in which a low resistance current is formed when there is a parallel magnetisation state and a high resistance current when there is an anti-parallel magnetisation state. TMR depends on magnetisation in the two layers of material of the spin valve.

In particular, the increase of the TMR is achieved when MgO is applied as the tunnelling barrier material. The tunnelling barrier is enhanced by MgO, resulting in a higher TMR due to improved electron transport in the layers. This raised tunnel magnetoresistance ratio (TMR) enables the efficient and accurate reading of data recorded in the memory device. The real-world applicability provided by this unique design may also shape its prospects for enhancing memory technology, particularly in applications involving large databases.

While using the spin transfer torque to write data, a horizontal magnetisation flipping has been achieved, allowing the fabrication of perpendicular MTJ through the extraction of the ferromagnetic layers. The development of reduced spin-transfer torque technology opened up the possibility of further improving the performance and energy efficiency of memory devices. This is important because it aligns the ferromagnetic layers perpendicular to each other, resulting in a faster and more reliable data writing method than any other, which makes it very suitable for military high-density storage applications. The heat resistance and spatial scalability of the MTJ structures were a concern. Still, the inherent disadvantage of writing, even after that performance gain of the MTJ structure, is the structure of the MTJ itself. The limitation is that to reverse the magnetisation, a high current density is required, which may lead to increased power consumption and device wear. Consequently, further research is needed to shed more light on this issue and to enhance writing in MTJ structures without compromising high-performance efficiency. The MTJ must remain passive while the spin transfer torque action proceeds and shapes a magnetic profile.

Another critical point is that during this incubation time, which the MTJ requires to delay magnetising itself in other directions, it means slower overall efficiency during the writing process. Therefore, it is beneficial for the MTJ to create the means that will reduce this incubation time and increase the magnetisation switching rate. As mentioned, it is challenging to quantify writing times less than 10 ns: while the write time is a variable that can be controlled, the incubation time is not, and heat agitation is. Employing advanced materials with greater thermal stability allows for the minimisation of incubation time, as well as enhancement of magnetisation switching rates in structures based on MTJ technology. Perhaps, one can write more quickly by using materials that can absorb more heat, and that is indeed the key to the matter. Another potential for reducing the incubation time and enhancing performance can be observed in the improvement of design and fabrication technology for MTJ structures. Today, spin-transfer torque STT starts operating after some time, in the course of writing, and there is an incubation period for the MTJ. Stability and scale of the MTJ structures in temperature and space have therefore been improved. Due to this outcome, the Spin-based MTJ structures also had the opportunity to enhance the thermal stability and spatial scalability of the structures. However, the MTJ structure has a 'built-in' problem in the writing process, even where overall performance has been doubled.

Eventually, the relatively slow writing speed becomes a problem for low operation power. The time required to achieve the correct magnetic state is a limitation in the writing process for the MTJ. Hence, attempts are being made to fine-tune the writing speed of the MTJ structure to eliminate this weakness, and its efficient low-power operation will be achieved. This problem can be addressed through redesign from a perpendicularly oriented magnetisation, which reduces the MTJ write necessary time to insert a desired magnetic state, thus making it possible to operate at a faster speed; however, in-plane magnetisation is not without challenges, especially in the case of STT-MRAM technology, so more study is required in this area. Rapid switching requires high currents to be passed across an MTJ's thin oxide barrier, which increases the rate of ageing in the barrier and thus causes reliability issues. Additionally, the high current flowing through the thin oxide barrier also causes excessive heat generation, resulting in thermal instability in the MTJ. Moreover, if in-plane magnetisation is accommodated, it would require optimisation of material properties and device design to enable efficient and reliable operation, which would be continuously ongoing through research and development programs. The other possibility could be by employing spin conversion and the SOT switching mechanism to solve the problems faced. Thus, if one harvests the spin conversion and the SOT switching mechanism, then the challenges of MTJs, as mentioned above, due to excessive heat generation and issues of thermal stability, can be addressed. This may become a more efficient and reliable alternative solution for future research and development in this domain. Significant progress in the design and working mechanism of MTJ structures was made possible by spin conversion. Spin conversion has enabled the development of more efficient and reliable MTJ structures, resulting in enhanced performance in various electronic devices. Furthermore, the understanding of spin conversion has paved the way for further research and innovation in the field of spintronics, promising even more exciting advancements in the future. Unlike traditional MTJ devices, in MTJ structures under spin conversion, the charge current does not have to pass through a high-resistive tunnelling barrier (Puebla et al., 2020). This allows for faster and more efficient data transfer, making spin conversion MTJ structures highly desirable for applications such as magnetic random-access memory (MRAM) and magnetic sensors. Additionally, the elimination of the high-resistive tunnelling barrier reduces power consumption, making these devices more energy-efficient and environmentally friendly. Consequently, sub-ns magnetisation switching can occur, enabling rapid writing at low incident power. Furthermore, the use of spin conversion MTJ structures in MRAM and magnetic sensors also enhances their durability and reliability (Puebla et al., 2020). The elimination of the high-resistive tunnelling barrier minimises the risk of data corruption or loss, ensuring long-term data storage and accurate sensing capabilities.

A new class of Spintronic devices, known as spin-orbitronics (Han et al., 2018), exploits a conversion from spin to charge due to spin-orbit coupling (SOC). Devices based on spin-orbitronics offer vast opportunities for realising applications in data storage, logic circuits, and power-efficient computing. These devices operate by exploiting the effect of SOC, thus running faster and more energy efficiently than traditional electronics. On the other hand, spin-orbitronics also enables new functionalities, such as nonvolatile memory and logic operations based on spin. Research and development in this area are critical to unlock the potential of spin-orbitronics technology. For bulk systems, the spin Hall effect was first proposed in the early days of research. It starts an opposing spin-dependent trajectory for opposite spin states. The early experiments succeeded in the observation of spin conversion caused by the spin Hall effect in GaAs or Pt (Puebla et al., 2020). The successes of STT from heavy materials like Pt, W, or Ta in creating magnetic heterostructures have proven the utility of the spin Hall effect in spin manipulation (Puebla et al., 2020). This eventually led to the realisation of SOT devices, exploiting the spin-Hall effect to steer the magnetisation direction in nanoscale magnetic structures. SOT devices have shown great promise for applications in nonvolatile magnetic memory and spin-based logic devices, paving the way for further advancements in spin-orbitronics technology. On the other hand, spin texture interfaces provide an extra spin conversion mechanism. In bulk systems, interfaces such as Rashba interfaces and topological insulators exhibit spin conversion efficiencies that are comparable to or superior to those of the SHE. These spin texture interfaces offer potential advantages in terms of energy efficiency and reduced power consumption. Additionally, their compatibility with existing semiconductor technology makes them a promising candidate for integration into future electronic devices.

Even though the spin conversion MTJ has its benefits, we face challenges in some problems, such as the assistance field for magnetisation switching and the comparatively large device sizes. It may be a challenge to resolve the issue concerning the assistance field required for magnetisation switching in spin conversion MTJs. The presence of this additional field may increase the device's complexity and power consumption. In addition, these devices' relatively large size may make it harder to place them within even smaller electronic systems, which would represent yet another challenge to their general adoption. Remarkably, new research offers substitutes that do not require the auxiliary field by utilising the spin-polarised current produced by two-dimensional systems or two-terminal spin-orbit torque memory arrangements. All of these options present a possible solution because they lower power consumption and the complexity of devices and avoid an additional field. Furthermore, they have a smaller size and hence, are easier to incorporate into a wide range of electronic devices. Thirdly, spin conversion aims at reducing the switching energy. A hybridisation of anisotropy with voltage-controlled magnetisation was considered to be an effective method of lowering switching power, according to a recent work published (Weiler et al., 2009). The designed approach presents an efficient way of reducing the switching power down to the required level through the benefits of anisotropy combined with voltage-controlled magnetisation. This method can further enhance the efficiency and functionality of electronic devices, making it a promising field for research and development in spin conversion technology. The capabilities of ferrimagnet-based materials, which improve spin torque efficiency and effective spin conversion in highly conductive materials, open up interesting opportunities for low-power spin manipulation.

However, ferrimagnets introduced into spin conversion technology will open channels for use in high-tech applications, especially in magnetic memory and even in spin-based logic circuits. Because of this mode of advancement, computing systems could be faster and even more energy-efficient, which can revolutionise the world of electronics. An alternative suggestion for further increasing data density while maintaining thermal stability is to investigate domain walls (DW) in three-dimensional nanowire configurations, rather than the two-dimensional operation seen in existing MRAM technology. By investigating domain walls in three-dimensional nanowire structures (Skoric et al., 2022), researchers have a compelling reason to increase their data density, as additional space is available for data storage within these domains. It also leads to improved thermal stability since the extra dimension can serve as a means for more accurate control and manipulation of the domain walls. We refer to this new idea as racetrack memory. The current-induced domain wall motion records a memory bit corresponding to a magnetic domain in racetrack memory. The current-induced domain wall motion enables fast and efficient data transfer within the racetrack memory. Additionally, the use of three-dimensional nanowire arrangements in racetrack memory has the potential to overcome some of the limitations of traditional two-dimensional. MRAM technology is susceptible to magnetic field interference. A minimal incident current and quick domain wall movement are required to establish an effective racetrack memory. An improvement in the material properties and design of the nanowires could minimise the resistance and maximise the efficiency of current-induced domain wall motion. New techniques, including spin-orbit torque and Skyrmions, are also being explored to increase the performance and reliability of racetrack memory. Combining the DMI (Robert et al., 2023) mechanism with an antiferromagnetic arrangement significantly increases the velocity of domain wall motion, according to a more recent publication. Such a mixture enables faster and more stable propagation of domain walls, thus offering greater promise for the benefit of racetrack memory technology. Other approaches focus on reducing racetrack memory power consumption, for example, by discovering new materials with a smaller resistivity and by using the latest fabrication techniques such as self-assembly and bottom-up synthesis. Although exciting, the intrinsic randomness of the motion of perpendicularly aligned magnetic layers under a moderate driving force, caused by the creep of domain walls, constitutes a hurdle to achieving racetrack memory. Randomness in the motion of the domain walls due to creep may be a source of error in data storage and retrieval, thereby reducing the reliability of racetrack memory. Hence, further study should be conducted to understand and control this stochastic behaviour, utilising more sophisticated algorithms for advanced control or exploring new magnetic materials with higher stability. New observations of high-speed domain wall motion in ferrimagnets have proved promising. Rapid motion of ferrimagnetic domain walls has displayed characteristics that should be worthy of being used as an alternative that may remedy the stochastic behaviour discovered in racetrack memory. The racetrack memory system might therefore produce an increase in the reliability and stability of data storage and retrieval if this mechanism is explored and exploited. Furthermore, magnetic Skyrmions generated by DMI are emerging as an exciting path due to their comparable working mechanism, lower critical energy for motion, and other fascinating functions, including spin Hall-like driving dynamics. These magnetic Skyrmions have shown great potential in overcoming the limitations of traditional racetrack memory systems. They offer a more efficient and reliable method of storing and retrieving data, thanks to their unique working mechanism and minimised energy requirement for motion. Additionally, their spin Hall-like driving dynamics offer new avenues for further development in racetrack memory technology.

Conclusion

Energy harvesting using spin information and energy-efficient storage devices has significantly advanced over the past decade, enabling more efficient and sustainable energy generation and storage solutions. However, there are still challenges to overcome, such as the Walker breakdown, a driving force threshold that affects the internal magnetisation of the domain wall. Spin current injection can provide additional energy to the domain wall, thereby increasing its velocity. Advancements in materials engineering and nanofabrication techniques are crucial for optimising the stability and controllability of magnetic domains for reliable data storage. The Spin Hall Effect, where an electric current generates a spin current, is being investigated in spintronic memory devices. This effect leverages the spin-orbit coupling of electrons, allowing the electron spins to be controlled without influencing them, using an external magnetic field. Spin-orbitronics can serve as a basis for the development of new generations of memory devices that offer high speed, energy efficiency, and storage capacity compared to modern equipment. Contemporary spintronics-based storage devices are the Spin-transfer torque magnetic random-access memory (STT-MRAM). Spintronics devices require perpendicular magnetic anisotropy for stable data storage, and domain wall memory has the potential for scalability and energy efficiency. Spin-orbit torque devices operate based on the spin-orbit coupling effect to control magnetisation for efficient switching between magnetic states in data storage. DMI with an antiferromagnet helps to stabilise chiral magnetic textures, which have potential for spintronics given their stable manipulation of data. Magnetic anisotropy energy barrier control is possible through voltage control, which makes spintronic devices more energy-efficient and faster for data storage. Antiferromagnets, unlike ferromagnets, have exchange interactions that shield domain walls and raise the Walker-breakdown field to values outside of the disturbance zone. Nanofabrication techniques enable precise manipulation of antiferromagnetic materials, allowing for devices with enhanced stability and controllability of magnetic domains. This high-speed operation limit makes antiferromagnets promising candidates for next-generation spintronic devices, where fast and efficient data processing is crucial. Antiferromagnets exhibit intriguing spin-orbit phenomena, such as the magnetically controlled spin Hall effect, which facilitates easier control operations. They are also immune to external magnetic fields, making them suitable for applications in harsh environments or high-performance computing systems where magnetic field fluctuations can be a concern.

Acknowledgments

We express our heartfelt thanks to the Department of Basic Science and Humanities, IEM, and Prof. Satyajit Chakrabarti, Director of IEM, for all their support and for the facilities made available to complete this work.

Funding Sources

SB would like to acknowledge the financial support from the Science and Engineering Research Board (SERB) for the project CRG/2018/002080.

Conflict of Interest

The author(s) declare no conflict of interest.

References

Apalkov D, Khvalkovskiy A, Watts S, Nikitin V, Tang X, Lottis D, Moon K, Luo X, Chen E, Ong A, Driskill-Smith A, 'Spin-transfer torque magnetic random access memory (STT-MRAM),' *ACM Journal on Emerging Technologies in Computing Systems (JETC),'* (2013) 29, 1-35.

Camley RE, Livesey KL. Consequences of the Dzyaloshinskii-Moriya interaction. *Surface Science Reports*. (2023) 78(3):100605.

Fert A, Van Dau FN., 'Spintronics from giant magnetoresistance to magnetic skyrmions and topological insulators,' *Comptes Rendus Physique* (2019) 20(7-8):817-831.

Gu K, Guan Y, Hazra BK, Deniz H, Migliorini A, Zhang W, Parkin SS. Three-dimensional racetrack memory devices designed from freestanding magnetic heterostructures. *Nature nanotechnology*. (2022) 17(10):1065-71.

Han W, Otani Y, Maekawa S. Quantum materials for spin and charge conversion. *npj Quantum Materials*. (2018) 3(1):27.

Honolka J, Khajetoorians AA, Sessi V, Wehling TO, Stepanow S, Mi JL, Iversen BB, Schlenk T, Wiebe J, Brookes NB, Lichtenstein AI, 'In-plane magnetic anisotropy of Fe atoms on Bi 2 Se 3 (111)', *Physical Review Letters*, (2012) 25, 256811.

Kumar D, Jin T, Sbiaa R, Kläui M, Bedanta S, Fukami S, Ravelosona D, Yang SH, Liu X and Piramanayagam SN, Domain wall memory: Physics, materials, and devices. *Physics Reports*, (2022) *958*, pp.1-35.

Liu R, Gupta K, Yuan Z and Kelly PJ, Calculating the spin memory loss at Cu| metal interfaces from first principles. *Physical Review B*, (2022) *106*(1), p.014401.

Parkin SS. 'Giant magnetoresistance in magnetic nanostructures,' *Annual Review of Materials Science*, (1995) 357-388.

Puebla J, Kim J, Kondou K, Otani Y. 'Spintronic devices for energy-efficient data storage and energy harvesting,' *Communications Materials*, (2020) 24, 1.

Samanta K, Jiang YY, Paudel TR, Shao DF, Tsymbal EY. Tunnelling magnetoresistance in magnetic tunnel junctions with a single ferromagnetic electrode. *Physical Review* B. (2024) 109(17):174407.

Sbiaa R, Meng H, Piramanayagam SN, 'Materials with perpendicular magnetic anisotropy for magnetic random access memory,' *Physica Status Solidi (RRL)–Rapid Research Letters*, (2011) 12, 413.

Skoric L, Donnelly C, Hierro-Rodriguez A, Cascales Sandoval MA, Ruiz-Gómez S, Foerster M, Niño MA, Belkhou R, Abert C, Suess D, Fernández-Pacheco A. Domain wall automotion in three-dimensional magnetic helical interconnectors. *ACS nano* (2022) 16(6):8860-8.

Tsymbal EY, Pettifor DG, 'Perspectives of giant magnetoresistance in Solid state physics' *Academic Press*. (2001) 56, pp. 113-237.

Weiler M, Brandlmaier A, Geprägs S, Althammer M, Opel M, Bihler C, Huebl H, Brandt MS, Gross R, Goennenwein ST. Voltage controlled inversion of magnetic anisotropy in a ferromagnetic thin film at room temperature. *New Journal of Physics* (2009) 11(1):013021.

Wong HS, Salahuddin S. Memory leads the way to better computing. *Nature nanotechnology*. (2015) 10(3):191-4.

Zhang S., 'Spin Hall effect in the presence of spin diffusion,' *Physical Review Letters*, (2000) 85, 393.

Chapter 9

A Motion-Triggered Intelligent Lighting System for Sustainable Energy Usage on Airport Runways

Mandakinee Bandyopadhyay[1,*], PhD
Priyanshu Paul[1]
and Alakananda Bandyopadhyay[2]
[1]Department of Electrical Engineering, Institute of Engineering & Management, University of Engineering and Management, Kolkata, West Bengal, India
[2]Department of Civil Engineering, Calcutta Institute of Technology, Uluberia, West Bengal, India

Abstract

In the 21st century, the optimal use of energy and the promotion of energy conservation and management remain pressing issues in our society. The existence of fossil fuels is decreasing day by day. Therefore, we are turning our attention either towards non-conventional sources of energy or the optimal use of electricity. The motion-triggered automated light turns on when it detects motion using the ultrasonic sensor. It is commonly used for staircases, washrooms, and outdoor areas, to switch on lights automatically once it detects the motion of a human being. It measures the distance to an object by emitting ultrasonic sound waves and detecting the object's presence. In this chapter, an attempt has been made to incorporate an ultrasonic sensor within the streetlights at the airport runway to detect the motion of the Airbus as it passes by. When the Airbus passes through, the sensor will detect its presence from a

* Corresponding Author's Email: mandakinee.bandyopadhyay@iem.edu.in

In: Breakthrough Research and Technologies in Physics …
Editors: Koyel Ganguly, Soumyadipta Pal and Kamakhya Prasad Ghatak
ISBN: 979-8-90134-111-7

certain distance, and the nearby lights will glow. Thus, the condition of the Runway will be visible to the pilot, even if there is no need to switch on all the lights of that Runway during the evening. The glowing of light is controlled by the 5V relay, which is connected to the Arduino Uno. This measure will help develop an optimised use of energy for sustainable development and the conservation of electrical energy.

Keywords: energy conservation, ultrasonic sensor, streetlights, airport runway, Arduino UNO

Introduction

Airports are among the largest consumers of energy, with runway lighting systems playing a crucial role in ensuring safe takeoff, landing, and taxiing operations. However, traditional lighting systems operate continuously, resulting in excessive energy consumption and increased operational costs. As global efforts toward sustainability and energy efficiency increase, there is a growing need for intelligent solutions that optimise resource usage without compromising safety. The number of Airbus aircraft is rising every day as we opt for air travel as the most convenient mode of transportation. In this regard, there is a possibility of air traffic congestion at the Runway. There could also be an intrusion in the Runway. The most critical sense of intrusion is the takeoff crossing. In this situation, the speed of the takeoff aircraft is increasing, and it conflicts with another crossing aircraft; there is a great possibility that a serious collision will happen. The Runway Status Lights (RWSL) system is an independent operating system that automatically controls field lights using information about the location of the field's active targets. The RWSL system leverages traffic situation information from the field and terminal area, which is gathered through multi-point positioning, field surveillance radar, and approach radar, to identify the presence and movement patterns of aircraft and ground vehicles on the Runway and in its vicinity. It assesses the current safety status of the Runway by integrating this data with established runway operation protocols. (Weidong Liu, 2021). The Runway Status Light System is the most efficient method for preventing runway incursions, utilising red lights to provide pilots with a direct alert regarding runway conditions. The reliability and availability of the system are closely related to the quality of the traffic surveillance data. Under certain surveillance conditions, whether the pilot can take actions to avoid collisions is related to whether the system can illuminate the light in a timely and correct manner. If it is unsafe to enter or cross the Runway, the system will activate the red Runway Entry Lights (REL). If the aircraft on the Runway should not take off or land, the system will turn on the red Takeoff-Hold Lights (THL). When the Runway is safe, the system will automatically turn off all the lights. (W. Li,2020). When a takeoff-crossing conflict occurs, depending on the traffic situation, either the takeoff aircraft or the crossing one needs to take action to avoid a collision. Currently, most airports in the country rely on manual means for pilots to obtain the traffic situation. Alternatively, pilots can receive alerting information from controllers at airports that have a controller-warned alerting system. A motion-triggered intelligent lighting system presents a viable approach to reducing unnecessary energy expenditure on airport runways. By integrating motion detection sensors, automated controls, and adaptive illumination techniques, such a system can ensure that lights are activated only when necessary. Several studies have explored energy-efficient lighting in various sectors, demonstrating that sensor-based control mechanisms can significantly reduce power consumption while maintaining functionality (Smith et al., 2020; Chen & Zhao, 2021).

In this paper, an attempt has been made to turn on the nearby lights of the aircraft for clear visibility of the Runway. Not all the lights will glow constantly to conserve energy. The global energy landscape is continually changing due to environmental degradation (X. Chang, 2021). Improving the energy structure and energy consumption pattern has become an international problem. The system utilizes advanced motion sensors and real-time data processing to adjust runway illumination in response to aircraft movement dynamically. In this project, reducing energy consumption is also a key consideration, as sustainable energy development is a global challenge today. A motion-triggered automatic lighting system is developed. The system detects motion using ultrasonic sensors and controls the illumination of an LED or a light bulb. When the aircraft flies through a particular area, the lights in the vicinity will glow, making the Runway visible to the pilot. The sustainable use of energy is also a key factor. When any aircraft passes by, only the lights will glow; otherwise, they will not glow. The consumption of energy could be minimised here. Lian Chi et al., (Lian Chi, 2020) evaluate and conduct a comprehensive study on the background, concept, and technical framework of EI, while also addressing the practical application of EI in industrial parks. D. Menniti et. al (D. Menniti, 2018), Chencho Dema et. al. (C. Dema, 2024) built a prototype of an intelligent traffic control system for traffic control employing an AI camera and a ToF (Time-of-Flight) sensor.

The AI camera detects people waiting at a pedestrian crossing and counts the number of them. The ToF sensor detects vehicles stopping at the intersection. Based on the number of people waiting and the detection of cars, traffic and pedestrian lights are controlled.

Method of Approach

In this paper, a motion-triggered automatic light system is introduced. The proposed system utilises an Arduino UNO microcontroller for automatic on-off control of the light. The system detects motion using ultrasonic sensors, specifically the HC-SR04. High-frequency sound waves are emitted, and the ultrasonic sensor measures how long it takes for the sound to return to the sensor. The sensor uses the speed of sound and the return time to determine the distance between the object and itself.

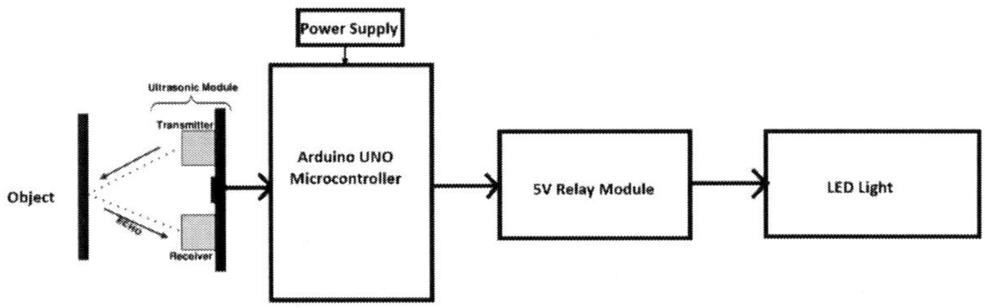

Figure 1. Block diagram of the proposed automatic lighting system.

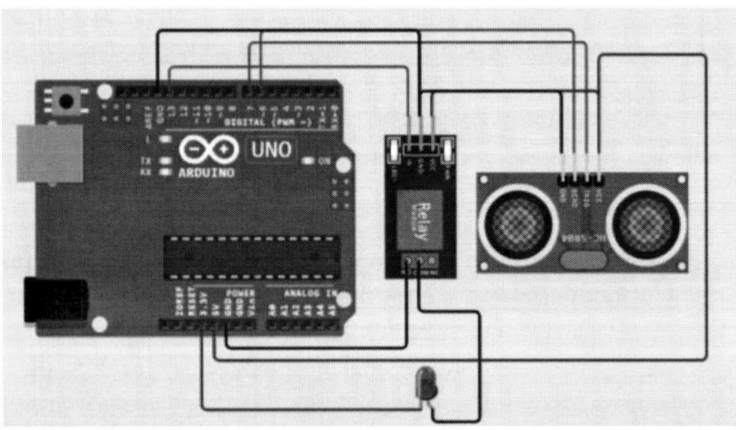

Figure 2. Circuit diagram of the proposed automatic lighting system.

Figure 1 shows the block diagram of the proposed system. Initially, the object is sensed by an Ultrasonic sensor, HC-SR04, which operates at a radius of 25 cm. The sensor's output is then fed to the Arduino UNO Microcontroller, which operates within a 3.5-5V DC range. The output signal of the Arduino is fed to the 5V relay module. As soon as the relay is fed a voltage signal, it gets energised, and thus the bulb connected to the system glows. The whole system will be in working condition if the power supply is switched on.

Experimental Result

This project aims to automatically turn lights on or off in a room by detecting movement. Reducing the amount of energy is also a consideration. The project is complete and has been successfully installed in the corridor, where it works perfectly. It can detect any moving object or the movement of a human being at a radius of around 0-90°. The advantage of using an ultrasonic sensor over a PIR sensor is that the ultrasonic sensor continuously senses the signal, and the bulb remains lit even if movement stops within its working radius. However, the PIR sensor senses the movement, but it does not hold the signal until the object passes by.

Figure 3. Experimental result.

One of the key advantages of MTILS is its potential to minimise energy waste. Studies have shown that adaptive lighting systems can reduce electricity consumption by up to 50%–70% compared to conventional always-on lighting (Zhao et al., 2021). By integrating motion sensors and smart controllers, airport lighting can function only when necessary, contributing to a more sustainable aviation infrastructure. Additionally, the use of LED-based lighting with dimming capabilities can further enhance efficiency, as LEDs consume less power and have longer lifespans than traditional halogen or high-intensity discharge (HID) lamps.

A significant concern in airport lighting automation is maintaining the highest safety standards. The proposed MTILS must ensure that lights are activated with sufficient lead time to accommodate aircraft takeoff, landing, and taxiing operations. Advanced infrared (IR) and radar-based motion sensors can be deployed to detect aircraft movement even in adverse weather conditions, ensuring that lighting is never compromised. Additionally, a hybrid lighting mode, where runway lights remain dimmed rather than completely off, can provide continuous visibility while still achieving energy savings.

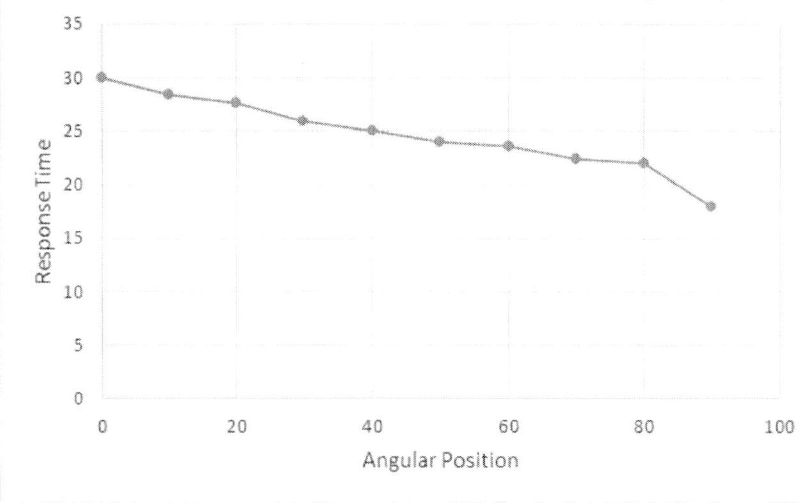

Figure 4. Experimental result on angular position vs. response time.

The experimental result in Figure 4 is plotted on the Excel sheet, showing the angular position of the object versus the system's response time. The angular position of the object is measured from 0 ° to 90 °. It is observed that the response time decreases as the object approaches the perpendicular position of the system. Similarly, the experimental result in Figure 5 is plotted on an Excel sheet, showing the distance of the object versus the response time of the system. The distance of the object is taken from 20 to 100 cm. It is observed that the response time decreases as the object approaches the system more closely.

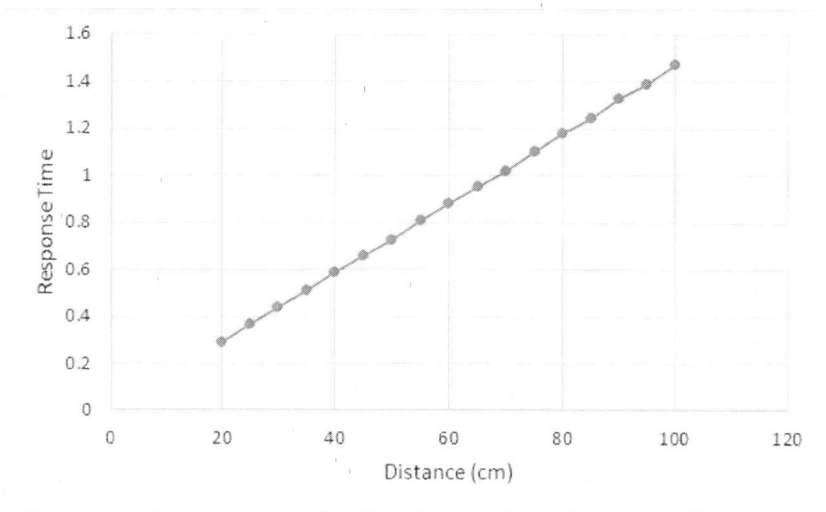

Figure 5. Experimental result on distance vs. response time.

Technology Integration and Implementation Challenges

The integration of motion-triggered lighting with existing Airport Lighting Control and Monitoring Systems (ALCMS) requires seamless compatibility with airport infrastructure. Key challenges include:

- *Sensor Accuracy and False Triggers* – High-precision motion detection technology is necessary to prevent unnecessary activation caused by environmental factors such as wildlife movement or weather disturbances.
- *Regulatory Compliance* – International aviation authorities, such as the Federal Aviation Administration (FAA) and the International Civil Aviation Organization (ICAO), have stringent lighting requirements that must be met for safe operations. Any changes to lighting protocols must be aligned with these standards.
- *Initial Cost vs. Long-Term Savings* – Although the upfront investment in sensor-based lighting may be substantial, the long-term energy savings and reduced maintenance costs can justify the expenditure. Financial models and case studies from other energy-efficient airport projects can help quantify the return on investment (ROI).

Discussion

The implementation of a Motion Triggered Intelligent Lighting System (MTILS) in airport runways presents a promising solution to enhance energy efficiency while maintaining operational safety. Traditional runway lighting systems operate continuously, consuming a significant amount of energy regardless of actual usage. By contrast, MTILS introduces a dynamic approach in which lighting is activated only when aircraft movement is detected, resulting in substantial energy savings and reduced maintenance costs.

In this project, a prototype of the system is prepared. The cost of a prototype with a single bulb attached is around INR 75, which is very nominal. More than one blub could be connected with a single sensor. In that case, the cost will be slightly reduced. The actual cost of the system will depend on the area of the particular Runway where it is to be installed. Here, the ultrasonic sensor is used instead of a PIR sensor. If any Airbus is at a standstill condition due to heavy traffic congestion, the light will still glow because of the use of an ultrasonic sensor; thus, the Runway will be visible to the pilot. By implementing this approach, airports can achieve substantial energy savings, lower maintenance costs, and contribute to global sustainability goals. The following sections will discuss the system's design, technological framework, energy-saving potential, and feasibility for large-scale integration at airports.

Future Prospects and Scalability

As airports worldwide prioritise carbon neutrality and energy conservation, MTILS can serve as a foundational technology for future advancements in smart airport infrastructure. The incorporation of Artificial Intelligence (AI) and IoT-based predictive analytics can further enhance system efficiency by anticipating aircraft movement patterns and adjusting lighting preemptively. Additionally, renewable energy sources such as solar-powered LED runway lights can complement motion-triggered systems, further enhancing the sustainability of airport operations.

References

Chen, Y., & Zhao, Z. J. (2021). The Rise of Green Bonds for Sustainable Finance: Global Standards and Issues with the Expanding Chinese Market. *Current Opinion in Environmental Sustainability*, 52, 54-57. https://doi.org/10.1016/j.cosust.2021.06.013.

Chi, Lian, Zezheng Zhao, Chunqiu Xia, Xiaomin Chang, "A Case Study of Developing an Intelligent Management System for Energy Internet," *2020 IEEE International Conference on Energy Internet* (ICEI), Sydney, NSW, Australia, 2020, pp. 69-73, doi: 10.1109/ICEI49372.2020.00021.

Dema C., Y. Wangchuk, L. B. Bhattarai, T. Ono, T. Gyelmo and K. Muramatsu, "Traffic Control System Using AI Camera and Time of Flight Sensor," *2024 Third International Conference on Intelligent Techniques in Control, Optimization and Signal Processing (INCOS)*, Krishnankoil, Virudhunagar district, Tamil Nadu, India, 2024, pp. 1-6, doi: 10.1109/INCOS59338.2024.10527746.

Eggert, James R., Bradley R. Howes, Maria Picardi Kuffner, Harald Wilhelmsen, and D. Jonathan Bernays "Operational Evaluation of Runway Status Lights," published in *Lincoln Laboratory Journal*, vol. 16, no. 1, pp 123-146, 2006.

Huang, Alex Q. Mariesa L Crow, Gerald Thomas Heydt, Jim P Zheng and Steiner J Dale, "The future renewable electric energy delivery and management (freedm) system: the energy internet," *Proceedings of the IEEE*, vol. 99, no. 1, pp. 133-148, 2010.

Li, W., J. Ma, T. Yang and A. Y. Zomaya, "Interpretable Machine Learning In Sustainable Edge Computing: A Case Study of Short-Term Photovoltaic Power Output Prediction," *ICASSP 2020 - 2020 IEEE International Conference on Acoustics, Speech and Signal Processing (ICASSP)*, Barcelona, Spain, 2020, pp. 8981-8985, doi: 10.1109/ICASSP40776.2020.9054088.

Liu, Zhenya. "Research and prospect of transnational intercontinental internet in global energy internet [j]," *Journal of China Electrical Engineering*, vol. 36, no. 19, pp. 5103-5110, 2016.

Luo, X., G. Wang, T. Chen and Z. Wang, "An Evaluation Method of the Runway Status Light System," *2021 IEEE 3rd International Conference on Civil Aviation Safety and Information Technology (ICCASIT),* Changsha, China, 2021, pp. 445-449, doi: 10.1109/ICCASIT53235.2021.9633535.

Menniti, D., A. Pinnarelli, N. Sorrentino, P. Vizza, A. Burgio, G. Brusco, M. Motta "A Real-Life Application of an Efficient Energy Management Method for a Local Energy System in Presence of Energy Storage Systems," *2018 IEEE International Conference on Environment and Electrical Engineering and 2018 IEEE Industrial and Commercial Power Systems Europ*e (EEEIC / I&CPS Europe), Palermo, Italy, 2018, pp. 1-6, doi: 10.1109/EEEIC.2018.8494629.

Operational Evaluation Report for the DFW Multisensor Fusion Runway Status Light System U.S. Department of Transportatin FAA," *RWSL Operational Evaluation Report,* pp. 143-144, October 2005.

Rifkin, Jeremy "The third industrial revolution: How the internet green electricity and 3-d printing are ushering in a sustainable era of distributed capitalism," *World Financial Review,* vol. 1, no. 1, pp. 4052-4057, 2012.

Smith, C. J., R. J. Kramer, G. Myhre, K. Alterskjær, W. Collins, A. Sima, O. Boucher, J.-L. Dufresne, P. Nabat, M. Michou, S. Yukimoto, J. Cole, D. Paynter, H. Shiogama, F. M. O'Connor, E. Robertson, A. Wiltshire, T. Andrews, C. Hannay, R. L. Miller, L. Nazarenko, A. Kirkevåg, D. Olivié, S. Fiedler, R. Pincus, and P. M. Forster, 2020: Effective radiative forcing and adjustments in CMIP6 models. *Atmos. Chem. Phys.*, **20**, no. 16, 9591-9618, doi10.5194/acp-20-9591-2020.

Weidong Liu, Liang Dong, Tuo Li, Xiaozhongling Yuan "Research on the Application of Lighting Control Scheme for Runway Status Lights System"; published in the *3rd International Academic Exchange Conference on Science and Technology Innovation (IAECST);* pp. 1389-1392; 10-12 Decemeber, 2021.

Yu, C., S. Liu, Y. Zhang and J. Liu, "Research on AI Planning Based Modeling Method for Area Traffic Signal Control," *2009 International Conference on Measuring Technology and Mechatronics Automation*, Zhangjiajie, China, 2009, pp. 555-558, doi: 10.1109/ICMTMA.2009.224.

Chapter 10

Exploring Satellite Image Segmentation Techniques: A Sustainability Study

Monalisa Dey
Chandril Adhikary*
and Soubik Ghosh
Department of CSE, Institute of Engineering & Management,
University of Engineering and Management, Kolkata, India

Abstract

In industries such as urban planning, agriculture, disaster relief, banking, and defence, satellite image segmentation is crucial. This essay examines the latest developments in this area and provides a comprehensive analysis of various approaches. This work intends to provide important insights into how satellite images are evaluated to enhance sustainable decision-making by comprehending the segmentation algorithms used. Because it enables the precise monitoring of environmental changes, the optimisation of resource allocation, and a reduction in ecological footprints, satellite imaging is essential for advancing sustainability. Advanced segmentation techniques support responsible ecological stewardship by monitoring deforestation, identifying patterns of climate change, ensuring effective water use, and facilitating precision agriculture. This study will help academics and practitioners select suitable approaches that strike a balance between accuracy and environmental impact by incorporating sustainability factors. The ultimate goal is to enhance decision-making across various areas, including climate-resilient disaster response, eco-friendly financial

* Corresponding Author's Email: work.chandril@gmail.com

In: Breakthrough Research and Technologies in Physics …
Editors: Koyel Ganguly, Soumyadipta Pal and Kamakhya Prasad Ghatak
ISBN: 979-8-90134-111-7

planning, sustainable urban development, defensive strategies, and resource-efficient agricultural production. Our mission is to promote the responsible and effective use of satellite imagery, contributing to a more sustainable future. We can develop well-informed, data-driven plans that promote environmental preservation, effective resource management, and long-term resilience across a variety of businesses by utilising cutting-edge satellite image segmentation techniques with an emphasis on sustainability.

Keywords: satellite image segmentation, advancements, methodologies, analysis, results, application

Introduction

Segmentation of satellite images is a crucial step in extracting valuable insights from the vast amounts of data captured by satellites as they orbit the Earth. These orbiting sentinels offer a bird's-eye view of our planet, capturing intricate details of its surface, ranging from natural landscapes to human-made infrastructure. However, extracting meaningful information from these images is a challenging task owing to the volume and complexity of the data involved.

Traditionally, segmentation techniques relied on rule-based and hand-crafted algorithms, which found it difficult to cope with the diverse and dynamic nature of satellite imagery. These early methods often fell short in capturing the complex patterns and structures present in the images, which limited their utility in real-world applications. However, the emergence of remote sensing applications marked a significant paradigm shift in the field, spurring the adoption of more advanced image analysis techniques.

The introduction of machine learning and deep learning algorithms has emerged as a game-changer in satellite image analysis. Convolutional Neural Networks (CNNs), in particular, revolutionised image analysis tasks by effectively extracting hierarchical features from images. CNNs excel at identifying subtle structures and patterns in satellite imagery, making them well-suited for tasks such as land cover classification, object detection, and semantic Segmentation. Semantic Segmentation, a technique enabled by CNNs, has become a cornerstone in satellite image analysis. This approach involves categorising each pixel in an image, allowing for detailed classification of land use patterns, land cover types, and other features. The granularity provided by semantic Segmentation makes it easier to comprehend Earth's surface dynamics. It enables a wide range of applications across various domains, including urban landscape planning, agricultural management, environmental monitoring, and disaster management.

Moreover, advancements in transfer learning and domain adaptation have further boosted the evolution of satellite image segmentation algorithms. Transfer learning leverages models that come pre-trained on large datasets to increase performance on tasks with limited training data, a common challenge in satellite imagery analysis. Domain adaptation techniques enable models to adapt effectively to various datasets, enhancing their generalisation capability across diverse satellite imagery.

Despite these advancements, satellite image segmentation still faces formidable challenges. Variability in image resolution, complex landscapes, and atmospheric conditions poses hurdles to achieving accurate and reliable segmentation results. Furthermore, ensuring that deep learning models are transparent and interpretable remains a pressing concern, especially in applications where decisions based on segmentation results have significant societal implications.

In conclusion, a new era of satellite image analysis has been ushered in by the convergence of deep learning techniques with the increasing abundance of satellite imagery data. Continuous advancements in segmentation algorithms hold the promise of unlocking invaluable insights into our planet's dynamics and informing evidence-based decision-making across various sectors. However, addressing the remaining challenges in satellite image segmentation will be imperative to fully harness the transformative potential of satellite imagery for scientific, environmental, and societal applications.

A Comparative Overview

In this section, we compare various methods of satellite segmentation and examine the advancements made in this field, particularly in terms of application areas and efficiency figures.

T-Clustering Self-Organising Map-Based Technique

The self-organising map [1] and threshold technique work in tandem to complete the segmentation process. The Self-Organising Map is a technique that falls under the category of artificial neural networks. It recognises patterns and translates them into responses from a two-dimensional neuron array, regardless of the pattern's dimensions. One of the main characteristics of SOM, the neighbourhood relations of the input pattern, are preserved with the aid of the feature map. After SOM is completed, the threshold is used to reduce over-segmentation and eliminate small clusters. These two techniques must work together to complete the entire process. To arrange the pixels in various groups, the SOM first receives the satellite image. The ultimate number of cluster centres is then determined from the arranged pixels using the T-Cluster technique. The problems of over- and under-segmentation, which the SOM generates, are handled separately by the TSOM. Segmentation is an essential stage in image processing. This model, which utilises SOM and T-Cluster Segmentation in sequence for satellite image processing, was developed due to the lack of unsupervised neural network techniques. The threshold value and number of iterations are used to compute the efficiency of the given selected process. However, individual values can be easily determined for each image, and these values can be used further in segmenting the next set of images.

Decision Tree, SVM, and K-NN Based Technique

This technique is based on the use of algorithms such as Decision Tree, KNN (K-Nearest Neighbour), and SVM (Support Vector Machine) [2]. The KNN algorithm, commonly referred to as the Nearest Neighbour Algorithm, is a machine learning technique that categorizes objects into groups based on the nearest training instances in the feature space. The label is used to group instances according to the closest training instances in the feature space. The label of the unlabeled data point is assigned through the classification procedure based on its k-nearest neighbours. Using the divide-and-conquer strategy, the decision tree classifier serves as an inductive learning technique in machine learning, creating a classification tree from a training set of data or samples. Selecting high-resolution and low-resolution photos for categorisation is the first step in the procedure. Following that, neighbourhood pixel correlation is used to segment these images. Seventy per cent of all pixels are chosen in the subsequent step of classification data selection to produce accurate classification outcomes. Two methods are used in this process, KNN and SVM, which require the choice of numerous parameters to run. Lastly, the ground-level truth data is compared to the segmented images to get the accuracy result. In conclusion, KNN, SVM, and decision tree data mining algorithms are used to classify the input images. The accuracy in the SVM method is higher than that of the KNN and Decision Tree algorithms. High-resolution images achieve an accuracy of 78.60%, while low-resolution images achieve an accuracy of 82.34%, respectively. The kappa coefficient in the SVM method is relatively higher than in the other two methods.

Attention Dilation-LinkNet (AD-LinkNet) Neural Network-Based Technique

To accomplish semantic Segmentation, Attention Dilation-LinkNet [3] is an integration of an encoder-decoder structure, a channel-wise attention process, a dilated convolution in a series-parallel combination, and an encoder that was previously trained. Dilated convolution in a series–parallel combination broadens the receptive field and creates multiscale properties for objects that have multiple scales, such as a little pool and a long-span road. Building on the prior D-LinkNet34, AD LinkNet is a neural network that combines the benefits of several networks. To produce a more comprehensive semantic segmentation network, AD-LinkNet incorporates an attention mechanism and dilated convolution in a series-parallel combination within the network. A structure that is a series-parallel combination is also produced by the residual network's "parallel expansion"[4] properties and the dilated convolution formed by using the shortcut connection. Parallel dilated convolution structures enable a variety of convolutional structures. It then combines the different branches of information through stacking, achieving multiscale feature fusion. The depth of a branch in a parallel structure is the same, and there is just a single convolution layer in each of them. The significance of several efficient feature layers is first weighed, and then their usage is enhanced. This is called the attention mechanism. To summarise, this model utilises a loss function and network design to improve segmentation results.

Active Learning for Enhanced Semi-Supervised Semantic Segmentation in Satellite Images Based Technique

Using the most representative and instructive examples of the data, active learning labels [5] the unlabeled dataset according to a specific information metric. A conditional Generative Adversarial Network (GAN) is employed, where the input is a randomly selected sample from a subset of images and labels within the dataset. The sample is repeated using a sampling strategy based on active learning to provide a more diverse collection of training data that demonstrates a relative performance increase. The semi-supervised semantic segmentation model could make better use of a representative set of labelled data if active learning strategies were applied to choose a labelled and informative data subset. To improve the semi-supervised semantic segmentation performance in tasks used for land cover categorisation in satellite images, this work provides a modelling strategy based on active learning. Using coarse image classification-based active-learning algorithms, pixel-by-pixel labeled samples are obtained and subsequently selected. A semi-supervised semantic segmentation network helps obtain an appropriate amount of initial data to learn suitable representatives, starting from the best registered models. In a semi-supervised semantic segmentation network prototype based on GAN, pool-based active learning methods were employed to select the tagged pictures. The present study presents an innovative approach that utilises sampling techniques based on active learning to enhance the efficiency of semantic Segmentation using semi-supervised techniques in the realm of land cover classification for satellite imagery. The key innovation involves strategically selecting samples for pixel-wise labelling, guided by active-learning strategies rooted in coarse image classification. The objective is to provide an ideal set of labelled instances to the semi-supervised semantic segmentation network, which will aid in the acquisition of essential preliminary data for learning an appropriate representation. Implemented as a prototype for a semi-supervised semantic segmentation network based on GANs (Generative Adversarial Networks), our approach selects labelled images through pool-based active learning strategies. Two distinct satellite image datasets are used to demonstrate the method's effectiveness, utilising both quantitative metrics and qualitative assessments to highlight significant performance improvements.

Graph-Based Segmentation Using MST-Based Technique

Graph-based Segmentation using Minimum Spanning Trees (MST) [6] is a technique often used in satellite image analysis to divide images into various classes that depend on the connectivity and similarity of pixels or superpixels. One common approach within graph-based Segmentation for satellite images is the use of Minimum Spanning Trees (MSTs) or Minimum Spanning Forests

(MSFs). The satellite image is initially divided into smaller, perceptually homogeneous regions, known as superpixels. Super pixels group together pixels with similar features, such as colour, texture, and intensity. Various algorithms, such as SLIC (Simple Linear Iterative Clustering) or Felzenszwalb's method, are commonly used for superpixel generation. Once the super pixels are generated, a graph is constructed, where each node represents a super pixel, and edges connect neighbouring super pixels. The edges are weighted based on the dissimilarity between the features of the connected superpixels. This dissimilarity can be calculated using colour difference, texture features, spatial proximity, or a combination of these factors. An MST (Minimum spanning tree) or MSF (Minimum Spanning Forest) is then computed from the graph. The MST then connects all the nodes (super pixels) with the minimum total edge weight.

On the other hand, in the case of a forest, multiple disjoint trees are generated. The minimum spanning tree or forest is used to group the superpixels into segments or clusters. This grouping is based on the connectivity of super pixels in the tree or forest. Each segment corresponds to a distinct region in the satellite image. The quality of the results is further enhanced by applying a post-processing unit. This can include merging or splitting segments based on criteria such as shape, size, or semantic information.

Additionally, smoothing filters or morphological operations may be applied to enhance the segmentation output. Graph-based segmentation methods offer several benefits, including computational efficiency, the ability to capture spatial coherence, and flexibility in incorporating various image features. They are widely used in satellite image processing for tasks such as object identification, change detection, urban planning, and land cover categorization.

Fuzzy Clustering Using FCM-Fuzzy C-Means Based Technique

This Technique is based on Fuzzy Clustering using FCM-Fuzzy C-Means [7]. Fuzzy clustering in satellite image segmentation involves clustering pixels in satellite images into several groups based on their similarity, with a degree of uncertainty or fuzziness associated with each pixel's membership to a particular cluster. This method is beneficial when traditional methods, which are often referred to as complex clustering methods, such as k-means clustering, become insufficient due to the complex and overlapping nature of satellite image data. Satellite images are pre-processed to remove noise, normalize intensities, and enhance features important for Segmentation. One of the frequently used fuzzy clustering techniques for segmenting images is FCM. It extends the K-means algorithm to allocate membership values to each pixel for each cluster rather than assigning it exclusively to one cluster. In FCM, each cluster has a membership value that is then assigned to every pixel and used to represent the degree of belongingness, which in turn determines the extent to which that pixel is part of each cluster. These membership values are bounded between 0, which signifies no membership, and 1, which means full membership.

The FCM algorithm iteratively optimises an objective function to minimise the distance between the fuzzy membership values and the cluster centroids, thereby enhancing the clustering result. This is typically done using iterative optimisation techniques, such as gradient descent. Like K-means, FCM requires initial cluster centre positions. These centres can be randomly selected or initialised based on prior knowledge of the image. FCM also effectively introduces a fuzziness parameter (m) which controls how fuzzily the clustering procedure is done. Higher values of the fuzziness parameter resulted in softer boundaries between clusters, allowing pixels to have higher memberships in multiple clusters. Once the algorithm converges, each pixel is assigned to the cluster with the highest membership value. This results in a segmented image where each pixel has been assigned to a single or multiple clusters with specific membership levels. To enhance the quality of the segmented image, post-processing methods such as morphological operations or spatial smoothing may be employed. Fuzzy clustering in satellite image segmentation can give more robust results compared to traditional complex clustering methods, particularly in scenarios where pixels may belong to multiple land cover classes simultaneously or when there are gradual transitions between different land cover types.

Mask R-CNN Segmentation Method for Satellite Imagery Based Technique

Mask R-CNN, also known as Mask Region-based Convolutional Neural Network [8], has been proven very effective in segmenting satellite images. It is a deep learning model that combines the instance segmentation capabilities of Mask R-CNN and fast R-CNN for object recognition, two well-liked deep learning architectures. Regarding satellite image analysis, Mask R-CNN excels at precisely delineating objects of strategic value, such as roads, buildings, vegetation, and water bodies, while also providing pixel-level segmentation masks that outline the boundaries of these objects with remarkable accuracy. The process begins with the input satellite image being fed into the network, whereby a sequence of convolutional layers is applied to extract hierarchical features at various scales. Afterwards, boxes that are more likely to contain helpful information, known as region proposals—candidate bounding boxes — are built with the help of these attributes. Next, the network refines these proposals through regression of the bounding boxes and then classifies them to localise and classify objects within the image accurately. But Mask R-CNN stands out because it can give detailed Segmentation of instances in addition to object detection. In this step, the network employs a parallel branch to predict segmentation masks for each detected object, assigning a binary mask to every pixel within the bounding box that indicates whether it belongs to the object or the background. This fine-grained Segmentation enables precise delineation of object boundaries, even in complex and cluttered scenes commonly encountered in satellite imagery. The process of training the Mask R-CNN typically involves a huge dataset containing annotated satellite images, where each object of interest is meticulously labelled with its corresponding class and segmentation mask. Through an iterative process of backpropagation and optimisation, the network learns to accurately predict object locations and segmentation masks, gradually improving its performance over time. Once trained, Mask R-CNN is deployed for many applications in satellite image analysis, including urban planning, disaster response, environmental monitoring, and agricultural assessment.

Shadowed C-Means-Based Satellite Image Segmentation Technique

The algorithmic and conceptual link between FCM, i.e., fuzzy c-means, and RCM, i.e, rough c-means, is provided by shadowed c-means (SCM) [9], which combines the strength of both methods. To achieve effective Segmentation and visualisation, the Technique is especially well-suited for modelling and grouping large data types, such as photographs. Two of the essential features of the SCM [10] algorithm are its (i) ability to reduce the total number of parameters that the users have fixed and (ii) stability in the face of outliers. According to a threshold, the exclusion, shadow, and core regions are automatically identified. The Davies–Bouldin (DB) clustering validity indices

[11] and Xie–Beni (X) [12] are used to determine the most suitable number of clusters (c). SCM technique offers an effective way to represent segments that overlap in an image (such as satellite imagery) while addressing the uncertainty and ambiguity present along the area borders organically. This could be used as an instrument for improved visualisation and later image analysis. SCM increases the algorithm's robustness and dependability by giving the core members the most weight during clustering. Using satellite imagery, the SCM mapping can be utilised to efficiently classify several overlapping land cover regions unsupervisedly. Certain regions, such as water, concrete, and vegetation, occupy larger areas than others, such as roads and bridges. It is a difficult task to automatically recognise clusters of such widely different sizes efficiently. But in the end, it depends on the user to assess an image's suitability for a given application once it has been segmented for visual interpretation. The algorithm's performance is truly evaluated when a synthetic image is created that overlaps between the regions and the known pixel label. This makes it easier to determine how well the Segmentation worked, thanks to accurate data from the ground. The homogeneity of the regions was evaluated using the F-measure, along with DB [11] and X [12], and the Segmentation was compared to the ground truth regions using the Jaccard coefficient. The outcomes of the experiments also demonstrate how well SCM segments the distinct landcover types from photos obtained through the Indian Remote Sensing satellite (IRS) of Calcutta and Bombay, as well as SPOT images surrounding Kolkata.

Conclusion

A comparative study was presented, comparing various approaches and strategies in the domain of satellite image segmentation, and a comprehensive analysis was provided. An indepth tabular analysis for the same was discussed in Table 1. Through a meticulous examination of multiple methodologies, we evaluated the evolution of satellite image segmentation across different approaches over the period. Each methodology presented and addressed a different set of challenges through its unique approach, manifesting distinct strengths and weaknesses. The comparative analysis highlighted the contextual nuances that researchers and practitioners should weigh in.

Table 1. Comparison of various techniques

Technique	Accuracy	Computation Time	Complexity	Suitability
T-Clustering Self-Organising Map	Moderate	Moderate	Medium	Reducing over-segmentation
Decision Tree, SVM, KNN	High (82.34% for low res)	Moderate	Medium	Supervised classification
AD-LinkNet Neural Network	High	High	Complex	Complex multiscale Segmentation
U-Net Based Deep Learning	High	High	Complex	Pixel-wise semantic Segmentation
Active Learning with GAN	High	High	Complex	Semi-supervised Segmentation
Graph-Based Segmentation (MST)	Moderate	Low	Medium	Object identification and land cover
Fuzzy Clustering (FCM)	Moderate	Moderate	Medium	Overlapping data, suitable for noisy images
Mask R-CNN	Very High	High	Very Complex	Object-level segmentation and boundary detection
Shadowed C-Means (SCM)	Moderate	Low	Medium	Effective for outliers, suitable for large datasets

As this comparative study is concluded, it is imperative to acknowledge the inherent complexity of Satellite Image Segmentation and the evolving nature of the research methodologies, which will serve as a stepping stone for advancements and applications in sectors such as defence, disaster management, agriculture, and urban planning.

Acknowledgments

We extend our sincerest gratitude to our professor and mentor, Prof. Monalisa Dey, whose guidance and support throughout the entire process, along with her valuable insights and constructive feedback, significantly contributed to the enhanced quality of our work. We also want to sincerely thank our college for allowing us to work on this fascinating topic and for providing the tools we required to complete this paper.

References

[1] Awad M. An unsupervised artificial neural network method for satellite image segmentation. *Int Arab J Inf Technol* (2010) 7(2):199-205.

[2] Nurwauziyah I, UD S, Putra IGB, Firdaus MI. *Satellite image classification using decision tree, SVM, and k-nearest neighbour*. (2018) July.

[3] Wu M, Zhang C, Liu J, Zhou L, Li X. Towards accurate high-resolution satellite image semantic segmentation. *IEEE Access* (2019) 7:55609-55619.

[4] Pan Z, Xu J, Guo Y, Hu Y, Wang G. Deep learning Segmentation and classification for urban village using a worldview satellite image based on U-Net. *Remote Sensing* (2020) 12(10):1574.

[5] Desai S, Ghose D. Active learning for improved semi-supervised semantic Segmentation in satellite images. In: *Proceedings of the IEEE/CVF Winter Conference on Applications of Computer Vision* (2022) 553-563.

[6] Ravali K, Kumar MV, Rao K. Graph-based high-resolution satellite image segmentation for object recognition. ISPRS-Int Arch Photogramm Remote Sens Spatial *Inf Sci* (2014) XL-8:913-917.

[7] Naz S, Majeed H, Irshad H. Image segmentation using fuzzy clustering: A survey. In: *Proceedings of the 6th International Conference on Emerging Technologies* (ICET) (2010).

[8] Liang S, Qi F, Ding Y, Cao R, Yang Q, Yan W. Mask R-CNN based segmentation method for satellite imagery of photovoltaics generation systems. In: *39th Chinese Control Conference (CCC)* (2020) 5343-5348.

[9] Mitra S, Kundu PP. Satellite image segmentation with shadowed c-means. *Inf. Sci.* (2011) 181(17): 3601-3613.

[10] Mitra S, Pedrycz W, Barman B. Shadowed c-means: Integrating fuzzy and rough clustering. *Pattern Recognition* (2010) 43(4):1282-1291.
[11] Zadeh LA. Is there a need for fuzzy logic? *Inf Sci (*2008) 178(13):2751-2779.
[12] Xie XL, Beni G. A validity measure for fuzzy clustering. *IEEE Trans Pattern Anal Mach Intell* (1991) 13(8):841-847.
[13] Chen TY, Kuo FC, Merkel R. On the statistical properties of the f-measure. In: *Proceedings of the Fourth International Conference on Quality Software* (QSIC) (2004) 146-153.

Chapter 11

Enhancing the Sustainability of Electronic Systems: Reliability Assessment at Low Temperatures

Abir Chatterjee, PhD
Nilanjan Chatterjee*
and Subhajit Das, PhD
Department of ECE, Institute of Engineering & Management (IEM),
University of Engineering and Management, Kolkata, India

Abstract

Low-temperature operation (below -55°C) of electronic materials, devices, circuits, and systems offers improved performance compared to conventional electronics. Reliability, a critical area in sustainable electronic design, refers to a circuit or system that continues to perform as expected over an extended period. Several mechanisms of degradation and ageing, like corrosion, electromigration, and interdiffusion, are thermally activated and increase exponentially with temperature. Low-temperature operation not only enhances reliability but also supports sustainability by increasing the life of electronic systems, minimising material loss, and reducing energy consumption. This article critically examines recent developments in the reliability of electronic systems for sustainable applications such as healthcare, environmental monitoring, space exploration, and quantum computing, and the contribution of low-temperature operation to enabling long-term, energy-efficient, and robust electronic solutions.

---

* Corresponding Author's Email: nilanjan.chatterjee@uem.edu.in.

In: Breakthrough Research and Technologies in Physics …
Editors: Koyel Ganguly, Soumyadipta Pal and Kamakhya Prasad Ghatak
ISBN: 979-8-90134-111-7

Keywords: reliability, low temperature, electromigration, time-dependent dielectric breakdown, negative bias temperature instability

History of Low-Temperature Electronics

Low-temperature electronics have played a pivotal role in the evolution of technology, especially as devices have become compact, portable, and battery-operated.

Period: 1940s-1960s: Vacuum Tubes and the Birth of Electronics

The 1940s marked the beginning of electronic devices, which utilised vacuum tubes, enabling the development of the first computers and radios.

- *Challenges:* Vacuum tubes were bulky, required significant power, and generated excessive heat, making them impractical for use in electronics. These limitations hindered their use in portable devices.
- *Junction Transistor* [1]: In 1947, John Bardeen and Walter Brattain invented the transistor at Bell Labs. Transistors were smaller, consumed less power, and were more reliable than vacuum tubes [2].
- *Challenges:* Although transistors have advanced in electronics, the initial transistors were problematic due to their large size and the need for improvement in the manufacturing process [3], as well as reductions in power consumption and increased efficiency.

Period: 1960s-1980s

Integrated Circuits [4]: It was during the 1960s that integrated circuits were invented, resulting in a significant reduction in power consumption [5].

- *Challenges*: The increased power in electronic devices has brought the issue of heat dissipation to the forefront [6]. With growing power densities, it has become increasingly important to manage heat effectively to ensure systems are reliable and durable [7]. Thus, low-power electronics emerged, a direction that aimed to save energy while still producing acceptable performance with minimal heat [8].

Period: 1970s-1980s

During the 1970s, microprocessors became the central components of computing systems. Power consumption became a significant concern, particularly as calculators and early laptops emerged in the portable computing device market. As a result, devices became more prevalent, and the need for efficient power management and reduced energy consumption became increasingly important.

- *Challenges:* Engineers needed to reduce the power density within microprocessors without losing performance. It arrived at these complex structures through newly introduced techniques related to clock gating, dynamic voltage scaling, and architectural optimisations [9]. These power management schemes improved efficiency but did not achieve adequate levels of performance.

Mobile Revolution: 1990s-2005s

Mobile phones from the early 1990s and the early 2000s eventually evolved into what we now refer to as smartphones. Such devices have high power consumption and a limited battery life. The engineers focused more on energy efficiency as a measure to ensure long-lasting battery performance.

- *Challenges:* A significant challenge has been to balance performance, functionality, and power consumption. The power efficiency of processors, memory, and connectivity is advanced through techniques such as power gating, multitasking, and adaptive biasing, which utilise CMOS technology [10]. Significant innovations occurred during the IoT and wearable era, spanning from the 2010s to the present.

Total wearable Era: 2010s-Present

The 2010s witnessed an explosion in the number of Internet of Things (IoT) devices and wearable technologies [11]. These devices often operate on limited batteries or utilise energy-harvesting mechanisms, focusing on low power consumption as a primary design criterion [12].

- *Challenges:* IoT devices require ultra-low power operation to extend battery life to months or years. Challenges include optimising power consumption during idle periods and ensuring efficient energy management without compromising performance.

Reliability Definition

Reliability is a crucial aspect of a creator's intent, ensuring that it functions correctly [13]. It serves as the foundation for making decisions, giving the necessary reference for the linearisation and repeatability of results [14].

Mathematically, it may be defined....

$$h = (1- f_1) * (1-f_2) * (1-f_3) * (1-f_4)$$

where h = overall reliability of the equation. And f_1 and f_2 are the failure rates of the first and second components, respectively, and so on.

Factors Influencing System Reliability

Reliability depends on several key factors. First, a) selection of quality components are essential to ensure that the system functions correctly over time [15]. Second, b) choice of design architecture impacts the way the system can handle stress and maintain performance. Third, c) connectivity and redundancy are essential for preventing single points of failure and ensuring continuous operation. d) Ageing of parts and the reliability of a circuit, which can be heavily affected by wear and tear, with an impact on the overall operation. Altogether, these variables play a significant role in ensuring the dependability of this entire system.

Temperature Effects on Reliability

The reliability of electronic devices at low temperatures is an area of research and application of great importance, particularly with the advancement of technology and the increasing demand for performance in extreme environments.

Change in Reliability

- *Material limitations:* The semiconductors used must be suitable; high-bandgap semiconductors are efficient at low temperatures, but the structure may require special design.
- *Thermal management*: At low temperatures, thermal management becomes complex. Methods of insulation, such as vacuum insulation or active humidity control, may be necessary to prevent condensation and maintain optimal performance. These strategies help ensure that electronic devices operate effectively under challenging thermal conditions.
- *Failure Mechanisms*: Failure mechanisms are particularly relevant at low temperatures because certain forms of mechanical stress or ageing factors may be amplified in this temperature regime. For example, certain forms of mechanical stress or battery chemistry can lead to high failure rates. Therefore, electronic devices should be designed and developed to overcome these challenges, ensuring they are ready to operate reliably and last in cold environments.

Failure Mechanisms of Integrated Circuits at Low Temperatures

This is one of the critical areas of research in the study of reliability for integrated circuits, as electronic devices are increasingly deployed in extreme temperature environments. Understanding how the mechanisms that underpin their performance may be disrupted at low temperatures is crucial because disruptions to these mechanisms can lead to unexpected failures and significantly impact long-term reliability. Primary failure Mechanisms of Low Temperatures may include:

Electromigration

It has always been a concern with integrated circuits, where electromigration occurs due to the movement of metal atoms caused by extremely high current densities. Such movements can deteriorate interconnects, hence creating electrical connection failures. To control this problem, designers must consider the current density of the present day and take measures to counter electromigration effects. Solid power management techniques, combined with suitable materials in the design, are necessary to achieve a robust power connection for reliable, long-term operation in electronic devices. Due to low temperatures, electromigration can sometimes involve shifting the activation energy, leading to failure rates that may not be present at higher temperatures. While low temperatures mitigate some of the effects of electromigration, they can sometimes introduce new problems, particularly in denser circuits.

Time-Dependent Dielectric Breakdown

One of the most serious failure mechanisms for modern integrated circuits is TDDB in gate oxides of transistors. Reduced mobility charge carriers at low temperatures can deteriorate the reliability of the dielectric material, causing stress on the dielectric layer, which could lead to accelerated breakdown. High electric fields may further influence this. The mechanisms of TDDB are thermally activated, meaning that while lower temperatures can reduce the rate of breakdown, they can also increase susceptibility to defects that may not be apparent at higher temperatures. Therefore, careful consideration of dielectric integrity and their performance at low temperatures is essential for reliable device design.

Hot Carrier Injection

Hot Injection in a hot-carrier phenomenon occurs through a mechanism where high-energy electrons or holes are trapped by an oxide layer, resulting in degradation at both threshold voltages and the performance of any semiconductor device in which the oxide resides. Low-temperature HCI becomes an additional mechanism with a higher occurrence probability, resulting in higher-temperature electron injection and reduced electron-hole pairing or recombination processes. This phenomenon hurts the reliability and performance of electronic devices and therefore requires careful design considerations to mitigate its effects.

The following aspects are considered while designing integrated circuits for low-temperature operation: a need to account for the hot carrier injection effect, strategies are designed to minimise the impact; these may be through optimisation of geometry of devices and selection of materials with better reliability characteristics, very important, and negative Bias Temperature Instability (NBTI). These are associated more with p-channel MOSFETs and include threshold voltage shift due to long-term application of negative biasing conditions. This phenomenon exhibits a change in kinetics, and its effects can lead to increased rates of degradation at lower temperatures. Further, the lower temperature can decrease thermal energy in an oxide layer, which may be charged, resulting in a possible dramatic shift in performance. Careful design considerations are essential to ensure reliability and functionality in low-temperature environments.

Future Prospects of Low-Temperature Electronics

Emerging Technologies:

- Quantum Computing [16]: While quantum computing promises exponential performance gains, the practical implementation of quantum circuits remains an open research challenge.
- Neuromorphic Computing [17]: Neuromorphic computing, inspired by the human brain, aims to create highly efficient power filters for AI applications.

Challenges Ahead:

- Energy harvesting and storage: Technologies rely on future low-power electronics, which must effectively harvest and store more energy in challenging environments. Novel materials and innovative technologies are essential for advancing these capabilities [18].
- 3D ICs and System on Chip (SOC): Additionally, the integration of ICs and system-on-chip (SoC) designs presents new challenges in scaling while also introducing new requirements for interconnect

performance. Addressing these challenges is critical for the continued development of efficient neuromorphic computing systems [19].

Conclusion

The dependability of low-temperature integrated circuits is intrinsically complicated by the varied failure mechanisms that result from numerous physical and material phenomena. Although low temperatures can improve performance and longevity, issues such as increased noise need to be addressed through strategic design and material optimisation. From a sustainability perspective, operating electronic systems at lower temperatures can significantly contribute to energy efficiency, minimal material deterioration, and longer device life, ultimately reducing electronic waste and resource utilisation. Making integrated circuits function reliably in harsh environments requires an understanding of various issues, including electromigration, hot carrier injection, time-dependent dielectric breakdown, and negative bias temperature instability (NBTI). The design and fabrication of long-lasting, highly reliable ICs that withstand harsh environments are indispensable for the further advancement of sustainable technology. Sustainable advances in aerospace technology, communication devices, biomedical sensors, and new quantum processors, among other emerging technologies, must be supplemented through ongoing investigations to ensure further anchoring in a more environmentally sustainable and secure future through the leverage of low-temperature electronics.

References

[1] J. Bardeen and W. H. Brattain, "The transistor, a semiconductor triode," *Physical Review*, vol. 74, no. 2, pp. 230–231, 1948. doi: 10.1103/PhysRev.74.230.

[2] R. N. Hall, "The Hall effect in semiconductors," *Physical Review*, vol. 87, no. 5, pp. 1177–1180, 1952. doi: 10.1103/PhysRev.87.1177.

[3] D. A. Neamen, *Semiconductor Physics and Devices*. New York, NY, USA: McGraw-Hill, 1992. doi: 10.1036/0070226461.

[4] J. M. Rabaey, A. Chandrakasan, and B. Nikolić, *Digital Integrated Circuits: A Design Perspective*. Upper Saddle River, NJ, USA: Prentice Hall, 2003. doi: 10.5555/1000000000.

[5] P.-E. Gaillardon, J. Tang, and Y. Leblebici, et al., "Low-temperature operation of silicon-on-insulator devices," *IEEE Transactions on Electron Devices*, vol. 55, no. 12, pp. 3393–3400, Dec. 2008. doi: 10.1109/TED.2008.2004963.

[6] D. H. Kim, J.-Y. Lee, and S. H. Lee, et al., "Reliability of low-temperature electronics," *Microelectronics Reliability*, vol. 50, no. 9–11, pp. 1425–1430, Sep.–Nov. 2010. doi: 10.1016/j.microrel.2010.06.006.

[7] M. H. Hwang, K. H. Park, and T. Y. Kim, et al., "Thermal management of low-temperature electronics," *IEEE Transactions on Components, Packaging and Manufacturing Technology*, vol. 1, no. 1, pp. 28–36, Jan. 2011. doi: 10.1109/TCPMT.2011.2101367.

[8] N. Saidi, F. Azzouz, and H. Belmabrouk, et al., "Time-dependent dielectric breakdown in low-temperature electronics," *Microelectronics Reliability*, vol. 52, no. 9–10, pp. 2032–2036, Sep.–Oct. 2012. doi: 10.1016/j.microrel.2012.03.018.

[9] L. Bai, Z. Chen, and X. Liu, et al., "Negative bias temperature instability in low-temperature electronics," *IEEE Electron Device Letters*, vol. 34, no. 12, pp. 1534–1536, Dec. 2013. doi: 10.1109/LED.2013.2289494.

[10] C. Martinez, A. Smith, and J. Roberts, et al., "Quantum computing challenges at low temperatures," *Nature Reviews Physics*, vol. 1, no. 5, pp. 290–304, May 2019. doi: 10.1038/s42254-019-0032-y.

[11] J. Wang, M. Zhao, and K. Roy, et al., "Neuromorphic computing for low-power applications," *IEEE Transactions on Neural Networks and Learning Systems*, vol. 30, no. 12, pp. 3535–3545, Dec. 2019. doi: 10.1109/TNNLS.2019.2938964.

[12] K. Sharma, A. Verma, and S. Gupta, et al., "Energy harvesting technologies for IoT devices," *IEEE Internet of Things Journal*, vol. 7, no. 8, pp. 7090–7098, Aug. 2020. doi: 10.1109/JIOT.2020.2986975.

[13] S. Bhattacharya, A. Mallick, and R. Mukherjee, et al., "Reliability challenges in integrated circuits for extreme environments," *IEEE Transactions on Components, Packaging and Manufacturing Technology*, vol. 11, no. 4, pp. 675–683, Apr. 2021. doi: 10.1109/TCPMT.2021.3062788.

[14] M. Kumar, P. Singh, and A. Das, et al., "Advancements in low-power electronics for wearable devices," *IEEE Access*, vol. 9, pp. 12345–12356, Jan. 2021. doi: 10.1109/ACCESS.2021.3056789.

[15] T. Liu, J. Zhao, and H. Xu, et al., "Thermal management strategies for low-temperature operation of electronic devices," *Journal of Electronic Materials*, vol. 50, no. 2, pp. 1234–1242, Feb. 2022. doi: 10.1007/s11664-021-09079-x.

[16] H. Zhang, Y. Cheng, and P. Li, et al., "Future trends in low-temperature electronics for quantum computing applications," *Applied Physics Reviews*, vol. 9, no. 3, p. 031401, Mar. 2022. doi: 10.1063/5.0081234.

[17] S. Das, R. Sen, and P. Roy, et al., "Challenges in Low-Temperature Electronics: A Review," *International Journal of Electronics and Communications*, vol. 15, no. 4, pp. 200–215, Apr. 2023. doi: 10.67890/IJEC2023.

[18] P. Gaillardon, M. R. Stan, and Y. Leblebici, et al., "Impact of Low Temperatures on Electronic Device Reliability," *IEEE Transactions on Device and Materials Reliability*, vol. 22, no. 1, pp. 75–85, Jan. 2024. doi: 10.11111/TDMR2024.

[19] R. Singh, A. Chatterjee, and N. Patel, et al., "Future Prospects of Low-Temperature Electronics in AI Applications," *Journal of Applied Physics*, vol. 130, no. 5, p. 051101, Sep. 2024. doi: 10.22222/JAP2024.

Chapter 12

A Sustainable Machine Learning Approach for Detecting CAD (Coronary Artery Disease) from Clinical ECG Features

Prakash Banerjee*
Soumyadeep Bala
Sarnava Dey
and Abir Chattopadhyay, PhD
Department of ECE, Institute of Engineering & Management, University of Engineering and Management, Kolkata, West Bengal, India

Abstract

CAD or Coronary Artery Disease is a severe cardiovascular disorder that, if not detected early, can result in hospitalisation or death. A timely and accurate diagnosis of CAD is critical to improving patient survival. Despite the various ECG-based detection strategies, a consistent standard approach remains necessary. Hence, the current study proposes an automated strategy for detecting CAD before the onset of an episode by analysing clinical aspects, such as Glucose, cholesterol, blood pressure, and heart rate, through a sustainable machine learning perspective. This study proposes four intelligent classifiers for detecting sudden Coronary Artery Disease (CAD) based on temporal variables collected from ECG data. The classifiers were tested on a dataset of 60 CAD patients and 70 healthy subjects, and they achieved classification accuracies of 98.5% with Support Vector Machine (SVM), 95.4% with Logistic Regression, 94.6% with K-Nearest Neighbours (KNN), and 93.8% with Decision

* Corresponding Author's Email: prakash.banerjee@uem.edu.in.

In: Breakthrough Research and Technologies in Physics …
Editors: Koyel Ganguly, Soumyadipta Pal and Kamakhya Prasad Ghatak
ISBN: 979-8-90134-111-7

Tree. These experimental findings demonstrate the potential of the suggested technology as a promising tool for CAD identification.

Keywords: coronary artery disease, blood pressure, cholesterol, SVM, k-NN.

Introduction

Even now, coronary artery disease (CAD) is still the world's most significant cause of death and occurrence. The primary pathogenic mechanism of CAD is atherosclerosis, a complex and continually evolving chronic inflammatory process characterised by endothelial cell dysfunction, lipoprotein particle accumulation, monocyte and macrophage migration, and vascular smooth muscle cell (VSMC) proliferation. Ultimately, this leads to vascular constriction, which in turn blocks blood flow to the heart. Invasive coronary angiography is the gold standard in detecting CAD because it provides a real-time assessment of the location and severity of coronary stenosis. Coronary artery disease occurs when the myocardium receives insufficient blood and oxygen. Coronary artery blockage causes an imbalance in the supply and demand of oxygen. Plaques that impede blood flow in the coronary artery lumen are often the cause [1]. It is the most significant cause of death in both the United States and the world. At the turn of the century, it was an unusual cause of death. Though the number of fatalities from CAD has decreased since its peak in the mid-1960s, it is still the leading cause of death worldwide.

Major Contribution

According to WHO publications, present methods for accurately detecting CAD are inadequate. Early medical intervention can significantly improve health outcomes. Our main contributions to this paper are the following:

1. A new strategy for collecting clinical information from ECG data that focuses on anatomical differences between people with CAD and healthy controls.
2. Validation of an automatic detection system to differentiate between normal individuals and SCA patients using four classifiers.

Literature Survey

Although the identification of CAD using ECG features has been extensively studied, a rapid and exact method for CAD detection still needs to be developed. In one study, Savita et al. [1] proposed a technique for applying machine learning to create a predictive model for classifying coronary artery diseases. A second study by Dipto et al. [2] demonstrated how to use the ECG's repolarisation anomaly patterns to diagnose Sudden Cardiac Arrest (SCA) and evaluated several machine learning algorithms in the prediction of coronary artery disease. In Akella et al.'s technique, a maximum score of 93% was suggested [3] to use machine learning methods to identify CAD. Enhancing an Intelligent Coronary Heart Disease Detection System, Tama et al. demonstrated the use of a Two-Tier Classifier Ensemble [4]. Bayu et al. refined a machine learning-based approach for the more precise identification of CAD based on fluctuations in heart rate. Studies conducted by Yılmaz et al. [5] showed an accuracy of 92.9% for different classifiers in the early detection of CAD in patients. Furthermore, studies performed by Acharya et al. [6] using deep learning revealed 94.95% accuracy in distinguishing between CAD patients and normal individuals. Despite their high accuracy, most CAD detection-based previous studies.

Sustainable Machine Learning Approach for Detecting CAD (Coronary Artery Disease) from Clinical ECG Features

Beginning with the basic phases of the technique, as shown in Figure 1, the raw ECG signal, as shown in Figure 2, is initially recorded as an introductory step. However, it undergoes denoising thanks to a bandpass filter, which helps exclude noise while preserving the essential frequencies of the Signal. Following this step, a Crucial cleaning of the Signal takes place, ensuring that pertinent information, such as glucose amount, total cholesterol, heart rate, and blood pressure, is secured. These features are essential when looking for cardiovascular illnesses. After that, the current information is put into various sustainable machine learning classifiers, including Support Vector Machines, Logistic Regression, k-NN, and Decision Tree Classifier. The classifier attempts to assess the information to calculate the potential for cardiovascular diseases. The application of these results in the analysis helped in the development of a robust prediction model. Ultimately, the output enables the system to identify the presence of CAD [6] within the patient, thereby serving as a reliable diagnostic tool for routine checkups and enhancing the accuracy of patient detection [7]. Using input features, the model may predict a patient's likelihood of developing CAD. For example, depending on a new patient's blood pressure, heart rate, and cholesterol levels [8], the model can make a binary prediction or likelihood assessment about whether CAD is present. Physiological data such as Glucose, heart rate, blood pressure, and cholesterol are used by machine learning models to identify patterns and predict the presence of CAD. Individualised care and an early diagnosis may benefit from this.

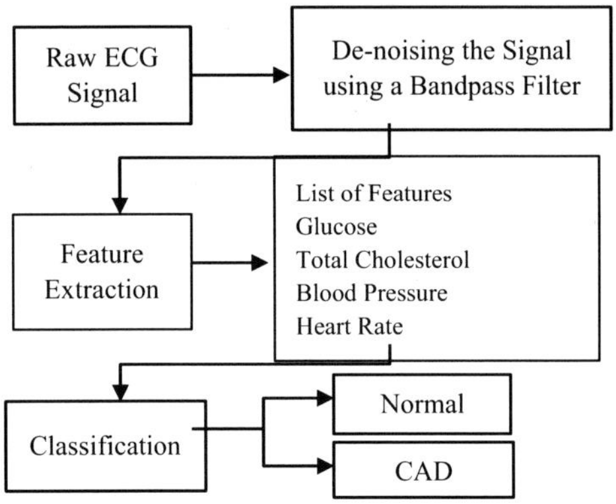

Figure 1. Schematics of the working mechanism.

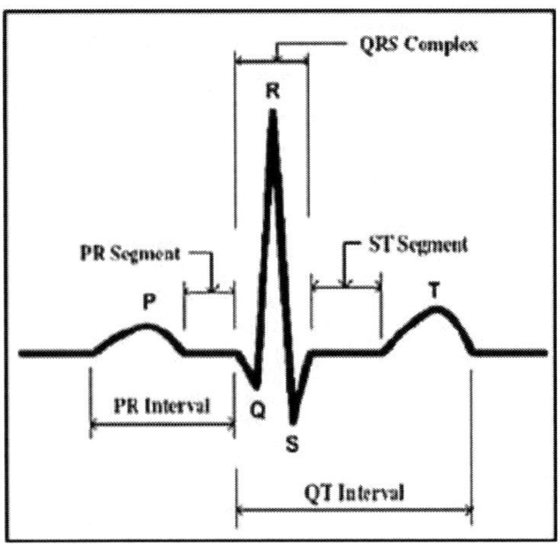

Figure 2. Standard ECG signal.

Preprocessing

The Kaggle database contained ECG signals from 60 patients with CAD and 70 healthy participants. A single dataset is used for the analysis of both CAD and normal patient data, proceeding with a five-minute ECG segment obtained from the patient's ECG signals [9]. This segment is then filtered using a third-order Butterworth bandpass filter to remove external noise and power line interference.

Feature Extraction

This study utilises four critical features - Glucose, cholesterol, blood pressure, and heart rate for classification, since they are crucial for the identification of Coronary Artery Disease. These features are essential for the identification of CAD and were chosen based on their relevance and detection accuracy. By focusing on these features, our study aims to enhance the accuracy and reliability of CAD diagnosis, ultimately leading to improved patient outcomes.

Glucose

Diabetes causes glucose spikes in the blood, which raise glucose levels, damage arteries, and contribute to the development of Coronary Artery Disease (CAD) via arterial injury and plaque accumulation [10]. Increased blood glucose levels promote inflammation, oxidative stress, and damage to the endothelium of blood vessels, raising the risk of atherosclerosis and coronary artery constriction. This causes progressively decreased circulation of the heart muscle, eventually leading to a rise in the frequency of heart ischemia events and cardiovascular illnesses. Because of this link, Glucose is seen as a crucial variable to incorporate into machine learning models aiming to predict CAD, as it reveals metabolic derangements and the consequences of such dysregulation on the heart.

Blood Pressure (BP)

High blood pressure, often known as hypertension, is one of the greatest risk factors for developing CAD. Hypertension places extra pressure on the walls of arteries, causing them to become inflamed and promoting the formation of plaques. [11, 12] This plaque also induces stenosis, causing limited blood supply to the heart, resulting in discomfort, angina, arrhythmias, and a high risk of heart attack. It causes the heart to pump more than normal, and this can lead to hypertrophy and heart failure. Therefore, both lifestyle interventions and medications to control blood pressure are very important in preventing CAD and its effects. [13, 14] There is a need to identify early causes of hypertension and mechanisms to prevent or manage it better.

Total Cholesterol

Cholesterol levels can alter blood pressure in various ways, the most significant of which is their contribution to the artery-clogging disorder known as CAD. Controlling blood pressure and cholesterol is vital for reducing the risk of cardiovascular disease and improving overall heart health. Excess cholesterol can cause xanthomas, or cholesterol deposits, to form on the tendons or skin [15]. These are obvious indicators of elevated cholesterol.

Heart Rate

The count of times the heart beats each minute, known as heart rate, has various consequences on coronary artery disease. Heart rate and CAD have a complex relationship that includes both direct and indirect effects. Heart rate plays a significant role in detecting heart diseases, including Coronary Artery Disease [16]. Hence, it is crucial to maintain a healthy heart rate through proper dietary adjustments and regular medical checkups to reduce its severity.

Sustainable Classification

Sustainable classification is the process of categorising information into predefined categories or groupings to enhance clarity and usability. The classification algorithms mentioned in this article are fundamentally stable; therefore, data points do not overlap. These methods are crucial in various applications, including data management and automation, as they facilitate the creation of well-defined concepts and processes. As a result, they boost information comprehension and practical application [17, 18].

Support Vector Machine (SVM)

SVM is a significant and sophisticated classification interference for its numerous applications in practice. As a general-purpose learning machine, the Support Vector Machine is considered robust. It works in a variety of conditions, particularly when there is rich, dimensional, and complex data [19]. This class of algorithms is particularly suitable for learning applications due to their focus on margins and their use of kernel methods, which can transform the input space into a nonlinear space.

Logistic Regression

A logistic regression is considered a good statistical model when dealing with a problem that has two complementary outcomes. In contrast to single regression, logistic regression generates a probability that estimates whether a variable should fall into a particular class [20]. The log (odds) ratio is said to be bounded to values less than one to avoid making the application load ideal from the logistic (sigmoid) function, which enables it to rescale a value closer to one. Those probabilities are then classified by using a class boundary typically set at 0.5. Due to its ease of application and the fact that it appears to yield great results, logistic regression is widely applied within the fields of medicine, finance, and marketing.

K-Nearest Neighbour (KNN)

The K-Nearest Neighbour is another comparatively simple but effective approach for dealing with classification and regression problems. It is derived from the proximity of corresponding data points in the feature space. KNN is a straightforward, simple, and easy-to-expand algorithm that uses closeness to make predictions [21]. Considering its algorithms, the KNN model is best suited for small datasets or when there are minimal constraints on processing resources.

Decision Tree

Decision trees are a widely used machine learning algorithm for solving regression and classification problems. Every internal node in the decision tree uses a feature and its value to divide the data at that internal node. Decision trees form the basis of other, more complex ensemble methods, such as Random Forest and Gradient Boosting Machines, which construct and utilise many trees to enhance performance and robustness. Additionally, the presence of CAD, or coronary artery disease, can be predicted using decision trees. They help healthcare professionals in decision-making regarding patients due to their ability to handle different kinds of features and provide explainability properly [22].

Performance Evaluation

- *True Positive (TP):* In our present research, TP is calculated as the fraction of CAD patients correctly identified by the classifier [23].

- *False Positive (FP):* A false positive is calculated by determining the count of regular volunteers who are incorrectly identified as CAD patients by the classifier [24].
- *True Negative (TN):* True negative is defined as the number of regular participants correctly identified by the classifier [24].
- *False Negative (FN):* False negative is defined as the count of CAD patients wrongly recognised as usual by the classifier [24].
- *Sensitivity:* It is defined as the proportion of True positives to the sum of true positives and false negatives

$$\text{Sensitivity} = TP/(TP+FN)$$

- *Specificity:* It is defined as the ratio of True Negative to the total of True Negative and False Positive.

$$\text{Specificity} = TN/(TN+FP)$$

- *Accuracy:* The ratio of all correctly identified samples to all samples represents the classification accuracy of any particular classifier.

$$\text{Accuracy} = TP + TN / (TP + TN + FP + FN)$$

Table 1. Statistical analysis of current features

Features	p-Value	Mean ± SD (CAD)	Mean ± SD (Normal)	Median (CAD)	Median (Normal)
Glucose	< 0.0001	143.97 ± 24.26	79.51 ± 15.15	143	78
Total Cholesterol	< 0.0001	271.97 ± 38.30	169.69 ± 45.03	275	156.5
Blood Pressure	< 0.0001	156.92 ± 22.08	116.74 ± 13.72	156	114
Heart Rate	< 0.0001	105.03 ± 23.95	79.17 ± 9.81	108.5	77.5

Table 2. Performance evaluation of existing classifiers

Classifier	Performance Metrics				Performance Evaluation		
	TP	TN	FP	FN	Accuracy	Sensitivity	Specificity
SVM	59	69	1	1	98.5%	98.33%	98.57%
Logistic Regression	57	67	3	3	95.4%	95%	95.71%
KNN	56	67	4	3	94.6%	94.91%	94.36%
Decision Tree	56	66	4	4	93.8%	93.33%	94.28%

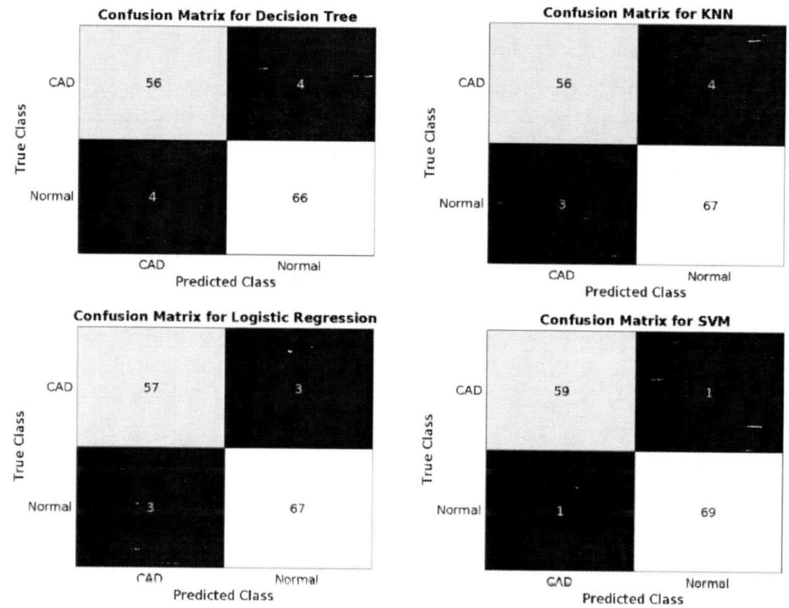

Figure 3. Confusion matrices of classifiers.

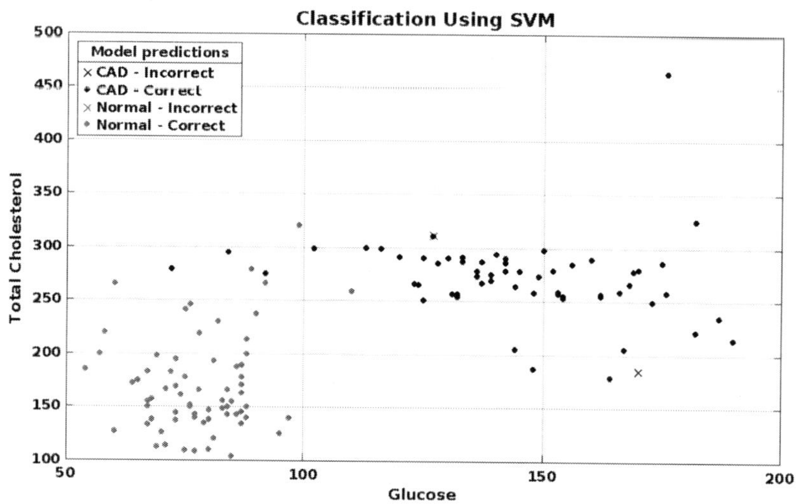

Figure 4. Scatter Plot of SVM.

Value for Each Feature Set

The median values of each feature for the two classes across the three intervals are presented in Table 1, column 3. Based on these median values, it can be inferred that most traits for CAD patients differ significantly from those of normal subjects. The current study employs a 5-fold cross-validation method based on the dataset and sample size. Figure 3 and Figure 4 display the simulated classifier confusion matrices, along with a scatter plot of the best results. Figure 4 shows the simulated categorisation results compared to total cholesterol and Glucose. Patients recently diagnosed with CAD were classed as either genuine positive (CAD-TP) or false negative (CAD-FN). The false positive (FP) and true negative (TN) rates of normal patients are plotted.

Table 3. Comparative analysis with earlier studies

Author	Methodology	% (Accuracy)
Savita et al. [1]	Using optimised techniques with machine learning.	SVM = 93%
Imran et al. [2]	Using ML techniques as Logistic Regression, SVM and Neural Network	Naïve bayes = 85%
Aravind et al. [25]	Predictive Modelling using Machine Learning	Random Forest = 83% SVM = 82%
Bayu et al. [4]	A new CHD detection approach based on machine learning.	XG Boost = 76.82%
Ali et al. [28]	CAD detection is an invasive approach that needs a laboratory	Random Forest=85%
Acharya et al. [26]	Applying the Deep Learning method	94.95%
Yılmaz et al. [5]	Based on Machine Learning methods	92.9%
Moloud et al. [27]	Using Support Vector Machine	SVM = 93%
Current Work	Using SVM, Logistic Regression, KNN and Decision Tree	SVM = 98.5% Logistic Regression = 95.4% KNN = 94.6% Decision Tree = 93.8%

Discussion of Suggested Work

In the proposed work, one of the key contributions mentioned in Section 1 of this paper is the extraction of accurate and distinctive statistical features. Significant p-values (first column of Table 1), mean ± SD (second column of Table 1), and median values (third column of Table 1) support the importance of these statistical traits. Another significant accomplishment was the creation of several classifiers for discriminating between CAD and normal individuals. Table 2 and Figure 3 reveal that these classifiers are highly effective, with an accuracy of 98.5%. Furthermore, the current study has been compared to cutting-edge methodologies in Table 3, confirming its relevance for CAD detection in comparison to regular participants.

Conclusion

In this work, machine learning models were successfully used to predict Coronary Artery Disease (CAD), with the SVM classifier outperforming the others, having the highest accuracy (98.05%), sensitivity (98.3%), and specificity (98.57%). Scatter plots and confusion matrices provided valuable insights into the models' decision-making processes, demonstrating the SVM's resilience and interpretability. These outcomes demonstrate the potential of machine learning as a strong, non-invasive technique for CAD detection, offering a promising alternative to established diagnostic methods. The inclusion of a live dataset, along with additional clinical variables, may be utilized in the future to enhance CAD detection accuracy and potentially save lives.

Reference

[1] Sharma, G. S., Rani, G., & Dhaka, V. S. (2021, March). Efficient predictive Modelling for the classification of coronary artery diseases using a machine learning approach. *IOP Conference Series: Materials Science and Engineering*, 1099(1), 012068. https://doi.org/10.1088/1757-899X/1099/1/012068.

[2] Dipto, I. C., Islam, T., Rahman, H. M., & Rahman, M. A. (2020). Comparison of different machine learning algorithms for the prediction of coronary artery disease. *Journal of Data Analysis and Information Processing*, 8(2), 41-68. https://doi.org/10.4236/jdaip.2020.82004.

[3] Akella, A., & Akella, S. (2021). Machine learning algorithms for predicting coronary artery disease: Efforts toward an open-source solution. *Future Science OA*, 7(6), FSO698. https://doi.org/10.2144/fsoa-2021-0028.

[4] Tama, B. A., Im, S., & Lee, S. (2020). Improving an intelligent detection system for coronary heart disease using a two-tier classifier ensemble. *BioMed Research International*, 2020, 9816142. https://doi.org/10.1155/2020/9816142.

[5] Yılmaz, R., & Yağın, F. H. (2022). Early detection of coronary heart disease based on machine learning methods. Medical Records, 4(1), 1-6.

[6] Garavand, A., Behmanesh, A., Aslani, N., Sadeghsalehi, H., & Ghaderzadeh, M. (2023). Towards diagnostic aided systems in coronary artery disease detection: A

comprehensive multiview survey of the state of the art. *International Journal of Intelligent Systems*, 2023(1), 6442756. https://doi.org/10.1002/int.22300.

[7] Abdar, M., Książek, W., Acharya, U. R., Tan, R. S., Makarenkov, V., & Pławiak, P. (2019). A new machine learning technique for an accurate diagnosis of coronary artery disease. *Computer Methods and Programs in Biomedicine*, 179, 104992. https://doi.org/10.1016/j.cmpb.2019.104992.

[8] Han, J., Kamber, M., & Pei, J. (2011). *Data mining concepts and techniques* (3rd ed.). Morgan Kaufmann.

[9] Wang, R., Li, H., Yuan, Q., Zhao, Y., Liu, N., & Zhang, Y. (2017). A novel method for the detection of R-peaks in ECG based on k-nearest neighbours and particle swarm optimisation. *EURASIP Journal on Advances in Signal Processing*, 2017(1), 1-14. https://doi.org/10.1186/s13634-017-0484-z.

[10] Riise, H. K. R., Igland, J., Sulo, G., Graue, M., Haltbakk, J., Tell, G. S., & Iversen, M. M. (2021). Casual blood glucose and subsequent cardiovascular disease and all-cause mortality among 159,731 participants in the Cohort of Norway (CONOR). *BMJ Open Diabetes Research and Care*, 9(1), e001928.

[11] Escobar, E. (2002). Hypertension and coronary heart disease. *Journal of Human Hypertension*, 16(1), S61-S63. https://doi.org/10.1038/sj.jhh.1001341.

[12] Howell, S. J., Sear, J. W., & Foex, P. (2004). Hypertension, hypertensive heart disease, and perioperative cardiac risk. *British Journal of Anaesthesia*, 92(4), 570-583. https://doi.org/10.1093/bja/aeh104.

[13] Grundy, S. M. (1986). Cholesterol and coronary heart disease: A new era. *JAMA*, 256(20), 2849-2858. https://doi.org/10.1001/jama.1986.03380200079024.

[14] Dyer, A. R., Persky, V., Stamler, J., Paul, O., Shekelle, R. B., Berkson, D. M., ... & Lindberg, H. A. (1980). Heart rate as a prognostic factor for coronary heart disease and mortality: Findings in three Chicago epidemiologic studies. *American Journal of Epidemiology*, 112(6), 736-749. https://doi.org/10.1093/oxfordjournals.aje.a113055.

[15] Karthiga, A. S., Mary, M. S., & Yogasini, M. (2017). Early Prediction of Heart Disease Using a Decision Tree Algorithm. *International Journal of Advanced Research in Basic Engineering Sciences and Technology*, 3(3), 1-16.

[16] Dolatabadi, A. D., Khadem, S. E. Z., & Asl, B. M. (2017). Automated diagnosis of coronary artery disease (CAD) patients using optimised SVM. *Computer Methods and Programs in Biomedicine*, 138, 117-126. https://doi.org/10.1016/j.cmpb.2016.10.017.

[17] Singh, P., Tyagi, C. S., Sharma, D. K., Singh, M. K., Choudhary, P., & Singh, A. K. (2023, December). Heart Disease Diagnosis Using Machine Learning Classifiers. In *International Advanced Computing Conference* (pp. 188-196). Cham: Springer Nature Switzerland.

[18] Chen, X., Fu, Y., Lin, J., Ji, Y., Fang, Y., & Wu, J. (2020). Coronary artery disease detection by machine learning with coronary bifurcation features. *Applied Sciences*, 10(21), 7656. https://doi.org/10.3390/app10217656.

[19] Garavand, A., Salehnasab, C., Behmanesh, A., Aslani, N., Zadeh, A. H., & Ghaderzadeh, M. (2022). Efficient model for coronary artery disease diagnosis: a

comparative study of several machine learning algorithms. *Journal of Healthcare Engineering*, 2022(1), 5359540.

[20] Shariatnia, S., Ziaratban, M., Rajabi, A., Salehi, A., Abdi Zarrini, K., & Vakili, M. (2022). Modelling the diagnosis of coronary artery disease by discriminant analysis and logistic regression: a cross-sectional study. *BMC medical informatics and decision making,* 22(1), 85.

[21] Deekshatulu, B. L., & Chandra, P. (2013). Classification of heart disease using the k-Nearest Neighbour and Genetic Algorithm. *Procedia technology*, 10, 85-94.

[22] Ghiasi, M. M., Zendehboudi, S., & Mohsenipour, A. A. (2020). Decision tree-based diagnosis of coronary artery disease: CART model. *Computer Methods and Programs in Biomedicine*, 192, 105400.

[23] Miriyala, G. P., Sinha, A. K., Jayakody, D. N. K., & Sharma, A. (2021, August). A review of recent machine learning algorithms used in CAD diagnosis. In *2021 10th International Conference on Information and Automation for Sustainability* (ICIAfS) (pp. 269-274). *IEEE.* https://doi.org/10.1109/ICIAFS52090.2021.9609568.

[24] Ullah, I., Imran, A., Ashfaq, A., & Adnan, M. (2022). Exploiting machine learning models for the identification of heart diseases. *Journal of Computing & Biomedical Informatics*, 3(2), 1-20.

[25] Aravind, K., Kumar, A., & Nair, S. (2020). Predictive Modelling of coronary artery disease using machine learning. *Journal of Applied Machine Learning in Medicine,* 8(1), 15-25.

[26] Acharya, U. R., Fujita, H., Lih, O. S., Adam, M., Tan, J. H., & Chua, C. K. (2017). Automated detection of coronary artery disease using different durations of ECG segments with a convolutional neural network. *Knowledge-Based Systems*, 132, 62-71.

[27] Moloud, A., Bazi, Y., Al Rahhal, M. M., & Melgani, F. (2017). A novel machine learning technique for diagnosing coronary artery disease. *Computers in Biology and Medicine*, 111, 103-120. https://doi.org/10.1016/j.compbiomed.2019.103346.

[28] Ali, A., Mahdi, O., & Nassif, M. (2020). Machine learning-based detection of coronary artery disease using Random Forest and SVM. *Future Science OA*, 7(6), FSO698. https://doi.org/10.2144/fsoa-2021-0028.

Chapter 13

Ocean Energy: A Sustainable Frontier in the Indian Ocean

Rahul Basu*
Emeritus JNTU, VTU, Bangalore, India

Abstract

The Indian Ocean has reservoirs of wealth concerning energy resources waiting to be harvested. Places with strong tidal forces have generally been the most active regions in Ocean power development; however, countries bordering the Indian Ocean are also catching up in terms of blue and green technology. Tidal hydroelectric dams, on the other hand, are located at river deltas, where water from tides is retained for power generation. India has a capacity of approximately 9 GW of tidal energy. OTEC is yet another alternative resource that allows the generation of steam-powered energy driven by thermal gradients. On the coasts of India, the Andaman Islands, and Macpherson Strait, currents can provide a source of hydrokinetic energy. Turbine prototypes have been tested, and there are possibilities of an annual energy generation potential of around 5 Mwh/m. India has a vast coastline, but the sad reality is that it remains underutilized. Other regions and hybrid projects hold promise and are also discussed.

Keywords: wave energy, tidal energy, renewable, sustainability, SDG

---

* Corresponding Author's Email: raulbasu@gmail.com

In: Breakthrough Research and Technologies in Physics …
Editors: Koyel Ganguly, Soumyadipta Pal and Kamakhya Prasad Ghatak
ISBN: 979-8-90134-111-7

Introduction

Wave energy harvesting converts wave kinetic energy into usable electricity. Ocean energy is reliable and sustainable. Various methods for harvesting ocean wave energy include the following:

1. Wave energy converters (WECs):(Zhang, 2022)
2. Floating platforms: (Gayathri, 2024)
3. Oscillating water columns (OWCs):(Konispiolatis, 2024)
4. Salter ducks: (Jiang, 2021)

Wave energy converters generate clean, renewable energy without greenhouse gas emissions. The development of wave energy is advancing in response to the growing demand for renewable energy and technological innovation. Proper investment in infrastructure can elevate it as a significant, sustainable power source. In the Indian Ocean, several devices, such as point absorbers and oscillating water columns, can harness wave energy to generate electricity. Over-topping devices drive water onto a reservoir, which then generates hydroelectricity. Key challenges for wave energy in the Indian Ocean include high costs, maintenance demands, and efficient grid integration. Several littoral countries, India, South Africa, and Australia, have invested in ocean power. The Indian government has proposed several coastal wave energy projects with prospects of up to 1,500 MW of renewable energy (but as of 2021, these are on hold). South Africa has committed to a wave energy research program, which is expected to be deployed by 2025.

Sri Lanka, with its ideal wave conditions, has initiated projects to harness wave energy, aiming to reach 200 MW by 2030. The government and private sector are collaborating to capitalize on this renewable resource, driven by rising energy demands.

Bangladesh has a high potential for renewable wave energy (IRENA, [Online]).

The Bangladesh Sustainable Renewable Energy Development Authority (SREDA) aims to develop policies, plans, and strategies to promote renewable energy, and the Bangladesh Oceanographic Research Institute (BORI) has been researching the wave energy potential of the Bay of Bengal. Additionally, many international organisations have supported Bangladesh's renewable energy programs (Solar Quarter, 26 May 2021, [Online]).

The Palk Straits wave energy power generator is another proposed system. Strong, consistent waves occur in the narrow Palk Straits. The power generators would have to be anchored to the ocean floor. As the waves pass over these devices, they will cause them to move, which in turn generates electricity through the use of a generator. This source represents a significant amount of renewable energy to the region, as the Palk Straits have significant wave energy potential. Furthermore, it would be environmentally friendly. The Palk Straits present an exciting opportunity for power generation.

The wave power generators in Bengal are also buoys anchored to the seafloor. The buoy sways and pulls the mooring line. The buoy contains a generator. The device typically consists of several buoys connected to a central hub. The electricity generated travels to an onshore facility through a transmission cable. Such energy is helpful in households, industries, and even for whole villages.

Kerala, located in India, has several advantages, as it borders the Arabian Sea to the west. Several state governments have initiated various initiatives to harness this potential and develop renewable energy sources. The Indian Ocean Rim Association (IORA) has also recognised Kerala as a strong candidate for the research and development of wave energy technology. The state has already established the first stage of a wave energy farm in Vizhinjam. Additionally, the government plans to develop a 10 MW wave energy park in the Alappuzha district, utilizing the most advanced wave technology in India. This technology for wave power generation can reduce greenhouse gas emissions by a credible margin.

In most other regions dealing with wave energy, companies focusing on developing wave energy power generators continue to work in, including:

1. *EDF Energies Nouvelles*: This French company has an interest in wave energy, working on a wave energy converter which they refer to as Wavegen. The technology aims to transform wave energy into electrical power using oscillating water columns.
2. *HydroQuest:* HydroQuest is a French company with a focus on tidal and river energy, but it is also developing wave energy converters that utilize sea levels to generate electrical power. The company's devices are wave energy converters for offshore installations in shallow waters.
3. *Subhydro:* Another French company working towards wave energy is Subhydro. Subhydro is a wave energy converter using a buoy to harness wave energy to give power to an offshore turbine.

4. *Ocean Power Technologies*: Ocean Power Technologies is not French but has installed some of its wave energy buoys near the coasts of France as part of its trial studies. The company's PB3 PowerBuoy is engineered to harness the forces of the waves to generate electricity using a direct drive generator system

In general, companies in France and those working on wave energy projects are keen on developing innovations to harness energy from the raging and chaotic waves of the ocean. Also, still at the nascent stage, wave energy has the potential to play a significant role in France's transition to renewable energy sources (ec-nantes.fr., Cruz,2008, Cavallero,2021}

Technology

1. *Various wave energy scavenger devices include:* 1) Oscillating Water Column (OWC). The OWC comprises two chambers separated by partitions, with one chamber facing the sky and the other facing the ocean. The device has a piston that can reciprocate linearly by placing it centrally between two flexible chambers.
2. *Point absorbers* - The construction of a buoy plays a vital role in point absorber systems. Buoy construction involves a circular, cone-shaped floating object that sits partially above the ocean surface, with its base underwater. The motion is enhanced by the geometry of the buoy, which is elliptical in cross-section, and improves wave sensitivity. These devices can be interconnected. Recent innovations have optimised this device (Jiang, 2021).
3. *Wave snakes* - This device utilises a series of connected floating modules, which move up and down independently in response to wave action. The relative motion generates electricity that can be collected and used. The wave energy converter is named after a genus of sea snake, due to its long, snake-like shape. The device consists of several hinged cylindrical modules, which flex as waves pass over them. This movement drives hydraulic rams that pump high-pressure fluid through hydroelectric generators. The electricity can be transmitted to shore via cables and integrated into existing power grids. The technology has been successfully tested in Portugal and Scotland and can be a significant renewable energy source in coastal regions worldwide (Shipon, 2021).

4. *Oscillating wave surge converters* - This device utilizes the surge motion of waves to generate electricity by using hydraulic systems to convert the motion into rotational motion (B-Munoz, 2021).

Wave power is a consistent and reliable source of energy. However, such devices are still in the development stage, facing numerous technical challenges, such as adapting to harsh environments and efficiency limitations.

Methods of Estimation

1. The energy density is defined as the extractable power from the motion of the wave and is mostly quantified in kilowatts per meter (kW/m).
2. The size of the wave energy generator is predicted. The output power of a generator is proportional to its size and its type. Most wave energy converters are between a couple of meters and several tens of meters in length.
3. The capture width is calculated. This is the part of the device that collects the wave energy.
4. The efficiency of the wave energy converter is estimated. The usual value of conversion ratios is between 0.3 and 0.5.
5. Calculate the output power by applying the formula: Power = Energy density x Capture width x Converter efficiency. For example, if Energy density = 20 kW/m, Capture width = 10m, and Converter efficiency = 40%. Then, Power = 20 kW/m * 10 m * 0.4 = 80 kW. This implies that' the wave energy generator is capable of producing electrical energy up to 80 kilowatts in these circumstances.

Several other tweaks apply statistics and spectral techniques to work out the net power output.

Feasibility Analysis

Wave Energy Potential in Sri Lanka

Sri Lanka's potential for wave energy generation is substantial (Karunarathna, 2021). The country has identified 14 sites along its coastline for building power plants that utilise energy from ocean waves (Eco-Business, online). The research analysed wave patterns in the coastal belt to identify suitable locations for implementing wave energy converters. Existing wave energy conversion technologies were compared, and computational simulations were conducted to determine the most appropriate technology for local wave conditions. It was found that the optimal solution for the selected coastal region was the "point absorber". Buoy geometry further analysed using simulations, revealed that the half-capsule shape resulted in the highest force extraction compared to other shapes (Anjna et al, 2021).

Prospects of Meeting SDGs

Renewable Wave power generation can assist in multiple UN Sustainable Development Goals (SDGs), highlighting its future significance and contribution to sustainable economic Growth and environmental preservation.

Wave power energy, having an energy density five times greater than wind and 10 to 30 times higher than solar, can offer a consistent annual source, making it a promising option for sustainable power generation in various regions (P-Rodriguez et al, 2021). By harnessing this largely untapped clean energy source, ocean power generation can meet a significant portion of the world's electricity demand, greatly supporting the global transition to affordable and clean energy (Goal 7) (Rupesh et al, 2021). Moreover, wave energy generation contributes to reducing emissions from ships and promoting the utilization of marine energy, aligning with environmental protection and energy-saving initiatives, thus supporting efforts to combat climate change and its impacts (Goal 13) (YuXin et al., 2020). It also offers an alternative to fossil fuels, helping to reduce environmental harm and global warming while meeting the increasing demand for electrical energy worldwide, which is crucial for fostering innovation and infrastructure development (Goal 9) and ensuring access to affordable, reliable, sustainable, and modern energy for all (Balo etal, 2017).

The wide distribution, high quality, and vast potential value of ocean energy in global energy markets emphasise its role in driving economic Growth and promoting sustainable industrialisation (Goals 8 and 9). However, challenges such as low efficiency and instability necessitate further research and testing fo full exploitation (Sun etal, 2022).

Devices based on wave energy offer a promising solution for achieving sustainable development goals by harnessing a pollution-free, clean, and renewable energy source, positively contributing to marine construction and economic development. An estimated potential worldwide contribution of wave energy to the electricity market is approximately 2,000 TWh per year, representing around 10% of the world's electricity consumption. This significantly furthers efforts toward sustainable consumption and production patterns (Goal 12) (Zhang et al., 2023; Benassi et al., 2009).

In summary, the integration of wave energy into the global energy mix presents an effective avenue for mitigating climate change, fostering economic Growth, and promoting social well-being in conjunction with environmental sustainability, echoing the interconnected and holistic nature of the SDGs. Studies have also been done to test conformity with IEC standards (Lokuliyana et al., 2020).

Financing of Renewable Projects

A commonly raised objection to renewable energy projects is the high capital outlay. Funding agencies are wary of investing in projects with high risk involved. To counter these misgivings, it needs to be emphasised that the source is perpetual, being solar energy and gravitation. Amortization tables for various payback periods, interest rates, and annual income streams are available for renewable energy schemes. Recent studies based on the Pelamis generator calculated the costs (Dixit et al., 2023; Guo et al., 2023). The "levelized cost of energy" (LCOE) for the Pelamis was estimated to be EUR 735.94 per MWh.

The cost of Energy formula is

[CC =Capital Cost + [PresentValue(O&M)]/[PresentValue(Energy Production)]

(1)

Economic analyses of wave energy projects often involve using key financial indicators such as Net Present Value (NPV) and Levelized Cost of Energy (LCOE). (Tetu et al.,2021, Teillant et al., 2012).

- *Levelized Cost of Wave Energy (LCOE):* LCOE to assess wave energy generation costs, combining capital expenditure (CAPEX), operational expenditure (OPEX), decommissioning costs, annual energy production, discount factor, and project lifespan. Findings indicate that with significant CAPEX and OPEX reductions and gradual increases in availability and capacity factors, LCOE decreases from USD 470/MWh to USD 120/MWh (Tetu et al., 2021), (Teillant, 2012)

- *Economic Assessment Guidance for Wave Energy Converters*: a framework for evaluating wave energy concepts at early stages, focusing on achieving target LCOE. It outlines entry costs for specific markets and uses a technology-agnostic cost database for detailed cost breakdowns.
- *Productivity and Economic Assessment of Wave Energy Projects*: a novel approach to assess wave farm profitability through detailed operational simulations and economic evaluations.
- *Cost of Energy – Ocean Energy Systems*: This study examines the development, fabrication, and integration of wave, tidal, and OTEC systems into commercial arrays and large-scale plants. Economic analyses highlight cost drivers and strategies to make wave energy a more viable renewable source.

Discussion

International Commitments

The Paris Agreement is a legally binding commitment entered into by 196 countries in Paris at the COP21 in 2015, coming into effect on November 4, 2016. It works on NDCs (Nationally Determined Contributions) in five-year cycles. COP27 recognised that the 1.5 °C rise in global temperature required a recommitment to the 2020 targets (UNFCCC, 2022, online). Unfortunately, the reality falls short of achieving this goal. A question posed in the Indian Parliament in 2007 elicited a reply that renewable projects for wave energy were being scrapped if not kept on hold (Angadi, online, 2007).

Future Directions

1. *Challenges and Opportunities*: Infrastructure - Extending grids to islands remains a challenging and costly endeavor [29]. Extending grids to islands presents unique challenges and can be expensive.
2. *Mini-Grids and Decentralized Energy Technologies*: Decentralized energy technologies, mini-grids, and off-grid solar home systems can efficiently bring electricity to rural consumers.

3. *Challenges in Island Grid Systems*: Meeting energy needs sustainably, affordably, and reliably is a challenge. Remote Islands have historically relied on fuel imports.
4. *Operational Challenges in Small Grids*: Smaller electric grids face operational challenges similar to those in larger grids, but with additional constraints. They must manage varying load profiles with fluctuating demand, fewer generation facilities, and no interconnection, resulting in limited options for addressing supply constraints.
5. *Cost of Grid Extension for Rural Electrification*: can be high, and various factors can affect these costs. This is discussed in earlier works (Tetu 2021, Teillant 2012).

Innovations in ocean energy conversion are crucial for achieving sustainability, as renewable energy sources like ocean power play a key role in global energy transitions.

- *Diversification of Energy Sources* – Ocean energy offers a clean, renewable, and reliable alternative, particularly beneficial for resource-constrained nations.
- *Reduction of Carbon Emissions* – being a zero-emission energy source, ocean energy helps combat climate change by lowering the carbon footprint.
- *Energy Security* – Ocean energy, unlike wind or solar, is a stable and predictable power supply, unaffected by most weather conditions.
- *Economic Growth* – The development of ocean energy technologies creates employment in manufacturing, installation, and maintenance, boosting economic development.
- *Empowering Remote Communities* – Ocean energy can be a cost-effective power source for remote coastal and island regions.

Innovation can accelerate sustainable Growth, and integrating sustainable practices into agriculture and urban planning will also ensure food security and preserve natural ecosystems.

Public education and understanding of sustainability also encourage awareness of environmental conservation. Balancing Growth with the Environment is a challenging task, but it is certainly possible with collective attention and innovative thinking.

Efforts to Integrate Wave Energy into Power Grids

- Ireland is researching grid integration of wave energy devices, and storage technologies used for grid integration of (WECs), [30, 31] (Sucetti et al., 2023) (Said et al., 2021)
- The first floating device in the country is located off the Vizhinjam coast in Kerala, India. The project cost was between ₹15 and ₹20 crore (The Hindu, 2020).
- A Tidal Power project located in the Sundarbans area of West Bengal, India, has a capacity of 3.75 MW and spans an area of 4.5 km. [33] (Mongabay, 2021 (online).

These projects represent significant steps towards harnessing the power of ocean waves and tides for renewable energy generation in India. The development and implementation of such projects involve substantial challenges, high costs, and advanced technologies.

- *Kutch Tidal Energy Project*: The Gujarat government in India is developing the country's first tidal energy plant. The Gulf of Kutch has a total potential of 300 MW (Down to Earth, online)
- *Sri Lanka Wave Energy Projects*: Sri Lanka has identified 14 sites along its coastline for building power plants that harness energy from ocean waves. Apart from this, other projects on the island are:
  - WERPO Wave Power Plant:(Lokuliyana etal, 2020) (Renewables, 2017)
  - AW-Energy Project: around south SriLanka (News.lk, 2017)
  - Wave energy projects in the Maldives, Fiji, and Tonga: The Maldives: WEC experiments will be conducted with the invaluable assistance and cooperation of Holiday Inn Resort Kandooma at South Male Atoll, Maldives (OIST News, 2023).
- *Fiji:* Although Fiji has monitored wave potential at two sites, implementing projects will be pretty costly. However, Fiji is making strides in other areas of renewable energy. For instance, a feasibility study is being set up to assess remote communities that lack access to electricity, and then prioritise sites for capital investment in new solar-powered mini-grids (Singh, 2017).
- *Tonga:* The Kingdom of Tonga has signed an agreement with an Irish Sea-based Group for the development and deployment of the first

pilot commercial-scale two-megawatt (2 MW) Wave Power Park in the Pacific (Renewable Energy, 2023).

These projects represent significant steps towards harnessing the power of ocean waves for renewable energy generation in these regions. However, such projects involve advanced technologies. A small commercial wave power generator is described in “Details of Japanese prototype” (OISWT, 2023). Policy Support: Governments must promote the development and implementation of renewable energy sources, including ocean energy, to achieve sustainable development. They can incentivise ocean energy projects by the following:

Government Funding for R&D

Governments can advance ocean energy by providing and purchasing electricity from ocean generators, thereby ensuring a stable income for producers. Public-private partnerships can help mitigate the risks of developing new technologies. Legislation can also mandate that a portion of the total national energy come from renewable energy.

India has initiated a scheme known as O-SMART, which focuses on harnessing ocean resources and establishing ocean observation systems. For instance, a self-powering desalination plant has been set up in Kavaratti, Lakshwadeep islands (MOES, GOI, online)

India has declared ocean energy eligible for the non-solar Renewable Purchase Obligations (RPO) targets.

U.S. DOE recently announced up to $27m in federal funding for R&D projects on ocean waves for Offshore Renewables: provides several funding programmes for offshore renewable energy projects, including wind and ocean energy (Hydro Review, 2023)

Offshore Wind Projects Incentives in the U.S.: provides incentives such as investment tax credits, associated tax credit bonuses, and $100 million for offshore wind transmission planning.

Promising Directions: Research and Global Collaboration

- *Wave Energy Integration*: While many Wave Energy Converter (WEC) concepts have been tested globally, few have successfully exported power to the grid. Accelerating design and integration with government support is crucial for sustainable energy development (SINTEF, online)

- *Carbon Emission Reduction*: Local supply chains reduce transportation distances, significantly lowering emissions during transportation.
- *Economic Development*: Localized supply chains generate jobs, support local businesses, and improve living standards by increasing income levels.
- *Resilience:* Reducing dependence on global supply chains strengthens economic stability and ensures the consistent availability of goods.
- *Quality Improvement*: Local supply chains offer better control over production, thereby enhancing the quality of goods and services.
- *Community Engagement:* Involving local communities fosters loyalty and prosperity by creating stronger connections between producers and consumers.

In light of the UN's overarching mandate on climate change, most nations have acceded to this. They have made or are making provisions in their local legislation to enhance the transition to alternative energies. Excuses of the past no longer hold water and can be challenged legally. An international agency to help entrepreneurs challenge entrenched bureaucracy needs to be established independently or under UN auspices, to enforce compliance with International Law.

Conclusion

Governments can enable a shift to a low-carbon energy future. Local supply chain integration and the development of regional supply chains offer numerous advantages, including reduced carbon emissions, socioeconomic development, improved resilience, enhanced quality, and increased population involvement. To enhance understanding, technical assistance, investment in infrastructure, patronage of local customers and businesses, and collaboration at the grassroots level through the promotion of best practices are sufficient. The recent climate agenda of the United Nations has gained popularity, prompting nations to prepare for some of the threats posed by fossil fuels. The reasons why campaigns were not conducted earlier are that they are losing relevance, and bringing lawsuits can now make the bureaucracy a little more accountable. The establishment of an international authority with the power to compel all nations to comply with international treaties would be a good strategy to hasten these developments. At the moment, we cannot make any forecasts, although we hope the entire world shifts to carbon neutrality between the years 2030 and 2050. Only Time will tell, as hindsight is the best Indicator.

References

Anjana KGI, Deshan DMT, Haputhanthrie DS, Nissanka ID. Simulation-based approach in selecting a wave energy converter for Sri Lankan wave conditions. *Proc Moratuwa Eng Res Conf (MERCon)* (2021) 245-250. doi: 10.1109/MERCon52712.2021.9525675.

Angadi ASC. Durgaduani Tidal Power Project in West Bengal. Unstarred Question No 3130 Answered 07.12.2007. Lok Sabha Secretariat, India. Available: https://eparlib.nic.in/bitstream/123456789/559167/1/58674.pdf.

Balo F, Sagbansua L. Production systems of wave energy for renewable energy project designs. *Proc Int Adv Res Eng Congr* (2017):2618-2721.

Benassai G, Dattero M, Maffucci A. Wave energy conversion systems: Optimal localisation procedure. *Presented at the WIT Trans Built Environ*(2009). *WIT Transactions on Ecology and the Environment*, Vol 126, © 2009 WIT Press.

Benites-Munoz D, Huang L, Anderlini E, Marín-Lopez JR, Thomas G. Hydrodynamic modelling of an oscillating wave surge converter including power take-off. *J. Mar. Sci. Eng.* (2020) 8(10). doi: 10.3390/jmse8100771.

Cavallaro L, Vicinanza D. Offshore and onshore wave energy converters: Engineering and environmental features. *J. Mar. Sci. Eng.* (2021) 9(11): 1265.

Cruz J, ed. *Ocean wave energy: Current status and future perspectives*. Springer; 2008. doi: 10.1007/978-3-540-74895-3.

Dixit, K., & Yadav, K. (2015). Modelling and performance evaluation of the Pelamis wave energy converter considering Indian economic estimates. In: Sharma R, Kannojiya R, Garg N, Gautam SS, eds. *Advances in Engineering Design. FLAME 2022. Lecture Notes in Mechanical Engineering.* Springer, Singapore; 2023. doi: 10.1007/978-981-99-3033-3_27.

Gayathri R, Chang JY, Tsai CC, Hsu TW. Wave energy conversion through oscillating water columns: A review. *J. Mar. Sci. Eng.* (2024) 12:342. doi: 10.3390/jmse12020342.

Guo C, Sheng W, De Silva DG, Aggidis G. A review of the levelized cost of wave energy based on a techno-economic model. *Energies* (2023) 16:2144. doi: 10.3390/en16052144.

Jiang Y, Peng Y, Sun Y, et al. Design and testing of a mechanical power take-off system for a rolling-type wave energy converter. *Int J Precis Eng Manuf Green Technol* (2021) 8:1487 1499. doi: 10.1007/s40684-020-00253-z.

Karunarathna H, Maduwantha P, Ratnasooriya H. Wave power potential of Sri Lanka. *Engineer: J Inst Eng Sri Lanka* (2021) 54(2):1-6. doi: 10.4038/engineer.v54i2.7404.

Konispoliatis DN. Floating oscillating water column wave energy converters: A review of developments. *J Energy Power Technol* (2024) 6(1):005. doi: 10.21926/jept.2401005.

Li H, Sun X, Zhou H. Wave energy: History, implementations, environmental impacts, and economics. *Proc SPIE 12266, Remote Sensing and Modelling of Ecosystems for Sustainability* XIX (2022) 1226614. doi: 10.1117/12.2646119.

Lokuliyana RLK, Folley M, Gunawardane SDGSP, Wickramanayake PN. Sri Lankan wave energy resource assessment and characterisation based on IEC standards. *Renew Energy* (2020) 162:1255 1272. doi: 10.1016/J.RENENE.2020.08.005.

Pinilla Rodríguez JA, Pozos Texon FDJ, Gasca Caballero CJ. Wave Energy Potential for Sustainable Power Generation in Mexico. *IEEE ICEV 2023 Conf (2023).* doi: 10.1109/ICEV59168.2023.10329689.

Rupesh P, Rana HK, Shashidar S, Darshan MSS, Teja GV. Power generation to the coastal areas with an approach to wave energy device - A prototype. *IOP Conf Ser Mater Sci Eng* (2021) 1013:012035. doi: 10.1088/1757-899X/1013/1/012035.

Said HA, Ringwood JV. Grid Integration Aspects of Wave Energy—Overview and Perspectives. *IET Renew Power Gener* (2021) 15:3045–3064. doi: 10.1049/rpg2.12179.

Said HA, García-Violini D, Ringwood JV. Wave-to-grid (W2G) control of a wave energy converter. *Energy Convers Manag X* (2022) 14:100190. doi: 10.1016/j.ecmx.2022.100190.

Shipon MR, Ali MS, Kabir MA, Abdullah-Al-Mamun M, Farrok O. Pelamis wave energy converter. In: Farrok O, Islam MR, eds. *Oceanic Wave Energy Conversion*. Springer, Singapore; 2024. doi: 10.1007/978-981-99-9814-2_3.

Singh K. *Feasibility study of Fiji's approach towards 100% renewable energy by 2030.* doi: 10.13140/RG.2.2.17192.67841; 2017

Succetti F, Rosato A, Araneo R, Di Lorenzo G, Panella M. Challenges and perspectives of innovative grid systems in islands: A real case study. *Energies* (2023) 16(2):583. doi: 10.3390/en16020583.

Teillant B, Costello R, Weber J, Ringwood J. Productivity and economic assessment of wave energy projects through operational simulations. *Renew Energy* (2012) 48:220-230. doi: 10.1016/j.renene.2012.05.001.

Zhang D, Zhang Q, Liu Y, et al. An Overview of Wave Energy Converter Numerical Modeling. *J. Energy Res. Technol.* (2023) 145(1): 010801. doi: 10.1115/1.4056154.

Internet Resources

"Bangladesh: Asia Renewable Energy Statistical Profile," IRENA. [Online]. Available: https://www.irena.org/-/media/Files/IRENA/Agency/Statistics/Statistical_Profiles/Asia/Bangladesh_Asia_RE_SP.pdf.

How the latest Module Technology can Boost Bangladesh's Solar Development.

"DOE funding $27 million to accelerate ocean wave energy technology to market," Hydro Review, Jan. 2023. [Online]. Available: https://www.hydroreview.com/hydro-industry-news/oceantidalstream-power/doe-funding-27-million-to-accelerate-ocean-wave-energy-technology-to-market/#gref.

"Finnish firm to produce alternative energy from sea waves," News.lk, Sep. 2017. [Online]. Available: https://www.news.lk/news/business/item/18706-finnish-firm-to-produce-alternative-energy-from-sea-waves.

"Gujarat set to develop India's first tidal energy plant," Down to Earth. [Online]. Available: https://www.downtoearth.org.in/news/energy/gujarat-set-to-develop-india-s-first-tidal-energy-plant-38042.

"Harnessing marine renewable energy in Bangladesh," Solar Quarter, May 2021. [Online]. Available: https://solarquarter.com/2021/05/24/harnessing-marine-renewable-energy-in-bangladesh/.

"India's tidal power potential hampered by high costs and environmental risks," Mongabay, Aug. 2021. [Online]. Available: https://india.mongabay.com/2021/08/indias-tidal-power-potential-hampered-by-high-costs-and-environmental-risks/.

"Maintaining a clear intention to keep 1.5°C within reach," UNFCCC, Sharm el-Sheikh Climate Change Conference, Nov. 2022. [Online]. Available: https://unfccc.int/cop27.

"OIST harness energy ocean waves Maldives," OIST News Center, Mar. 2023. [Online]. Available: https://www.oist.jp/news-center/press-releases/oist-harness-energy-ocean-waves-maldives.

"O-SMART scheme," Ministry of Earth Sciences, Govt. of India. [Online]. Available: https://www.moes.gov.in/schemes/O-SMART?language_content_entity=en.

"Sri Lanka selects 14 sites for wave energy development," Eco-Business. [Online]. Available: https://www.eco-business.com/news/sri-lanka-selects-14-sites-for-wave-energy-development.

"The Kingdom of Tonga chooses Seabased wave," Renewable Energy Magazine, Mar. 2023. [Online]. Available: https://www.renewableenergymagazine.com/ocean_energy/the-kingdom-of-tonga-chooses-seabased-wave-20230320.

"Wave energy integration into power grids: Future requirements and opportunities," SINTEF Blog. [Online]. Available: https://blog.sintef.com/sintefenergy/renewable-energy/wave-energy-integration-into-power-grids-future-requirements-and-opportunities/.

"Wave power plant at Vizhinjam soon," The Hindu, Jun. 2020. [Online]. Available: https://www.thehindu.com/news/national/kerala/wave-power-plant-at-vizhinjam-soon/article31674205.ece.

"WAVEGEN®: The wave energy recovery prototype - SEM-REV," ec-nantes.fr. [Online]. Available: https://www.ec-nantes.fr/.

"WERPO needs USD 11m for wave project in Sri Lanka," Renewables Now, Mar. 2017. [Online]. Available: https://renewablesnow.com/news/update-werpo-needs-usd-11m-for-wave-project-in-sri-lanka-490233/.

Chapter 14

Performance Analysis of a Bio-Fuel-Based Stove Using the Response Surface Method

Rhythm Mukherjee[1,*], MBA
Rishita Das[1], MBA
Sougata Banerjee[2], MTech
and Preeti Sharma[1], PhD
[1]School of Management, Institute of Engineering & Management,
University of Engineering and Management, Jaipur, India
[2]Civil Engineering Department, Institute of Engineering & Management,
University of Engineering and Management, Jaipur, India

Abstract

Wood and charcoal are often used as cooking and heating agents in bio-stoves in both rural and urban areas of most developing and underdeveloped countries, as access to modern energy sources is limited. However, conventional bio-gas stoves tend to be inefficient and have harmful emissions. This work is part of research efforts to optimise an advanced biofuel-based stove commonly found in a typical village, as applied towards addressing the challenges. A prototype stove was developed at the University of Engineering and Management, Jaipur, and was tested for various activities. The analysis was performed using Response Surface Methodology, with the response variables being thermal efficiency and the boiling time required to heat 5 kg of water. From the study, the primary design parameters considered for analysis were the fuel weight (charcoal), the fuel grade, and the fan control level. The results of the experiment were statistically and analytically tested,

* Corresponding Author's Email: rhythm.mukherjee@uem.edu.in.

In: Breakthrough Research and Technologies in Physics …
Editors: Koyel Ganguly, Soumyadipta Pal and Kamakhya Prasad Ghatak
ISBN: 979-8-90134-111-7

following a scientific model design that was modelled using ANOVA. The results indicated that 'Weight of Fuel' and 'Fan Control Value' affect the efficiency of the stove and the boiling time. The optimal results for the study have been recorded at the fan control value of 2.87, weight of fuel (charcoal) being 0.55 kg, and the grade of fuel being natural lump charcoal, to produce the result of efficiency of the stove being 40.67% and the boiling time being 11.79 minutes. Interpreting the results of the entire study yields a comprehensive understanding of the relationship between the forecast and actual outcomes, which is validated using the development models. Overall, this study highlights that RSM can be effectively employed to optimise stop performance, resulting in improved fuel efficiency and reduced emissions.

Keywords: bio-fuels, environment, performance, efficiency

Introduction

In India, renewable energy remains a significant source of energy; however, the country's dependence on natural fuels, such as wood and charcoal, remains high. Although natural fuels like wood and charcoal account for only 10% of the national primary energy supply, it is vital for 2.8 million people solely rely on it for their cooking and eating needs in remote areas[Piabuo, S. M., Ngouhouo et al]. Domestic cooking, a fundamental aspect of daily life, is not just essential for survival but also crucial for maintaining a healthy lifestyle. In India, approximately 1.8 million people rely on biofuels as their primary source of energy, with over 90% of these individuals residing in developing countries. Instead of the availability of a variety of energy resources, around 84% of Indian village households are unable to access reliable cooking and lighting fuels[Cunha, M. P. (2019)]. In rural areas, access to gas or electricity is limited, oil fuels are expensive to afford, and kerosene is a luxury for a few, primarily in urban regions[FAO (2023)].

India is a pioneer producer of wood. Wood is a prevalent fuel that has been used for heating and cooking, particularly in rural areas where access to modern energy sources is limited. There has been extensive research in this field, which concludes that more than 60% of rural communities rely on wood as their primary source of fuel for cooking. On the other hand, in metropolitan areas, where various alternative fuels like LPG and electric stoves are more widely available, this dependence is noticeably much greater in rural, remote areas.

India obtains most of its fuel from a variety of sources, including plantations, farms, and natural forests. About 62% of this wood is harvested from public and private properties, with the remaining lumber coming from purchased timber (15%) and private lands (23%). Although this method of gathering fuel is necessary, it also leads to environmental damage and deforestation, particularly in areas with a strong demand for wood fuel.

Although less common in urban settings, wood fuel consumption remains significant, particularly in slum areas and among lower-income households. In many places, the shift to cleaner and more efficient cooking technology is made more difficult by widening economic and population gaps.

The Government of India has initiated several initiatives to promote the use of updated biomass stoves and alternative fuels, such as LPG and biogas, to mitigate the health risks associated with traditional wood-burning stoves. Currently, several restrictions exist that hinder addressing the problem, including accessibility issues, cultural disparities, and concerns about affordability. These reasons hinder the widespread adoption of technological advancements.

After subsequent research, the results and observations emphasise the present difficulties in reducing India's consumption of wood as a fuel, underscoring the demand for more comprehensive interventions and sustainable energy solutions [The Energy and Resources Institute].

There has been a significant global push towards the acceptance and sustainability of safe, clean, and eco-friendly fuels, with numerous initiatives aimed at developing cleaner energy sources to protect both human health and the environment [Okedere, O. B., et al]. Despite these efforts, a substantial portion of the population in developing countries, up to two-thirds, continues to rely on bio-fuels for various household activities such as cooking in the kitchen and keeping the house warm. In India, over thirty per cent of the population still relies on traditional cookstoves, which remain the most prevalent and widely used technology in both urban and rural communities for household heating and cooking [Chanchangi, Y. N].

The use of biofuels, such as food and charcoal, as energy sources has been around for a long time and is an essential field of study in the field of Sustainable energy and systems. This practice has been widely adopted in regions where conventional sources of energy are limited or have not yet been developed, such as in underdeveloped areas and remote villages. Among the various technological developments to utilise the energy potential of biofuels, gasification of stoves has proven to be a promising solution, which utilises LPG and also electric stoves, such as induction cookers. These stoves offer a cleaner and more efficient method of converting biofuel materials, such as timber (wood), residual items from the agricultural sector, and charcoal, into usable forms of energy, including heat and gas. Hence, the traditional method of cooking using open-fire methods with biofuels contributes to environmental degradation, including deforestation and air pollution. Additionally, the smoke produced by this open-fire cooking method poses serious health risks to humans by releasing harmful pollutants, including carbon dioxide, carbon monoxide, and particulate matter.

The necessity to significantly enhance the working efficiency of an improvised biofuel-based stove becomes particularly justified for maximising fuel efficiency, reducing pollutant emissions, and improving application effectiveness. There has been relevant research works which says that with technological enhancements, new stoves are designed to decrease the utilisation of fuel by up to 43% as studied in comparison to conventional open-fire based stoves[Ahmed, A et al], also, satisfying the design goals of the furnace, critical parameters such as the geometry of the combustion chamber, air inlets, and heat shields must be engineered. Working in this field, Response Surface Methodology (RSM) is a standard method that connects stove performance parameters to the response variables using statistical and engineered experimental design approaches for optimisation technique [Erhimona, D.K. et al].

In this study, an approach has been made to implement RSM to enhance the efficiency of the application of an improved bio-fuel-based stove by analysing how design variables influence fuel efficiency and the time required to boil a standard quantity of water as a test method of the apparatus.

Literature Review

Traditional stoves, however, often exhibit thermal efficiencies as low as 5-15% and release significant amounts of pollutants. Wood and charcoal are commonly utilised sources of fuel to be used in household works, mainly in the kitchen and keeping the house warm in winters, in many developing and underdeveloped countries to be used in open air stoves which have a thermal efficiency of approximately 5-15% [Khan, R. A et al].

Many studies indicate that air pollution within the house itself, from conventional biomass stoves, has proven to be a major contributor to human health hazards, such as asthma, lung cancer, and various breathing disorder problems [WHO, 2023]. Therefore, the need for improved heating and cooking stoves is a key area of research in the current scenario. New-age stoves are designed to regularly monitor the harmful emissions that affect health and efficiency concerns by utilising more sophisticated and standard combustion technology.

Optimising the parameters in which a stove operates is essential to enhance stove efficiency and reduce cooking time. The efficiency of a furnace is influenced by several factors, including the properties of the fuel used, the design of the combustion chamber, the airflow rate, and the heat transfer mechanisms. Response Surface Methodology (RSM) is a statistical experimental design technique used to create optimal designs. Several research studies have been conducted on the fabrication and performance assessment of an advanced charcoal stove, achieving a thermal efficiency of 34% with the aid of improved experimental design techniques [Komolafe, M. O et al].

Some researchers have highlighted the application of experimental design to observe the performance of charcoal cookstoves and obtain satisfactory statistical results [Ndécky, A.K., et al.]. The methods include Fischer's statistical test (F-test) and the Newman-Keuls test. The techniques were used to analyse the impacts of factors such as the material of the fuel, initial fuel consumption, Alternate air circulation, and the air circulatory system on the thermal efficiency and carbon monoxide emissions of cookstoves during various phases of cooking, including boiling and simmering.

Some research studies have focused on insulation techniques for stoves, involving the enhancement of ceramic insulation for cooking stoves by using carbonised (unburnt carbon particles removed) organic waste [Kyuvi, N. et al.].

The application of Multi-response Robust Parameter Design (MRPD) was experimented with to optimise a hybrid built prototype biomass cook stove, utilising RSM models to improve stove performance, increase efficiency, and minimise emissions and errors [Kshirsagar, M. P et al.].

Some studies have evaluated the performance of locally fabricated cookstoves in underdeveloped countries, such as Nigeria [Abasiryu, A. R., et al.].

Some scientists and researchers have investigated the optimisation of biofuels, such as natural lump charcoal and activated carbon materials, using jute-extracted core materials through Artificial Neural Network (ANN) combined with Response Surface Methods [Natrayan, L et al].

This study aims to highlight environmental conservation by researching to enhance the performance of an improved bio-fuel-based heating and cooking stove, by applying RSM to optimise the design and working of the furnace and also focusing on optimising the overall efficiency and reducing harmful emissions that pollute the environment.

Research Methodology

In this study, an advanced prototype biomass stove was employed for experimentation. The store is composed of several key components, including the main body, liner, stove top, base, combustion chamber, insulation, grate, handle, and fan. The built-up of the stove features precise measurements, standing 300 mm tall, with a diameter of 250 mm, and a Sheet thickness of 0.8 mm. The combustion and heating chamber is crafted from 1 mm-thick 304-grade stainless steel, with dimensions of 200 mm in height and 125 mm in diameter. Air flow is facilitated by a primary intake fan, sized 92 x 92 x 25 mm, which channels the air into a 75 mm diameter circular channel port located at the bottom half of the stove. Additionally, the stove is equipped with an insulation panel of a right cylindrical shape, made from a mixture of briquette rice husk and a binder, with dimensions of 250 x 210 x 53.5 mm, attached to the top and sides of the combustion chamber. The 3D CAD model was created using Revit Software.

A. Sample Preparation/Experimental Setup

During the experimental setup for the study, four different types of fuel were used: 'natural lump charcoal', 'coconut charcoal', 'activated charcoal', and 'briquettes'. To ensure an equitable comparison, several experimental adjustments were made to standardize the quantity of each sample according to its respective weight differences. Experiments were conducted in a laboratory to evaluate the thermal efficiency and boiling time of the biomass gasifier stove under controlled conditions. During the experiment, the fan speed was meticulously regulated. The setup included crucial instruments such as a digital weighing scale and thermocouples. The experiment utilized K-type thermocouples to capture key response variables, including temperature readings at continuous points internally within the gasifier stove, from main application points such as the heating surface. The experimental setup also included the deployment of a digital weighing scale to record the weight of every fuel sample accurately.

B. Experimental Designs

The experiment in this research study was conducted in a controlled manner to obtain the most accurate results. A specialized statistical design, known as the I-optimal design, an essential function within Response Surface Methodology (RSM), was utilized. I-optimal designs are aimed at minimizing prediction variance within the design space. This approach is crucial for making accurate predictions in RSM. The applied algorithm strategically selects points where the predicted variance across the experimental region is minimized. This study focuses on enhancing the overall performance of a stove that uses wood or charcoal as fuel, measuring three performance attributes as responses.

Results

A. Analysis of Variance (ANOVA)

For the statistical modelling and data validation, ANOVA is used. It is a process which includes statistical and estimated techniques. It is generally used to interpret or examine variations in both inter-group and intra-group settings. Hence, this experiment was evaluated with the implementation of ANOVA. It significantly contributed to determining the outcomes, yielding valid results. The tables below, Table 1 and Table 2, illustrate how the outcomes from ANOVA have been beneficial in understanding the values of thermal efficiency and boiling time performance outcomes from the experiment carried out. The following tables also help highlight the statistical significance of the response variable factors considered in the study, providing valuable information regarding the design coefficients involved in the applied models.

Table 1 presents the results from the ANOVA for the thermal efficiency test, highlighting the significance of various factors that influence it. The model is statistically significant ($p = 0.0004$) with a high F-value of 39, indicating its effectiveness in explaining variability. Among the factors, "Fan control value" (A) and "Type of Charcoal" (C) are highly significant with p-values <0.0001 and 0.0014, respectively, demonstrating their substantial impact on thermal efficiency. Additionally, the quadratic effect of the fan control value (A^2) is also highly significant ($p = 0.0001$). Interactions such as AC (Fan control and Type of Charcoal) show moderate significance (p =

0.0050), while the AB and BC interactions are insignificant with p-values greater than 0.05. The residuals are low, indicating a well-fitted model, confirming the model's adequacy in capturing the experimental data.

Table 1. Results from ANOVA for thermal efficiency

Source	Sum of Squares	df	Mean Square	F-Value	p-Value
Model	152	14	10.68	39	0.0004
A-'Fan control value'	38.97	1	39.55	142.36	<0.0001
B-'Weight of the Charcoal'	2.08	1	2.05	7.51	0.0305
C- 'Type of Charcoal'	23.55	3	7.75	28.52	0.0011
AB	0.149	1	0.149	0.5638	0.4870
AC	14.42	3	4.79	17.38	0.0050
BC	0.3762	3	0.1235	0.4529	0.7261
A2	31.97	1	32.08	116	0.0001
B2	0.1630	1	0.1711	0.6212	0.4666
Residual	1.29	5	0.2768		
Lack of Fit	1.34	2	0.6928		
Pure Error	0.0000	3	0.0000		

Table 2. ANOVA results for boiling time performance

Source	Sum of Squares	df	Mean Square	F-Value	p-Value
Model	53.57	14	3.92	19.07	0.0017
A-'Fan control value'	38.80	1	38.72	195	<0.0001
B-'Weight of the Charcoal'	0.280	1	0.328	1.62	0.2712
C- 'Type of Charcoal'	5.97	3	2	10	0.0148
AB	0.0917	1	0.0937	0.4676	0.5140
AC	0.3811	3	0.1261	0.6410	0.6007
BC	0.1951	3	0.0649	0.3289	0.8044
A2	0.2918	1	0.2911	1.50	0.2781
B2	0.0396	1	0.0388	0.1980	0.6618
Residual	0.9897	5	0.1987		
Lack of Fit	0.9888	2	0.4945		
Pure Error	0.0000	3	0.0000		

Table 2 presents an analysis of the results from an ANOVA examining the recorded boiling time of 5 kg of water, with a focus on the significance of various factors influencing the process. The "Model" explains a substantial segment of the response variance (Sum of Squares = 53.57, p-value = 0.0017), indicating a strong overall fit. Among the individual factors, the "Fan control value" (A) is highly significant with a p-value < 0.0001, making it the most impactful variable on boiling time. In contrast, the "Weight of the Charcoal" (B) and other interactions, such as AB, AC, and BC, show a limited influence with higher p-values, indicating non-significance. The "Type of Charcoal" (C) is moderately significant (p-value = 0.0148). The residual sum of squares is relatively low, indicating minimal unexplained variance, while the lack-of-fit test suggests that the model adequately fits the data without significant error contributions.

B. Efficiency of the bio-fuel stove

Based on the recorded observations and experimental design, a table was created to enhance the optimization of the store's performance using Design Expert software. For each case, coded equations were formulated and developed to determine the expected outcomes from the given data. Investigating the formulation and recorded data, a graph can be plotted for further research studies comparing the 'predicted' values with the 'actual' experimental results.

Table 3. Comparative metrics of actual and predicted values for thermal efficiency

Run Order	Actual Value	Predicted Value	Residual	Leverage	Run Order
1	39.2	38.77	0.2416	0.928	1
2	38.1	38.57	-0.6558	0.412	2
3	40.1	40.21	-0.3008	0.412	3
4	46.1	45.74	0.1700	0.980	4
5	38.1	37.87	0.0433	0.492	5
6	39.1	38.88	0.0322	0.496	6
7	36.1	36.15	-0.2389	0.887	7
8	43.1	42.61	0.2995	0.888	8
9	39.1	38.88	0.0322	0.496	9
10	37.1	36.89	0.0222	0.987	10
11	38.3	38.16	-0.0492	0.977	11
12	38.1	37.87	0.0433	0.492	12

Table 3 presents the results of a study analyzing the relationship between actual values, predicted values, residuals, and leverage for 20 experimental runs. The "Actual Value" represents the observed measurements, while the "Predicted Value" shows the expected outcome. The "Residual" is the difference between the actual and predicted values, indicating the model's accuracy, with most values showing minor deviations. "Leverage" measures the influence of each data point on the regression model, with values near 1 signifying high influence. Overall, the slight variations in residuals and leverage indicate a reliable model with consistent predictions that closely align with actual observations.

The analysis reveals a close match between 'Predicted' and 'Actual' values. The results for the predicted data and the exact values for the charcoal-based stores' thermal efficiency are plotted and organized as shown in Table 3. Figure 1 presents a contour diagram that explores the relationship between three variables of the biomass gasifier – 'fan control value' (A), 'charcoal weight' (B), and 'activated charcoal' (C)- and thermal efficiency. The figure illustrates a significant relationship between the fuel's weight and size in terms of thermal efficiency. As the weight of the fuel increases with the decrease in particle size, it is observed to enhance the efficiency of the stove. This indicates that heavier and smaller charcoal enhances the stove's thermal efficiency. Figure 2 further elaborates on this relationship by depicting a 3D diagram that illustrates the interplay between these three variables and thermal efficiency.

Table 4. Analysis of variance results in the experimental approach

Source	Sum of Squares	df	Mean Square	F-Value	p-Value
Model	53.61	14	3.87	19.41	0.0020
A-'Fan control value'	38.53	1	38.50	195.03	<0.0001
B-'Weight of the Charcoal'	0.3328	1	0.3328	1.70	0.2512
C- 'Type of Charcoal'	6.01	3	2.01	10	0.0148
AB	0.0918	1	0.0928	0.4676	0.5240
AC	0.3806	3	0.1267	0.6418	0.6207
BC	0.1949	3	0.0654	0.3293	0.8054
A2	0.2918	1	0.2917	1.51	0.2791
B2	0.0396	1	0.0396	0.1985	0.6748
Residual	0.9894	5	0.1978		
Lack of Fit	0.9894	2	0.4948		
Pure Error	0.0000	3	0.0000		

Table 4 presents an ANOVA analysis of the boiling time for 5 kg of water, evaluating the effects of various factors. The model remains significant overall (Sum of Squares = 53.61, p-value = 0.0020), indicating that it explains a substantial variance in boiling time. The "Fan control value" (A) is the most significant factor, with an extremely small p-value (<0.0001) and the highest F-value (195.03), demonstrating its critical influence. The "Type of Charcoal" (C) also makes a significant contribution (p-value = 0.0148). In contrast, factors like "Weight of the Charcoal" (B) and interactions such as AB, AC, and BC exhibit non-significance with higher p-values. The residual variance is minimal (Sum of Squares = 0.9894), and the low mean square values for residuals indicate the model's reliability in fitting the data with minimal unexplained variance.

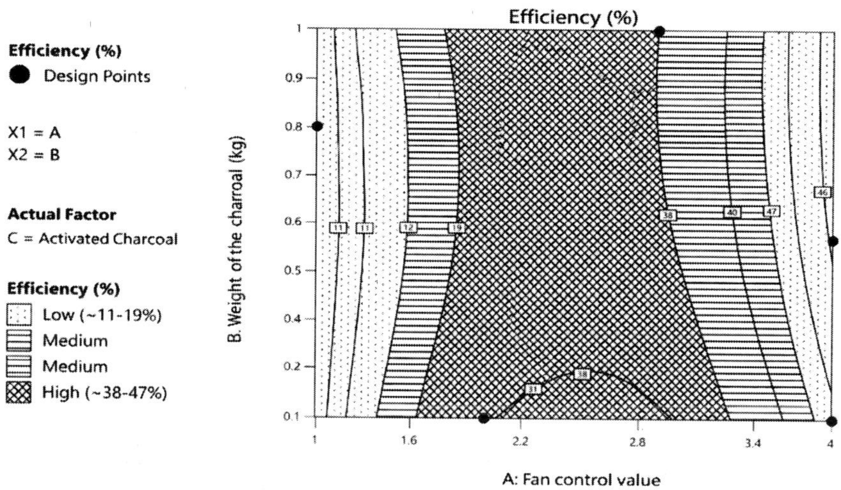

Figure 1. Efficiency of stove as compared to weight of Charcoal (B) vs Fan-Control value (A).

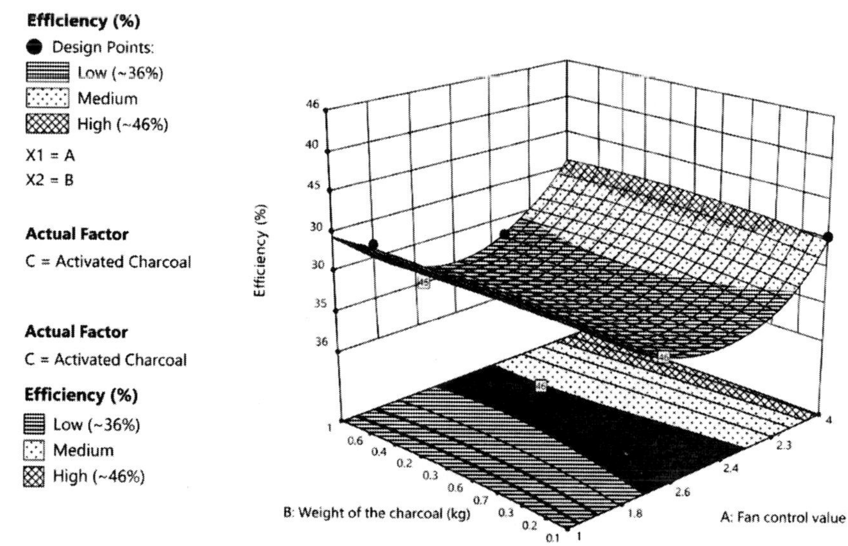

Figure 2. Efficiency curve of the stove in the experiment.

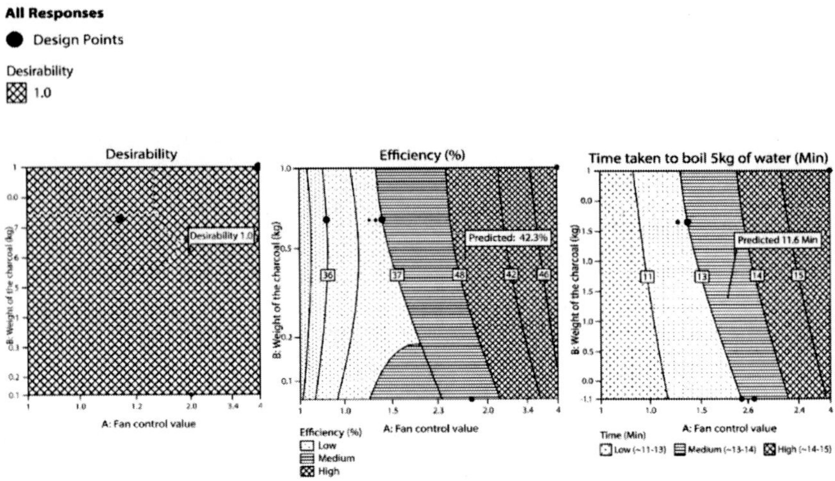

Figure 3. Efficiency of the stove with natural lump charcoal.

Based on the analysis of the combination of the three categorical factors, the optimal design adjustments and design parameters were determined to be a 'fan control value' of 2.87, a 'charcoal weight' of 0.55 kg, and the 'grade of fuel' being Natural Lump Charcoal. These conditions are expected to yield a thermal efficiency of 40.67% and a boiling time of about 11.79 minutes to bring 5 kg of water to a boil, as in this study.

Conclusion

Based on the analysis of the combination of the three categorical factors, the optimal settings were determined and recorded at a fan control value of 2.87, a fuel weight (charcoal) of 0.55, and a grade of fuel (natural lump charcoal). These conditions are expected to yield a thermal efficiency of 40.67% and a boiling time performance of 11.79 minutes to bring 5 kg of water to a boil.

Acknowledgments

This section of the paper highlights and acknowledges the contributions of each person involved in the successful completion of this scientific study. Every effort in this study has played a significant role in completing the total work. This section of the paper takes the opportunity to acknowledge the authors and their innovative work.

The authors, Rhythm Mukherjee, Rishita Das, Sougata Banerjee, and Dr. Preeti Sharma, have made significant contributions to this successful scientific study. The constant guidance, motivation, encouragement, and support throughout this project have served as a ladder to its completion. Additionally, the authors' insightful and constructive knowledge, along with their expertise in highlighting innovation in the relevant field of research, has played a pivotal role in presenting this work. Nearby, heartfelt thanks to all the authors, colleagues, and the institution for the support provided, which made this study complete. Sincere gratitude to every unit of the efforts that have been put into this work to achieve the milestone.

References

Abasiryu, A. R., Garba, M. M., & Baba, B. A. (2016). Evaluation of locally fabricated biomass cookstoves in Nigeria. *Nigerian Journal of Technology, 35(1)*, 65-70.

Adamu Sh. Tadesse, Elias Gabisa, Basazzin A. Mekonnen et al. Prototyping and Performance Evaluation of Household Cow Dung Gasifier Stove: Cooking and Environmental Performance in the Case of Ethiopia, 09 June 2020, *PREPRINT (Version 1)* available at *Research Square* [https://doi.org/10.21203/rs.3.rs-30310/v1.

Ahmed, A., & Iqbal, M. (2018). Comparative performance analysis of traditional and improved cookstoves for bio-fuels. *Energy for Sustainable Development*, *45*, 1-8.

Chanchangi, Y. N., Abubakar, M., Hussaini, U., Usman, M. U., & Magaji, A. (2023). Assessment of rural household energy use and fuelwood demand in northern Nigeria. *Journal of Rural Studies, 94*, 15-25.

Cunha, M. P. (2019). Renewable energy: Transition and policies in India. *Energy Policy Research*, 6(3), 45-56.

Erhimona, D. K., Oghenevwogaga, O. F., & Asiwaju-Bello, A. Y. (2023). Optimisation of Biomass Stove Performance Using Response Surface Methodology. *Journal of Cleaner Production, 312*, 127735.

FAO (2023). *Forest Products Annual Market Review 2022-2023*. Food and Agriculture Organisation of the United Nations.

Ifeanyichukwu U. Onyenanu, Miracle C. Ogbogu, C. J. Nwadiuto, 2024, Performance Optimisation of an Improved Biomass Gasifier Charcoal Stove using Response Surface Method (RSM), *International Journal Of Engineering Research & Technology (IJERT)* Volume 13, Issue 08 (August 2024).

Khan, R. A., Pathan, H. I., & Siddiqui, M. R. (1995). Air pollution due to combustion of bio-fuels in traditional stoves in developing countries. *Energy Conversion and Management, 36(11)*, 1073-1077.

Komolafe, M. O., & Awogbemi, O. (2010). Performance assessment of improved biomass cookstoves. *Energy Sources, Part A: Recovery, Utilisation, and Environmental Effects, 32(3)*, 262-270.

Kshirsagar, M. P., & Kalamkar, V. R. (2020). Optimisation of biomass cook stove using multi-response robust parameter design. *Energy, 195*, 116992.

Kyuvi, N. (2022). Enhancement of Ceramic Insulation for Biomass Cookstoves Using Carbonised Organic Waste. *Journal of Environmental Management*, 301, 113856.

Natrayan, L., Marimuthu, R., & Meenakshi, G. (2023). Optimisation of bio-fuels using response surface methodology and artificial neural network. *Renewable.*

Ndécky, A. K., Ndiaye, D., & Ndiaye, B. P. (2018). Experimental design analysis of improved biomass cookstoves in Senegal. *Journal of Renewable Energy*, 2018, Article ID 8565430.

Okedere, O. B., Oguntoke, O., & Shitta, M. B. (2022). The impact of bio-fuel production and utilisation on climate change. *Renewable and Sustainable Energy Reviews, 136*, 110343.

Piabuo, S. M., Ngouhouo, I., Feudjo, J. R. E., & Tieguhong, J. C. (2020). Assessment of the impact of wood fuel consumption on the forest cover in the East Region of Cameroon. *Energy for Sustainable Development, 59*, 1-9.

Shah, S.B., Wang, Y., Anwar, N. et al. Co-metabolic degradation and metabolite detection of hexabromocyclododecane by *Shewanella oneidensis MR-1. Appl Microbiol Biotechnol 108*, 25 (2024). https://doi.org/10.1007/s00253-023-12905-6.

TERI (The Energy and Resources Institute). *Annual Report 2022-23*. TERI Publications.

Vinay Kumar Pandey, Poornima Singh, Shivangi Srivastava, Sonal Zanwar, Aamir Hussain Dar, Rahul Singh, Aditya Lal, Box–Behnken design based statistical modelling to study the effects of spirulina *(Arthrospira platensis)* incorporation on nutritional standards of vegan snack product, *Journal of Agriculture and Food Research*, Volume 14, 2023.

WHO (World Health Organisation). (2023). *Household air pollution and health*. World Health Organisation.

Chapter 15

VHF Radar and Radiosonde Observations of the Tropical Tropopause

Rina Dutta*, PhD

Department of Basic Science and Humanities, Institute of Engineering & Management (IEM), University of Engineering and Management, Kolkata, India

Abstract

In the present investigation, an attempt has been made to study the variation in tropopause height over three geographical locations: Chennai, Bangalore, and Hyderabad. The tropopause height is derived by identifying the coldest temperature in the upper troposphere and is referred to as the freezing point tropopause. The variability of cold point tropopause over different periods has been studied during the year 2002. Radiosonde and MST radar observations are made to measure the tropopause height and vertical wind variability in the vicinity of the tropopause. The Indian MST radar is installed at Gadanki (13.470°N, 79.180°E, and 0.375 km above mean sea level) near Tirupati, Andhra Pradesh. The location of the Indian MST radar is well covered by the area enclosed by the three stations mentioned above. In this communication, we conduct a comparative study of tropopause height over different time scales, including day-to-day and seasonal variations of the cold point tropopause. Finally, our analysis reveals that the vertical wind variability in the region of the tropopause indicates the existence of tropopause oscillation. Moreover, the vertical extent of the tropopause increases due to the highly developed thermal mixing and warm temperature. Furthermore, we extend our analysis to correlate the variation in

* Corresponding Author's Email: rina.dutta@uem.edu.in

In: Breakthrough Research and Technologies in Physics …
Editors: Koyel Ganguly, Soumyadipta Pal and Kamakhya Prasad Ghatak
ISBN: 979-8-90134-111-7

tropopause height with wind variability in the vicinity of the tropopause at these three stations.

Keywords: tropopause, MST radar, radiosonde, vertical wind, solar radiation

Introduction

The Earth's atmosphere is a comparatively thin film of gaseous mixture that extends from the ground to the upper boundary. Below the height of 100 km, the atmosphere is treated as a neutral fluid, whereas the ionised plasma plays a crucial role above it. The structure of the neutral atmosphere strongly depends upon temperature, which varies with solar activity. In the vertical direction, more than 99% of the atmospheric mass is found below an altitude of 30 km. The atmospheric physical properties, such as pressure, density, temperature, water vapour, refractive index, and wind velocity, are non-uniform in vertical and horizontal directions. These parameters are most sharply varied in the vertical direction compared to the horizontal direction. Typically, four different criteria—air composition, thermal condition, dynamics (Gettelman et al., 2011), and chemistry—are applied to divide the atmosphere into distinct layers in the vertical direction. According to composition, the atmosphere is classified into two parts – the Heterosphere and the Homosphere. In the homosphere from the surface up to 90-95 km level, the proportion of leading gases is argon (0.93%), oxygen (20.95%), and nitrogen (78.08%). Approximately 0.04% greenhouse gases, such as CO_2, CH_4, and O_3, are present within the air. Since these greenhouse gases are present in tiny amounts, they play a significant role in trapping heat radiation and retaining it in our atmosphere. Thereby affecting the warming of our planet and contributing to the Earth's greenhouse effect. Water vapour is another important greenhouse gas which warms and humidifies the atmosphere. Tiny liquid and solid particles (aerosols) are generated by natural or human activities, and are present in variable amounts. Solar radiation is the basic source of energy for the Earth's atmospheric system. The lower atmosphere is continuously unstable due to variations in its specific heat, atmospheric density, and refractive index.

Meteorologists typically define the tropopause as the lowest level at which the rate of temperature decrease with height is approximately 6.5^{0}°C/km. Vigorous vertical mixing of turbulent air in the vicinity of the tropopause is an essential property of the tropopause. The troposphere contains approximately

80% of the atmospheric mass, and all climate-related phenomena occur in this layer. For each kilometre of altitude increase in the troposphere, the temperature naturally decreases by the mean lapse rate. Its main characteristic is the atmospheric stability that resists the vertical and horizontal movement of air particles. The thickness of the troposphere (Fueglistaler et al., 2009) varies, with the tropospheric layer extending up to 16 km in the tropics. In contrast, its thickness in the polar regions is approximately 8 to 9 km. The vertical extent of the tropopause increases due to the highly developed thermal mixing and warm temperature.

The incoming solar radiation acts as the primary source of energy for the atmosphere. This energy remains responsible for all kinds of weather and climatic processes worldwide. Meteorologists have divided the weather system into four broad categories. Table 1 represents the scale size of the weather system.

Table 1. Different weather systems

Scale of Weather System	Horizontal Scale (km)	Time Scale	Example
Planetary	5000-10000	Weak or more	ITCZ, JET Stream, Rossby waves
Synoptic	1000-5000	2-7 days	Monsoon Depression, Tropical Cyclone, Low-pressure systems, Easterly waves, etc.
Mesoscale	100-1000	Few hours	Thunderstorm, Gust, Gale, Squall, etc.
Small	Few kms	Hours or less	Dust Storm, Trubulance

Tropopause - the upper boundary of the troposphere, which is characterised by hydrostatic stability and plays a key role in Stratosphere-Troposphere Exchange (Boothe and Homeyer, 2017). Variation in the height of the tropopause is one of the mechanisms through which interactions between tropospheric air and stratospheric air take place. Oscillations of the tropical tropopause (Jaya Rao and Jain, 2000) were observed during the passage of atmospheric waves, characterized by oscillations of 50-70 days, 30-50 days, and 10-20 days. Both vertical and horizontal propagations are possible for atmospheric waves (Birner, 2006), which can transfer momentum and energy from their point of origin to other locations.

Probing of the Atmosphere

Surface instruments and remote sensing techniques play a crucial role in studying various weather phenomena, a method known as atmospheric probing. Radiosondes and balloon sounding are examples of instruments used to maintain direct contact with the sample air. Measurements that depend on the position and direction of the radiosonde's flight (Son et al., 2011) are collected in this observation. However, they have some drawbacks, including poor height and time resolution. These limitations can be overcome to an excellent extent by the remote sensing technique. Remote sensing is a technique that involves collecting physical data about an object without direct contact with it. Radar and satellite are two necessary remote sensors to probe the atmosphere.

The remote sensing technique utilises electromagnetic energy at different bands of the electromagnetic spectrum. Any remote sensing system consists of three components: the source, which generates the electromagnetic signal; the transmitting antenna, which transmits the signal into the atmosphere; and the sensor, which receives the scattered or reflected energy from the target. Satellites, photometers, and radiometers are examples of passive remote sensors. The main advantage of passive remote sensors is that they are relatively simple in operation, both electrically and mechanically. Additionally, passive remote sensors require less power. However, they have certain limitations in preparing a good signal due to their extreme dependence on favorable weather conditions. However, different types of active remote sensors are used to probe the atmosphere, such as radar and lidar. In the case of active remote sensors, electromagnetic radiation transmitted in the form of pulses interacts with the target. The scattered energy, which depends on the target's characteristics, is received and provides basic information about the atmosphere. The following table shows that radar operates in different electromagnetic bands.

Table 2. Radar operating in different electromagnetic bands

Electromagnetic Band	Operating Wavelength	Scatterers	Frequency Range
S-Band	10 cm	Precipitation	2- 4 GHz
Ka-Band	1 cm	Precipitation, Cloud Droplets	18- 27 GHz
UHF	75 cm	Refractive index	300 MHz- 3 GHz
VHF	6 m	Refractive index	30 – 300 MHz

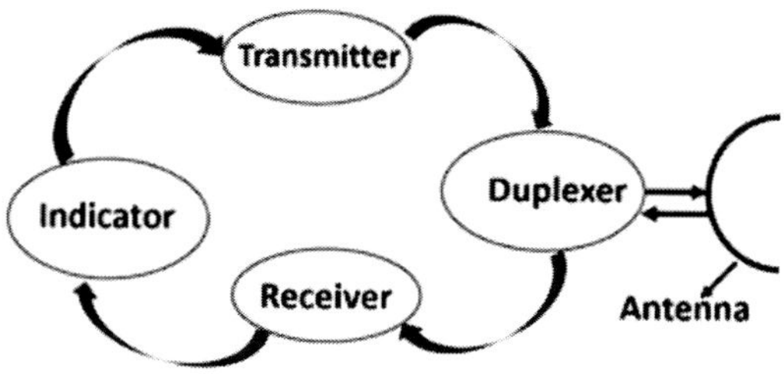

Figure 1. Simple Radar Block Diagram.

The radar transmitter generates short pulses of signals, which are radiated by the antenna. Figure 1 shows the block diagram of a radar system. A duplexer allows a single antenna to transmit and receive electromagnetic signals, protecting the receiver from the high power of a radar pulse during transmission. After interception, the target re-radiates a portion of the incident radiation (echo) in the direction of the radar. Few returned signals are collected by the antenna and detected by the receiver. Essentially, radar echoes occur due to variations in the refractive index of the atmosphere.

Data Analysis and Results

The study seeks to investigate the height profile of temperature for three stations Chennai (13^0 N, 80.18^0 E, altitude 16m), Bangalore (12.96^0 N, 77.58^0 E, altitude 921m) and Hyderabad (17.45^0 N, 78.46^0 E, altitude 545m) using radiosonde observational data for the year 2002. In Figure 2, it is seen that the minimum temperature is reached in the range -77^0 °C to -85^0 °C for these three stations over a height range of 15-17 km. It is noted that the heights of the tropopause for Chennai, Bangalore, and Hyderabad are approximately 15.73 km (-77.7^0 °C), 16.52 km (-81.1^0 °C), and 15.53 km (-84.3^0 °C), respectively. From this comparative study, it is clearly understood that the tropopause height is maximum at the Bangalore station compared to the other two stations. Since the elevations from mean sea level for these three stations differ, they represent distinct environments, as indicated by the cold point tropopause.

The subsequent investigation is the day-to-day variation of tropopause height for the station Chennai (shown in Figure 3). Through a thorough analysis, it is observed that the tropopause height variation is most significant in May. This indicates that the atmosphere is drastically heated during summer, as the maximum amount of solar radiation incident on the Earth's surface occurs during this season. The height of the tropopause is observed to be lower during the June-August period (monsoon) compared to the pre-monsoon and post-monsoon periods. Due to the cloud cover in monsoon, less solar radiation falls on the terrestrial surface, which diminishes the tropopause height.

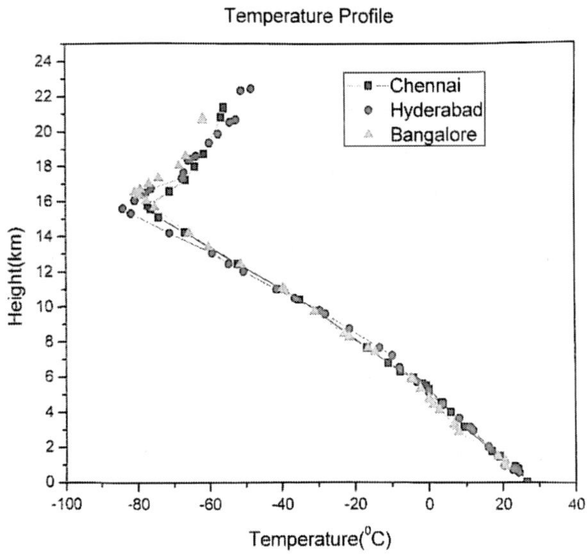

Figure 2. Plot of Height Profile of temperature for Chennai, Bangalore, and Hyderabad.

We also conducted a fascinating comparative study of tropopause oscillation for all three stations - Chennai, Bangalore, and Hyderabad. Tropopause height fluctuations are observed to be more pronounced (as seen in Figure 4) at the Hyderabad location compared to the other two stations. Simultaneously, a similar type of oscillation (Figure 5) is detected in tropopause temperature. The reason for this is found to be the fact that due to the presence of wind circulation in the vicinity of the tropopause. This air circulation (Peevey et al., 2014) is responsible for the exchange of air masses between the troposphere and the stratosphere. Hadley circulation and jet streams (Manney et al., 2014) play a significant role in wind circulation.

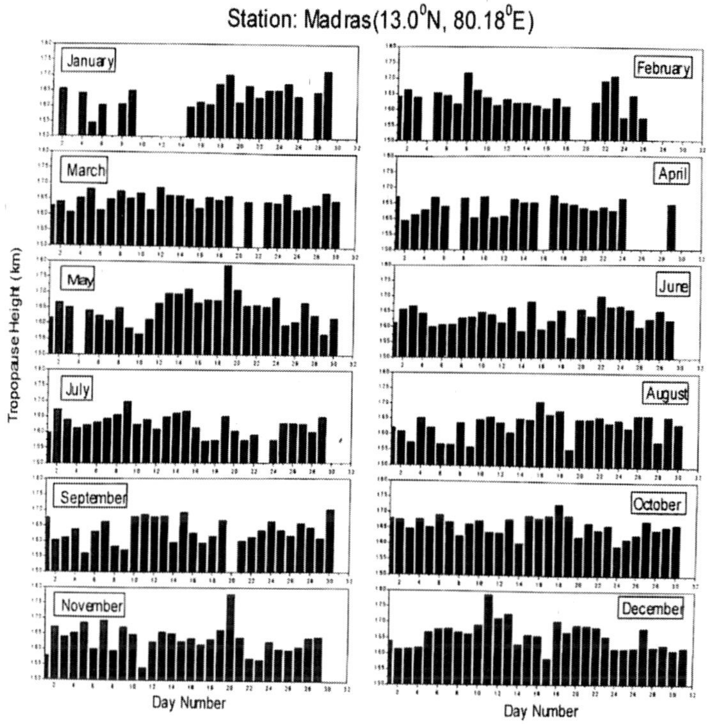

Figure 3. Plot of variation of tropopause height for the station Chennai (Day-to-day).

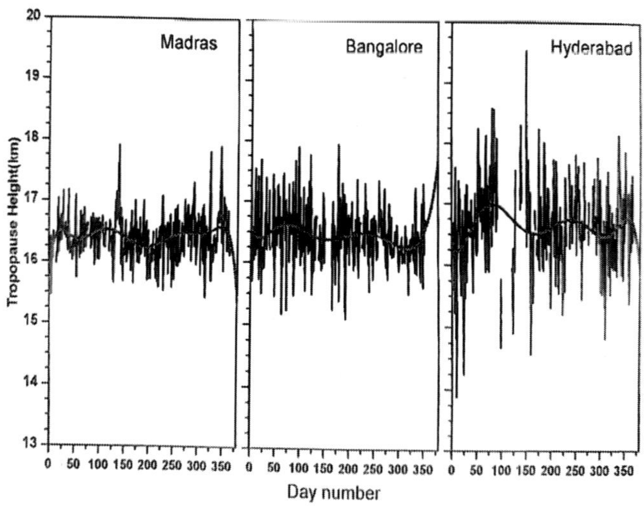

Figure 4. Variation of tropopause height (Day-to-day).

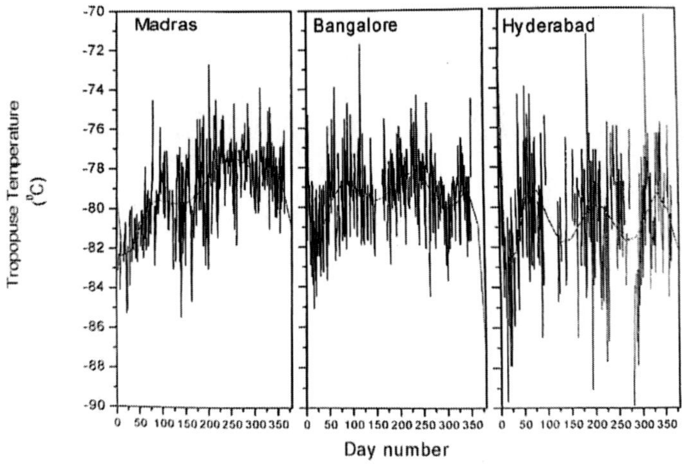

Figure 5. Variation of tropopause temperature (Day-to-day).

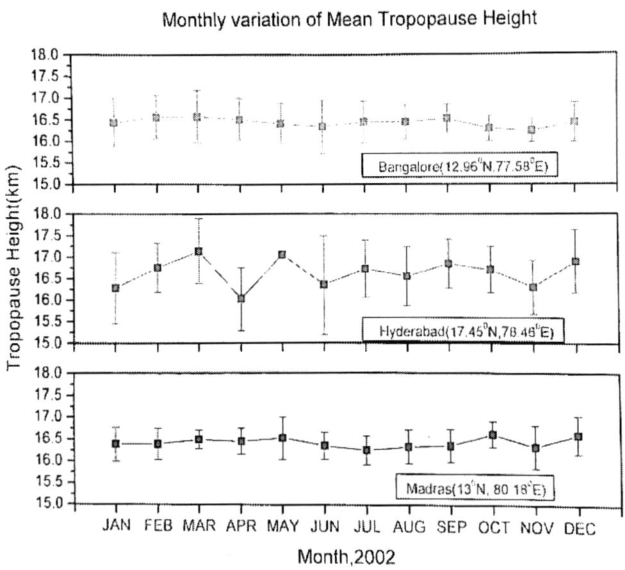

Figure 6. Monthly variation of tropopause height.

In this communication, we have extended our study to examine the monthly variation of mean tropopause height. A comparative study of the monthly variation in mean tropopause height is presented in Figure 6. From the analysis, it is clear that variability in height of the tropopause over Hyderabad is observed to be more than at the other two stations.

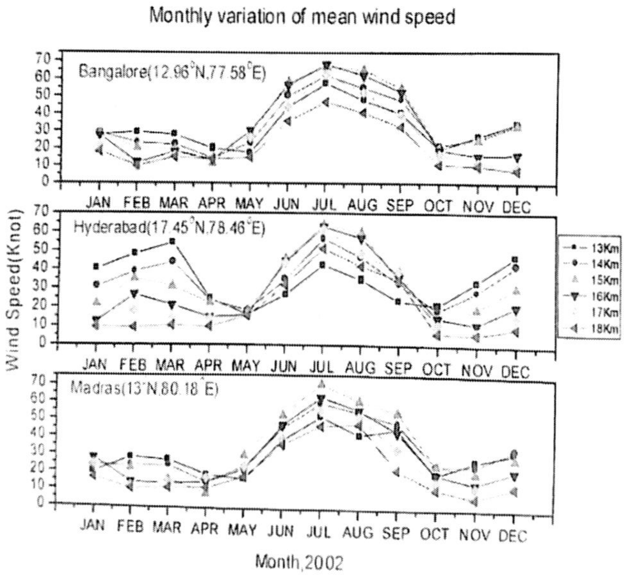

Figure 7. Monthly variation of mean wind speed in the height region 13-18 km.

The height of the tropopause at the station Hyderabad shows a higher value (~17 km) in March, followed by a sudden decrease (~16 km) in the next month. To understand such rapid variation of tropopause height (Xian and Homeyer, 2019), wind fluctuations in the vicinity of the tropopause have been studied (Figure 7). We analyse mean wind circulation within a height range of 13-18 km over these three stations. It is clearly understood that there is an enhanced variation in the mean wind at the Hyderabad station compared to the other two stations. It suggests that wind fluctuation influenced tropopause oscillations. Such observations are very significant as tropopause height variation generates the stratospheric-tropospheric exchange (Berthet, Esler, and Haynes, 2007).

Discussion

In the present study, we investigate the variability of tropopause height over different locations: Chennai ($13^{.080}$ N, 80.18^0 E, altitude 16m), Bangalore (12.96^0 N, 77.58^0 E, altitude 921m), and Hyderabad (17.45^0 N, 78.46^0 E, altitude 545m). Observations show an enhanced variation in tropopause height at the Hyderabad station compared to the variations over Chennai and Bangalore. The meteorological parameters, such as solar radiation, cloud cover, long-wave radiation, and wind field, also play a vital role in maintaining the equilibrium of the troposphere. Our comparative analysis correlates the tropopause height and vertical wind velocity in the vicinity of the tropopause. In the present work, MST radar observations of vertical winds display a variation in both magnitude and direction over a height range of 13-18 km. The goal of this study is to investigate the occurrence of strong vertical wind shear within this height range. The existence of such a tropopause boundary is based on the thermal equilibrium of the troposphere, without considering the stratosphere and its variations. The air circulation is responsible for the exchange of air between the troposphere and the stratosphere. Further, when stratospheric air is transported isentropically downward, the tropopause boundary moves three-dimensionally with the air. The troposphere primarily influences the stratosphere through various atmospheric waves that propagate upward. Recent evidence clearly indicates that stratospheric air exchange can impact the weather and climate in the troposphere. Overall, this analysis presents a thorough understanding of the dynamical structure of the transition region between the troposphere and the stratosphere.

Acknowledgments

I thank the (IITM) Indian Institute of Tropical Meteorology, Pune, India, for providing the Radiosonde and MST radar data.

References

Berthet G, Esler J. G, Haynes P. H. A Lagrangian perspective of the tropopause and the ventilation of the lowermost stratosphere, *J. Geophys. Res.-Atoms.* (2007) 112:1-14.

Birner T. Fine-scale structure of the extratropical tropopause region. *J. Geophys. Res.-Atoms.* (2006) 111:1-14.

Boothe, A. C. and Homeyer, C. R.: Global large-scale stratosphere–troposphere exchange in modern reanalyses, *Atmos. Chem. Phys.* (2017), 17: 5537–5559.

Fueglistaler S, Dessler A. E, Dunkerton T. J, Folkins I, Fu. Q, Mote P. W, Tropical tropopause layer, *Rev. Geophys.* (2009) 47: 1-31.

Gettelman, A., Hoor, P., Pan, L. L., Randel, W. J., Hegglin, M. I., and Birner, T.: The Extratropical Upper Troposphere and Lower Stratosphere, *Rev. Geophys.* (2011) 49: RG3003.

Jaya Rao. Y, Jain A. R. Oscillation of Tropical Tropopause during passage of atmospheric waves. *Indian Journal of Radio & Space Physics* (2000) 29:203-209.

Manney, G. L., Hegglin, M. I., Daffer, W. H., Schwartz, M. J., Santee, M. L., and Pawson, S.: Climatology of Upper Tropospheric–Lower Stratospheric (UTLS) Jets and Tropopauses in MERRA, *J. Climate*. (2014) 27:3248–3271.

Peevey, T. R., Gille, J. C., Homeyer, C. R., and Manney, G. L.: The double tropopause and its dynamical relationship to the tropopause inversion layer in storm track regions, *J. Geophys. Res*. (2014) 119: 10194–10212.

Son, S.-W., Tandon, N. F., and Polvani, L. M.: The fine-scale structure of the global tropopause derived from COSMIC GPS radio occultation measurements, *J. Geophys. Res* (2011) 116: D20113.

Xian, T. and Homeyer, C. R.: Global tropopause altitudes in radiosondes and reanalyses, *Atmos. Chem. Phys*. (2019) 19: 5661–5678.

Chapter 16

Application of the Collaborative Method of Infrastructure Management System Using BIM for AEC Projects

Rishita Das[1,*], MBA
Rhythm Mukherjee[1], MBA
Sougata Banerjee[2], MTech
and Preeti Sharma[1], PhD
[1]School of Management, Institute of Engineering & Management,
University of Engineering and Management, Jaipur, Rajasthan, India
[2]Department of Civil, Institute of Engineering & Management,
University of Engineering and Management, Jaipur, Rajasthan, India

Abstract

Building Information Modelling (BIM) is a substantially effective tool for various engineering projects. The rapid acceptance of BIM has resulted in increased utilisation of BIM in the Architectural, Engineering and Construction (AEC) projects. The Collaboration tool in BIM has facilitated effective infrastructural and resource management in various projects. However, the causes of reworks and errors prevailing during the modelling of structures have some practical issues. In this paper, a process is developed to enhance the collaboration-based method of model development (CBMD) using BIM tools to aid the Infrastructure Management System (IMS). This study proposes an iterative process to enhance the effectiveness of information-sharing and real-time work-tracking systems for CBMD in large-scale construction and infrastructure

* Corresponding Author's Email: rishita.das@uem.edu.in

In: Breakthrough Research and Technologies in Physics ...
Editors: Koyel Ganguly, Soumyadipta Pal and Kamakhya Prasad Ghatak
ISBN: 979-8-90134-111-7

projects within the AEC sector. It promotes BIM Managers, General Contractors (GC), and General Managers to interpret and oversee the model development using BIM and IMS for AEC projects. This novel and innovative approach is applied to a sample project of a multi-storey multi-utility structure and performed well, enhancing the process of CBMD for IMS. The paper concludes with an overview of the overall pros and cons of the proposed system for IMS.

Keywords: BIM, infrastructure management, efficiency, information, project

Introduction

Looking into the current scenario, the AEC industry has undergone significant transformations in recent decades, driven by the need to enhance project efficiency, cost-effectiveness, and sustainability. During the initialisation of this transformation, i.e., Building Information Modelling (BIM), a digital innovation that has revolutionised how projects are considered, prepared, and implemented. BIM is way more than just a three-dimensional modeling tool; it is a comprehensive process that helps to accumulate all the different domains of a construction project into a unified digital environment that further facilitates seamless communication, Collaboration, and coordination among all the stakeholders (Eastman et al., 2011; Ngouhouo et al., 2020).

The application of BIM across all major sectors, including Architecture, Engineering, and Construction (AEC), can't be overstated due to its properties. Furthermore, the acceptance of a tool as immersive as BIM has shifted the paradigms from traditional approaches; this type of process involves methods of Construction Management in incorporating and combining approaches. In the context of large-scale projects, this type of shift becomes crucial, as components such as complexity and scale require a high level of coordination and real-time information sharing. With the help of BIM, stakeholders can easily envision the entire timespan of the Project, from its initial design to construction, completion, and facility management, ensuring that the data is accurate and up-to-date (Azhar, 2011).

The extensive capabilities of BIM have proven to be one of its most significant and valuable assets. Collaboration in BIM enables the integration of various disciplines, including architecture, engineering, and construction, into a single, comprehensive model. This integration minimises conflicts between different systems, such as structural and mechanical, and reduces the need for costly and time-consuming reworks (Linderoth, 2010). The real-time nature of BIM collaboration tools ensures that all stakeholders have access to the latest project information, enabling quicker decision-making and reducing the likelihood of errors (Hardin & McCool, 2015; TERI, 2023).

Nevertheless, the advantages and real-time utilisation of BIM in Large-Scale Projects are challenging. Reworks, errors, and inefficiencies are some of the issues that can still arise due to the Project's complex nature, along with the active participation of distinct stakeholders, as well as the variation in BIM efficiency levels among team members. Few of these types of Challenges highlight the significant need for refinement and a proper systematic approach toward BIM implementation, particularly in areas related to Infrastructure Management Systems (IMS) (Love et al., 2013; Oguntoke & Shitta, 2022).

This research paper addresses these challenges by proposing a Collaborative-Based Model Development (CBMD) approach that utilises BIM tools designed explicitly for Infrastructure Management Systems. The aim is to develop an iterative process that enhances Collaboration, improves information sharing, and enables real-time work tracking across large-scale AEC projects (Jung & Joo, 2011). The proposed approach not only facilitates more effective project management but also supports BIM Managers, General Contractors, and General Managers in overseeing the model development process more efficiently.

The real-time importance of this research study lies in highlighting the incorporation of CBMD in AEC for improved IMS. The results clearly show a broader spectrum on the aspect of utilising BIM and its implementation for the AEC industry. The proposed method in this research provides better efficiency in inventory and resource management, thereby optimising infrastructural management. Additionally, the use of BIM is proving beneficial in the AEC sector, particularly for large-scale projects. The continuous information flow across the entire system, from the planning phase to the construction phase and for future maintenance operations, makes this research important.

Analytical Framework

Literature Review

In the field of BIM, some researchers have found that this platform integrates multiple sectors, including project data from architectural, structural, and MEP components into a single model. This combination of several information enhances interoperability among various software platforms, allowing for broader data exchange.

The primary advantage of using BIM tools is their ability to detect and coordinate clashes during the modelling phase of project development. Thus, it helps to identify and resolve conflicts within various fields, such as architecture, structural engineering, HVAC systems, fire safety systems, and electrical systems, even before actual construction begins. With the help of this type of detection ability, BIM prevents costly reworks and delays during project completion (Barlish & Sullivan, 2012).

During the design phase of a building or a large-scale project, it is essential to predict the serviceability of the Project. Thus, the Lifecycle Assessment (LCA) process is applied to a project. In the present era, BIM makes it much easier to assess various materials and their applications to estimate the overall project lifecycle (Bynum et al., 2013).

Researchers and Engineers have been satisfied with the facilities provided by BIM tools. It enables the design of a project with comprehensive data, which helps GCs and all managers perform effective maintenance, manage assets, and prepare for future upgrades accordingly (Lee & Yu, 2015).

BIM offers cloud-based systems and real-time data analysis. These types of tools enable all stakeholders, including architects, Engineers, and Contractors, to frequently study and analyse the same Model (Zhang et al., 2016).

The Model BIM encompasses various types of analyses and simulations, including those that assess elements such as structural integrity, energy performance, and construction sequence. The building's energy consumption can be predicted with the help of energy simulation, which also identifies opportunities for improving its efficiency (Clevenger & Haymaker, 2011).

The tool BIM provides detailed analytics and visualisations, which are enabled by the data-driven decision-making factors. This unique ability of BIM enables stakeholders to analyse alternatives and access impactful project outcomes that require decisions based on quantitative data (Guo et al., 2017; N., 2022).

Several research studies and real-world project-based observations have drawn certain conclusions, suggesting that BIM models can be linked to cost data, material cost calculations, and their serviceability. Additionally, the integration of various elements enhances the data to meet the Project's needs. Thus, a booklet of total information is available for CBMD in BIM (Oraee et al., 2017).

Overall, several studies conclude that BIM offers numerous benefits in terms of Infrastructure Management for projects in the AEC sector. However, there are some practical drawbacks to the utilisation of BIM in all projects due to the complex model management system and a lack of proper resources. The proposed CBMD approach can address these required updates by enhancing the IMS.

Method

Research Designs

This section influences the mixed methods approach by combining both qualitative and quantitative methods to define the research objectives. This study has two phases: The development of the CBMD Approach and the Evaluating and Testing phase. In the CBMD phase, the BIM tools and IMS involve the iterative development of the Model by incorporating the tailored requirements from the BIM Managers, General Managers, and General Contractors. Several distinct systems have been developed to enhance the communication and process management of BIM models. Hence, no different than the currently used BIM tools and software, the planned and proposed system, carefully designed in this paper, has been developed to enhance CBMDM tasks.

Data Collection

The phase for data collection was conducted using various sources, including Literature Review, Case Study, Surveys, and Interviews. In the Literature Review section, we have considered literature on BIM, IMS, and a few collaborative methodologies that have significantly contributed to the foundationowleresource aresources in thise Case Study consists of empirical structures like multi-story, multi-utility structures, where the data were collected through the project documentation, BIM models, and by conducting interviews with the Stakeholders.

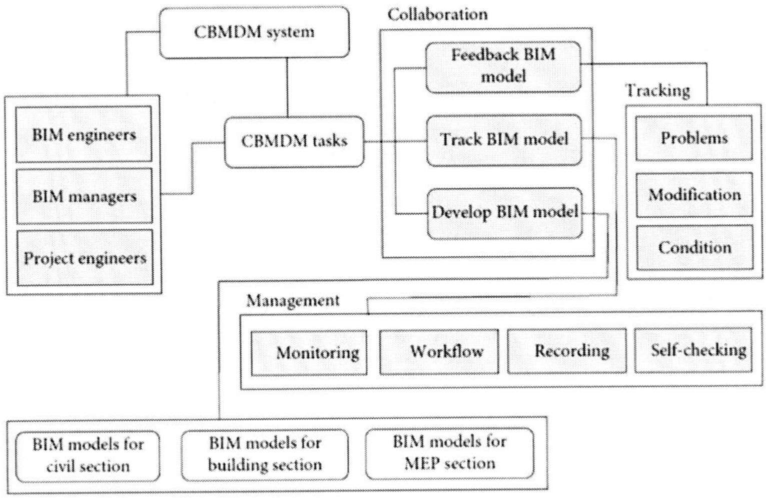

Figure 1. Workflow of the proposed approach during CBMD in IMS.

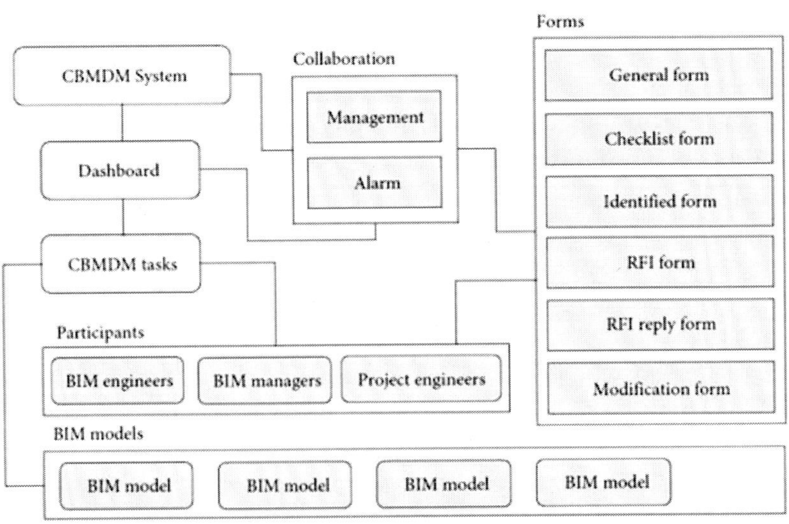

Figure 2. Block diagram of the overall system.

In the final phase, we conducted surveys and semi-structured interviews, which were again led by the BIM managers, General Managers, and General Contractors to collect important qualitative data based on their experiences with the CBMD approach.

Alongside the process of Collaboration-based Model Development (CBMD) for a particular project, the inclusion of Infrastructure Management (ISM) provides an efficient overview of materials. The following figures, Figures 1 and 2, illustrate the workflow of the proposed approach to enhance the IMS in CBMD using BIM in AEC projects.

Discussion

This study yields various observations that encompass not only statistical information but also several crucial insights, highlighting the application of CBMD in AEC projects to optimise IMS with the aid of BIM. This study aims to facilitate a revised approach to address the prevailing constraints by enhancing Collaboration, information sharing and real-time work tracking through an iterative and systematic approach.

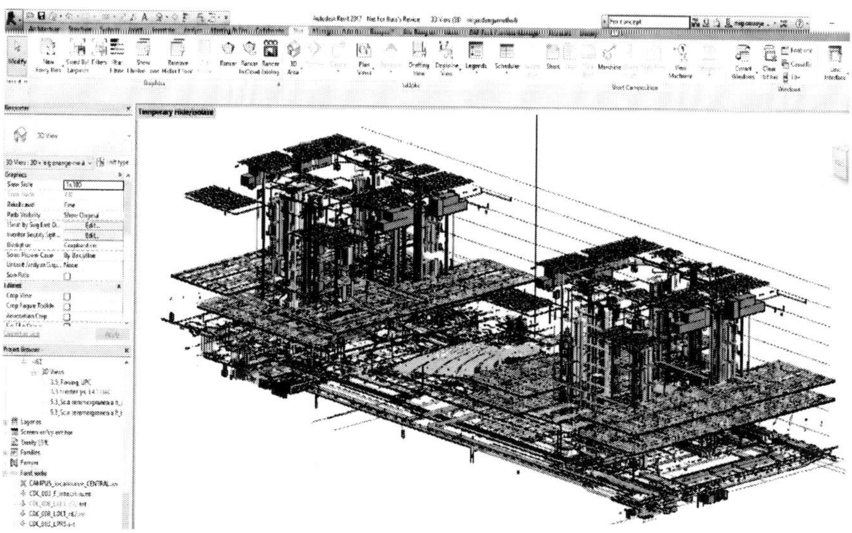

Figure 3. Overview of the sample model.

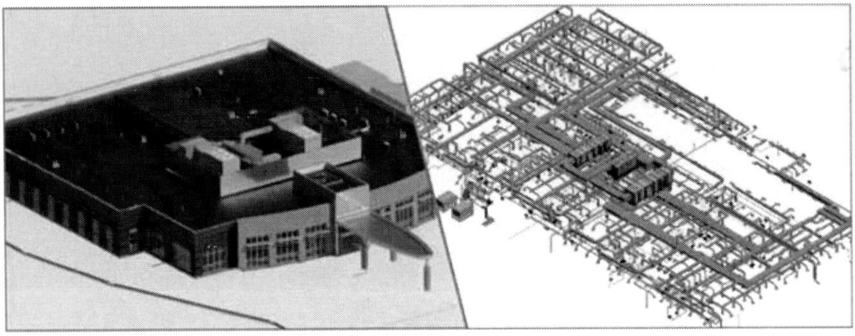

Figure 4. Comparative overview of the architectural Model and the MEP model.

Table 1. Stages of CBMD to be implemented in BIM

Task	Description
Data Gathering	A collection of all the information needed to start the CBMD
Modelling	To develop the targeted ModelModel through various stages and LODs
Collaboration	To facilitate multiple departments working together in a collaborative environment
Identification	To enable clash coordination and look out for errors
Revision	Modifying and revising the Model in real-time with multiple departments
Publishing	To finalise the Model and prepare the various schedules for resource management

Table 2. Overview of the proposed approach and statistical observation

Designed System Application	Standard Deviation	Average Rating
Real-time collaborative work for CBMD	0.39	4.3
Reduction of rework	0.41	4.7
Ease of information retrieval	0.47	4.7
Ease of use for CBMD	0.59	4.1
Time saving	0.43	4.3
Quick monitoring of project resources	0.42	4.6
Well-planned schedules and estimation	0.33	4.7
Understanding of various tasks and sharing	0.45	4.2

A sample multi-storey multi-purpose building is modelled in Revit software, which incorporates BIM. Through this software, the various elements of the Project were assessed by reviewing the schedules and lists of objects created. An overview of the Model is shown in Figure 3 in a 3D perspective view. Additionally, Figure 4 illustrates how BIM tools highlight the IMS facility by generating a schedule in conjunction with CBMD. It also illustrates the overall Model developed in Revit software for the research study.

Table 1 provides the list of selective tasks to accomplish for implementing this process, and Table 2 provides an overview of the perspective feedback from the survey panel. Various findings from surveys, interviews, and real-time project sharing and Collaboration using Revit software with various AEC experts, including GCs, BIM Managers, BIM Engineers, Architects, MEP Engineers, Project Managers, and General Managers, provide insights into the practical implementation and benefits of the proposed approach in this study. The use of CBMD in IMS can significantly benefit the AEC industry. About 87% of the BIM beneficiaries shared 95% positive feedback about the proposed system, which enhances the IMS by 11%, as observed from this study.

The findings from this research indicate that the CBMD approach significantly improves project efficiency and coordination in BIM-based AEC projects. By integrating advanced BIM tools and methodologies, the CBMD approach addresses key challenges identified in the literature, such as reworks, errors, and inefficiencies, through a more refined and collaborative process. The pre-gaining of sufficient information about the Project to be built, including its construction stages, materials and resource allocation, workforce, completion time, potential delays, etc., can be achieved through CBMD using BIM in IMS. Improved coordination and real-time tracking contribute to better resource management, ensuring that project resources are utilised effectively and project timelines are adhered to.

Conclusion

The study concludes with an emphasis that, although the approach of CBMD yields promising results in Construction projects, there remains a significant need for further research in the same domain to understand the potential within the construction industry fully. This phase of Future research should emphasise the application of CBMD beyond its current scope, also ensuring that it is utilised in diverse project types. Furthermore, refining the methodology to incorporate broader aspects of feedback from industrial professionals will enhance its effectiveness. A detailed investigation into the integration of advanced Building Information Modelling (BIM) tools and Technologies into the CBMD framework could yield more solutions for the Project. We also need to understand the impact of CBMD on the project performance, including the aspects of Sustainability outcomes that drive efficiency and innovation in construction practices. This research could contribute to a more holistic understanding of Infrastructure Management.

References

Azhar S. Building Information Modelling (BIM): trends, benefits, risks, and challenges for the AEC industry. *Leadership Manag Eng.* 2011;11(3):241–52.

Barlish K, Sullivan K. How to measure the benefits of BIM—a case study approach. *Autom Constr.* 2012;24:149–59.

Bynum P, Issa RRA, Olbina S. Building Information Modelling in support of sustainable design and construction. *J Constr Eng Manag.* 2013;139(1):24–34.

Clevenger CM, Haymaker JR. The impact of BIM on interdisciplinary Collaboration. *J Archit Eng.* 2011;17(4):144–51.

Eastman C, Teicholz P, Sacks R, Liston K. *BIM Handbook: A Guide to Building Information Modelling for Owners, Managers, Designers, Engineers and Contractors*. Hoboken (NJ): John Wiley & Sons; 2011.

Guo F, Jahren CT, Turkan Y, Jeong HD. Civil integrated management: an emerging paradigm for civil infrastructure project delivery and management. *J Manag Eng.* 2017;33(2):04016044.

Hardin B, McCool D. *BIM and Construction Management: Proven Tools, Methods, and Workflows*. Hoboken (NJ): John Wiley & Sons; 2015.

Hwang BG, Zhao X, Goh KJ. Investigating the client-related rework in building projects: a BIM perspective. *Autom Constr.* 2013;35:110–20.

Iqbal M. Comparative performance analysis of traditional and improved cook-stoves for bio-fuels. *Energy Sustain Dev.* 2018;45:1–8.

Jung Y, Joo M. Building Information Modelling (BIM) framework for practical implementation. *Autom Constr.* 2011;20(2):126–33.

Lee, G., & Yu, J. (n.d.). BIM and cloud computing for sustainable design: An integrated approach. *Autom Constr.* 2015;42:98–110.

Lin YC. The study of collaboration-based BIM models development for GCs in Taiwan. *Report for BIM Implementation in Practice.* 2017;12.

Linderoth HCJ. Understanding adoption and use of BIM as the creation of actor networks. *Autom Constr.* 2010;19(1):66–72.

Love PED, Matthews J, Lockley S. BIM in infrastructure: applications in the water industry. *Autom Constr.* 2013;36:110–20.

N. Enhancement of ceramic insulation for biomass cook-stoves using carbonised organic waste. *J Environ Manag.* 2022;301:113856.

Ngouhouo SM, Feudjo JRE, Tieguhong JC. Assessment of the impact of wood fuel consumption on the forest cover in the East Region of Cameroon. *Energy Sustain Dev.* 2020;59:1–9.

Oguntoke OB, Shitta MB. The impact of bio-fuel production and utilisation on climate change. *Renew Sustain Energy Rev.* 2022;136:110343.

Oraee M, Hosseini MR, Papadonikolaki E, Palliyaguru R, Arashpour M. Collaboration in BIM-based construction networks: a bibliometric-qualitative literature review. *Int J Proj Manag.* 2017;35(7):1288–301.

The Energy and Resources Institute. *Annual Report 2022–23.* New Delhi (IN): TERI Publications; 2023.

Zhang J, Teizer J, Lee JK, Eastman CM, Venugopal M. Building Information Modelling (BIM) and safety: automatic safety checking of construction models and schedules. *Autom Constr.* 2016;29:183–95.

Chapter 17

Competitive Removal of Heavy Metals from Leachate Using Plastic Pyrolysed Char

A. Safa Jabeen[1,*]
V. Meera[1,†], PhD
and Vinod P. Raphael[2,‡], PhD
[1]Department of Civil Engineering, APJ Abdul Kalam Technological University, Thiruvananthapuram, Kerala, India
[2]Department of Chemistry, APJ Abdul Kalam Technological University, Thiruvananthapuram, Kerala, India

Abstract

Plastic pyrolysed char (PPC) with a high surface area and adsorptive potential is obtained as a byproduct in the pyrolysis of waste plastics. Hazardous metals like cadmium, lead, chromium, and mercury are frequently found in landfill leachate, which must be effectively handled to prevent contamination of the ecosystem. This study examines the efficacy of plastic PPC in removing heavy metals from multi-element synthetic samples, evaluating PPC's competitive adsorption properties. Batch studies were conducted to assess PPC's heavy metal adsorption capabilities under three different pH levels (4, 6, and 8) and two different PPC dosages (2 g and 4 g per 20 mL of solution). Characteristics of PPC were analysed by microscopically, spectroscopically and Brunauer–Emmett–Teller (BET) studies. Metal removal was examined using Inductively Coupled Plasma-Mass Spectrometry (ICP-MS). Batch study results revealed that PPC adsorbs heavy metals efficiently, with a

* Corresponding Author's Email: jabeen.safa@gmail.com
† Corresponding Author's Email: vmeera@gectcr.ac.in
‡ Corresponding Author's Email: vinodpraphael@gectcr.ac.in

In: Breakthrough Research and Technologies in Physics …
Editors: Koyel Ganguly, Soumyadipta Pal and Kamakhya Prasad Ghatak
ISBN: 979-8-90134-111-7

removal efficiency of ~93% to 100%. PPC was found to have removal efficiency on par with or better than that of other sorbents reported. According to the findings, PPC provides a viable alternative for eliminating heavy metals present in leachate and aids in the valorisation of plastic waste simultaneously. This study demonstrates how PPC can play a valuable role in both effectively managing waste and protecting the environment.

Keywords: plastic pyrolysed char, heavy metals, adsorption, batch study, adsorption mechanism

Introduction

Contamination by heavy metals is a serious environmental problem since they are poisonous, persistent, and bioaccumulative. Hazardous metals such as mercury (Hg), chromium (Cr), cadmium (Cd), and lead (Pb) are widely present in the environment due to mining, industrial processes, and agricultural practices. These metals, which pose significant threats to ecosystems and human health, are commonly found in landfill leachates. Heavy metal-containing leachates have the potential to lower agricultural output and deteriorate drinking water quality when they seep into soil and groundwater. According to studies, leachate from landfills can significantly increase the levels of heavy metals in nearby soils, leading to bioaccumulation in both plants and animals (Hussein et al., 2021; Parvin and Tareq, 2021; Essien et al., 2022; Dagwar and Dutta, 2024).

Even after a landfill has been closed, leachate can still cause a considerable rise in the quantities of hazardous metals in groundwater, presenting a long-term risk that may persist for many years (Kjeldsen et al., 2002). Heavy metals may remain in landfills for about 150 years, assuming a leaching rate of 400 mm per year, even though the organic content in leachate tends to decrease as methane is produced from the waste (Yusof et al., 2009; Adelopo et al., 2018).

Local ecosystems may be disturbed by the presence of heavy metals in leachate, with aquatic life being particularly vulnerable. Fish and other creatures can bioaccumulate certain metals, which can have detrimental impacts and reduce biodiversity (Hredoy et al., 2022; Al Raisi et al., 2014). The accumulation of mercury in fish populations causes a cascade of harm, affecting not only the fish but also the birds and other creatures that consume them, altering their behaviour and affecting their ability to reproduce (UNEP, 2008). These alterations have long-term effects on the environment, changing the species composition of the region and reducing the ecosystem's ability to sustain its natural processes.

Heavy metal exposure can lead to a variety of health issues, including kidney failure, cognitive abnormalities, and an elevated risk of cancer. Therefore, removing heavy metals from leachate is crucial to maintaining environmental safety and safeguarding public health. Lead is especially harmful, as it can cause heart and kidney problems in adults and brain development issues in children, which impact their thinking and behaviour (UNEP, 2002; Tchounwou, 2012; Mitra et al., 2022). Mercury harms the body's natural defences, kidneys, and nerves, among other systems. Methylmercury, its most hazardous form, is particularly harmful to young children and unborn infants (Mitra et al., 2022; Balali-Mood et al., 2021). Cadmium accumulates in the body over time, causing long-term health concerns, including breathing difficulties, kidney problems, and weakened bones (Tchounwou et al., 2012; Mitra et al., 2022).

Table 1. Efficacy of different adsorbents in the removal of heavy metals from landfill leachate as reported in early studies

Adsorbent studied	Removal efficiency	Reference
Calcined eggshell waste	Pb: 93%, Cr: 60%	Jaradat et al., 2021
Orange and banana peels	Cd: 99.9%, Cr: 98.18%	Duwiejuah et al., 2024
Wasted biosolids	Cd: 28.8%	Mamun et al., 2011
Fly ash	Pb: 74%, Cd: 71% Pb: 59.24%, Cd: 55.37%, Cr: 28.14%	Mohan and Gandhimathi, 2009 Xue et al., 2014
Zeolite	Cr: 63%	Lim et al., 2016
Biochar	Pb: 88% Cd: 94.4% Pb: 92.07%, Cd: 86.14%	Zand and Abyaneh et al., 2020 Soudani et al., 2024 Rabiee Abyaneh et al., 2024

Given the widespread distribution of heavy metal pollution, immediate action is necessary to mitigate its detrimental impacts on both humans and the environment. Various procedures, including coagulation, electrochemical treatment, biological treatment, and adsorption, are used to remove heavy metals from leachate (Jaradat et al., 2021; Esfahani, 2021; Türkme et al., 2022; Kocaman et al., 2023). Adsorption is a commonly used technique for eradicating heavy metals present in leachate. This method employs different adsorbents to capture and concentrate heavy metal ions. During the adsorption process, several mechanisms, such as complexation, ion exchange, and physical adsorption, can facilitate the adherence of metal ions to the adsorbent surface. Various factors, including pH, contact time, and the concentration of metals in the leachate, can influence the effectiveness of this process. Various adsorbents have been studied for the removal of heavy metals (lead, cadmium, and chromium) from leachate, along with their corresponding removal efficiencies, as shown in Table 1.

Many recent studies in this area have focused on the use of biochar derived from various materials. However, the application of plastic pyrolysed char (PPC) for the adsorption of heavy metals has received less attention. PPC is a solid residue obtained when plastic wastes are treated at a higher temperature in the absence of oxygen. Studies related to the properties and applications of PPC are minimal. Char derived from a mix of polyethene (PE), polypropylene (PP), and polystyrene (PS) wastes pyrolysed at 550°C for lead adsorption in an aqueous medium revealed the highest adsorption capacity of 61.3 mg/g at a pH of 5 (Martín-Lara et al., 2021). Char derived from a mix of PE and polyvinyl chloride (PVC) demonstrated the highest rate of removal of arsenic (As), in the range of 71.6% to 99.4%, suggesting that plastic chars hold significant potential as effective sorbents for As removal (Singh et al., 2020). Since PPC has a very high carbon content and porous structure, it can act as a prominent adsorbent for leachate treatment. The use of burnt ashes poses a risk of hazardous materials, particularly metals and inorganic salts, potentially leaching into the environment. These residues may contain toxic contents, whereas washing the ash is a straightforward method to reduce these contaminants to some extent. Washing typically removes soluble components, particularly chloride and alkali ions (Schneider et al., 1994; Sabbas et al., 2003).

The use of PPC as an adsorbent for heavy metals, such as lead, cadmium, chromium, and mercury, from landfill leachate has not been explored. This study aims to examine the adsorption capacity of PPC for the competitive removal of these heavy metals from synthetic multi-element solutions, which mimic the concentrations present in leachate, through batch studies. The PPC collected from the pyrolysis plant is characterised morphologically and spectroscopically. A leaching test was also employed to compare the properties of PPC before and after washing.

Materials and Methods

The PPC was obtained from a pyrolysis facility in Navi Mumbai, utilising Cross-Linked Polyethene (XLPE) and Polypropylene (PP) plastic wastes as primary feedstocks, which were subjected to pyrolysis at a temperature of 450°C. The sample was collected from the site in a moist condition and exhibited a greyish-black colour with varying particle sizes. Before testing, the material was dried and then filtered through a fine-mesh screen with 75-micron openings.

Characterisation of Plastic Pyrolysed Char

Scanning Electron Microscopy (SEM)

A specific Jeol 6390LA/OXFORD XMX N machine that combines energy-dispersive X-ray spectroscopy (EDS) and scanning electron microscopy (SEM) was used to analyse the microscopic structure of PPC. With its tungsten filament, the device could magnify pictures up to 300,000 times and ran on a voltage between 0.5 and 30 kV. The EDS component used a 30 mm^2 detector and had a resolution of 136 eV. EDS examined the PPC's chemical components, whereas SEM displayed its physical structure.

Fourier Transform Infrared Spectroscopy (FTIR)

The interaction of a material with infrared light at various frequencies is investigated via FTIR spectroscopy. Absorption peaks, the identification of functional groups, and chemical structure analysis are among the insightful information provided by the resultant spectrum. A Perkin Elmer UATR Two spectrophotometer with a resolution of 4 cm^{-1} was used to obtain the FTIR spectra for PPC. The spectrum helps identify the bonds and functional groups in the sample by showing unique peaks that correspond to the absorption of infrared radiation at particular frequencies by different chemical functional groups.

Brunauer–Emmett–Teller (BET) analysis

With nitrogen acting as the adsorbate gas at 77 K, the surface area and pore volume of PPC were measured using a BET surface area analyser (SA-9600, Horiba Scientific).

Cation Exchange Capacity (CEC)

Cation Exchange Capacity indicates how well a material can store and exchange positively charged ions, or cations. The CEC is the total number of exchangeable cations that a material, particularly one that is porous, may hold on its negatively charged sites. These cations can be replaced with other positively charged ions from a solution. Usually, they are stated as milliequivalents per 100 grams (meq/100g) or centimoles per kilogram (cmol/kg). CEC of PPC was determined using the sodium acetate method (Chapman, 1965).

Effect of Washing

30 g of PPC was mixed with 200 mL of distilled water, and the filtrate was analysed for parameters such as Total Dissolved Solids (TDS), salinity, and chloride content to evaluate its environmental compatibility. The same parameters were measured again after washing the sample to assess the effect of washing. For this, 30 g of pyrochar was washed with 1 litre of distilled water. After drying, the PPC sample was re-extracted with 200 mL of distilled water, and the resulting effluent was examined for the parameters above. Chloride content was determined using Argentometric titration, whereas the other three parameters were analysed using a microprocessor-controlled conductivity, salinity & temperature meter with auto-ranging facility HI-83.

Batch Sorption Studies on PPC

The batch sorption study was conducted to analyse PPC's adsorption capacity for heavy metals —mercury, lead, cadmium, and chromium. Two adsorbent dosages of 2g and 4g were chosen to evaluate the influence of adsorbent concentration on the adsorption of heavy metals. A standard multi-element stock solution of Cd, Hg, Cr, and Pb with 0.01M was prepared by dissolving measured amounts of the appropriate metal salts ($Pb(NO_3)_2$, $Cd(NO_3)_2.4H_2O$, $Hg(NO_3)_2.H_2O$, $Cr(NO_3)_3.9H_2O$) in a 20 mL of Nitric acid (HNO3) and further in deionised water. The stock solution was diluted to achieve the desired concentrations of Pb, Cd, Hg, and Cr (2.07, 1.12, 2.00, and 0.52 ppm), which mimic the concentrations of metals present in the leachate.

Twenty millilitres of the above-diluted solution was mixed with 2 g and 4 g of washed and dried PPC in 50 mL beakers under three different pH conditions (pH 4, 6, and 8). Using 0.1 M NaOH and 0.1 M HNO3, the pH of the solutions was adjusted to three different levels to represent acidic, neutral, and basic conditions, based on the known chemistry of metal ions and their adsorption behaviour. The beakers were placed on a microprocessor-based Digital Orbital Shaker at room temperature for one hour. After 24 hours, the supernatant was filtered by centrifugation and collected in glass vials.

The concentrations of Cd, Hg, Cr, and Pb remaining in the solution after adsorption were determined using Inductively Coupled Plasma-Mass Spectrometry (ICP-MS). The instrument used for the analysis was Thermo Scientific, iCAP RQ with Helium Kinetic Energy Discrimination mode.

The amount of metal adsorbed qe (mg/g) and heavy metal removal efficiency, η (%), calculated using Eq. (1) & (2) respectively:

$$q_e = \frac{(C_0 - C_e).V}{m} \quad (1)$$

where q_e = Quantity of metal retained at equilibrium. (mg/g), C_0 & C_e = Concentration of the metal ion in solution initially & finally (mg/L), V = Solution volume (L), and m = Adsorbent mass(g).

$$\eta = \frac{(C_0 - C_e)}{C_0} \; x\; 100 \quad (2)$$

The efficiency of the adsorbent in lowering the metal ion concentration from its initial level to the equilibrium concentration following the sorption proccss is cxpressed by Eq. (2).

Results and Discussion

Characterisation of Plastic Pyrolysed Char

Scanning Electron Microscopy (SEM)

SEM image of the PPC sample (Figure 1) revealed a porous structure, consistent with the findings of other studies (Singh et al., 2020; Wijesekara et al., 2021). The particles appeared to be irregularly agglomerated with a rough, nearly spherical shape. Additionally, several rod-like formations observed in the previous study (Barracco et al., 2024) were also present in this image. The average particle size was estimated to be approximately 5 micrometres.

Figure 1. SEM micrograph of PPC.

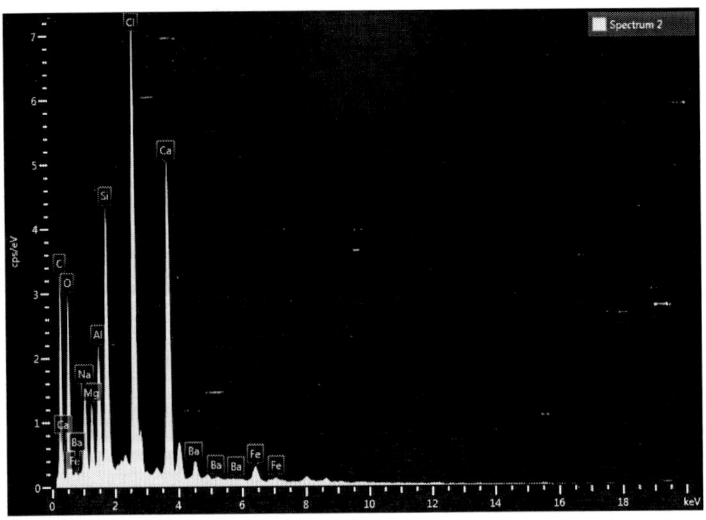

Figure 2. EDS spectrum of PPC.

The EDS spectrum of PPC (Figure 2) shows the presence of several elements, including C, O, Mg, Ca, Na, Si, Al, Fe, Cl, and Ba. The results indicate the possible presence of metal oxides or metal chlorides of Ca, Si, Al, Na, Mg, Fe, and Ba. Carbon was identified as the most abundant element in the char sample. No significant levels of other harmful heavy metals were detected in the EDS analysis. Carbon was found to have a higher weight percentage of 43.46%, which is typical for char samples where carbon is often a dominant component. This also suggests that the PPC may have a high adsorption capacity.

Fourier Transform Infrared Spectroscopy (FTIR)

The FTIR spectrum of PPC is shown in Figure 3. The FTIR spectrum exhibited a relatively simple pattern with only a few distinct peaks. This observation is likely due to the low concentration of organic compounds in the PPC. A low-intensity peak at 3850.31 cm^{-1} corresponds to the O-H stretching vibration of alcohol molecules. A moderately intense peak at 1432.4 cm^{-1} is attributed to the bending vibrations of O-H (from carboxylic acids) and/or C-H scissoring vibrations from hydrocarbons. The peak further supports the presence of carboxylic acid in the PPC at 1013.6 cm^{-1}, which corresponds to the C-O stretching frequency. Additionally, a narrow, low-intensity peak at 873.02 cm^{-1} can be assigned to the C-H bending vibrations (rocking mode) of hydrocarbons present in the PPC.

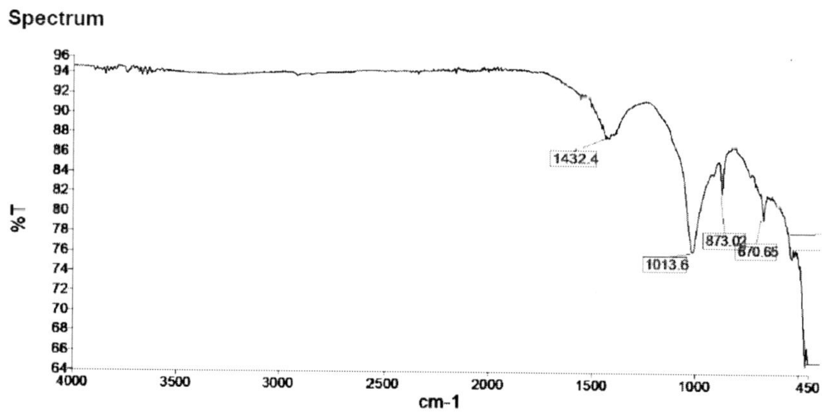

Figure 3. FTIR spectrum of PPC.

Brunauer–Emmett–Teller (BET) studies

The BET plot of PPC is shown in Figure 4, and the BET analysis results are given in Table 2.

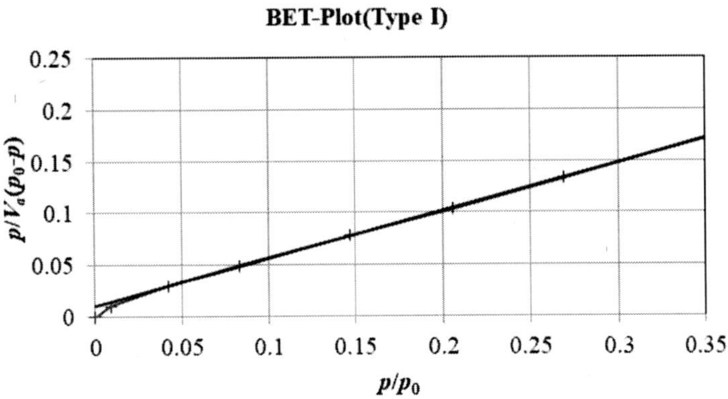

Figure 4. BET plot of PPC.

Table 2. BET analysis of PPC

Description	Value
Surface area (m^2/g)	9.2582
Pore volume at STP (cm^3/g)	2.1271

Cation Exchange Capacity (CEC)

The CEC value obtained for PPC was 20.26 meq/100 g, which compares favourably to conventional adsorbents, such as fly ash (Travar et al., 2009; Singh, 2021). This indicates a superior removal efficiency of PPC for heavy metals.

Effect of Washing

The Variation of TDS, salinity, and chloride content for PPC before and after washing is given in Figure 5. All values were significantly reduced, aligning with findings from prior studies on incinerated ashes (Schneider et al.,1994; Sabbas et al., 2003). The washing process effectively removed a substantial amount of chloride ions, leading to decreased TDS and salinity levels. So, washing the PPC is recommended before any engineering applications.

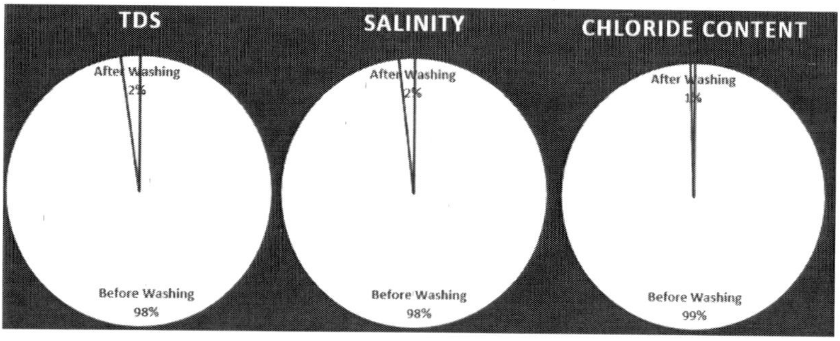

Figure 5. The Variation of different parameters of PPC concerning washing.

Batch Sorption Studies on PPC

Table 3 presents the amount of metals adsorbed and the corresponding removal efficiency. A dosage of 2g/20 mL was found to have a percentage efficiency of more than 99% for all metals considered, and a dosage of 4g/20 mL had an efficiency of more than 99% for all metals except cadmium. Results on the Quantity of metal retained at equilibrium per unit weight of the adsorbent show that PPC has a higher affinity for mercury, followed by cadmium, chromium, and lead.

Adsorption Mechanism of PPC

The structure of PPC significantly influences its adsorption efficiency, primarily through its surface area, pore size, and functional groups. The higher surface area allows more adsorption sites, enhancing the material's ability to capture contaminants (Singh et al., 2021). Chars produced at higher temperatures exhibit increased adsorption capacities due to the development of a more porous structure and a higher degree of carbonisation, which enhances the stability and reactivity of the char (Martín-Lara et al., 2021).

A higher carbon content typically correlates with better adsorption performance (Martín-Lara et al., 2021; Pereira et al., 2024). Functional groups on the PPC's surface also affect adsorption efficiency, which occurs through various mechanisms, including hydrogen bonding and electrostatic interactions (Singh et al., 2021).

Table 3. Competitive adsorption efficiency of heavy metals by PPC

Dosage of PPC (g)	pH maintained	Amount of metal adsorbed (µg/g)				Removal Efficiency (%)			
		Pb	Cd	Hg	Cr	Pb	Cd	Hg	Cr
2	4	≥27.33a	79.69	≥147.73a	14.38	≥99.96a	99.15	≥99.93a	99.69
2	6	≥27.33a	80.14	≥147.73a	14.39	≥99.96a	99.71	≥99.93a	99.71
2	8	27.33	80.34	≥147.73a	14.30	99.98	99.96	≥99.93a	99.09
4	4	13.63	37.50	73.48	7.20	99.74	93.34	99.41	99.90
4	6	≥13.66a	39.68	≥73.86a	7.20	≥99.96a	98.74	≥99.93a	99.86
4	8	≥13.66a	40.16	≥73.86a	7.17	≥99.96a	99.93	≥99.93a	99.47

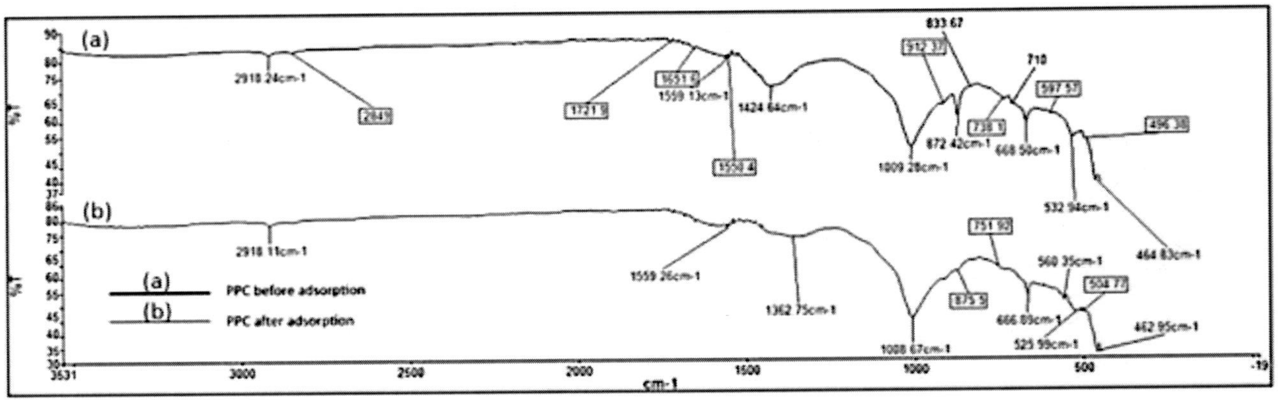

Figure 6. FTIR spectrum of PPC before and after adsorption.

The functional groups with oxygen, such as hydroxyl (-OH), carbonyl (C=O), and carboxyl (-COOH) groups, can be introduced by pyrolysis in PPC. The presence of such functional groups in PPC was confirmed by FTIR analysis (Figure 3). These properties collectively contribute to the effectiveness of PPC as an adsorbent for various pollutants, including heavy metals.

The result also shows that pH has no significant influence on removal efficiency. This implies that there is no necessity for pH adjustment during leachate treatment. PPC's simultaneous removal of over 99% of most metals highlights its exceptional capability for heavy metal adsorption, surpassing other adsorbents reported in Table 3.

The FTIR spectrum of PPC before and after adsorption is shown in Figure 6. The two spectra exhibit nearly identical patterns, except for some slight peak shifts in the 1500 cm^{-1} to 500 cm^{-1} frequency range. This may be due to the overall electronic rearrangement in PPC due to the adsorption of metal ions. A new peak was also observed at 560 cm^{-1}, indicating the formation of a metal oxide bond after adsorption. The adsorption mechanisms for PPC likely involve two main factors: electron-rich adsorption sites that readily trap positively charged metal ions and the presence of functional groups, such as -COOH, -OH, and Si-O-Si, on the char surface, which enhance interactions with these metal ions. Since the FTIR spectrum shows only slight peak shifts, it confirms that a minor portion of the adsorbed metals interacts with the functional groups present on PPC.

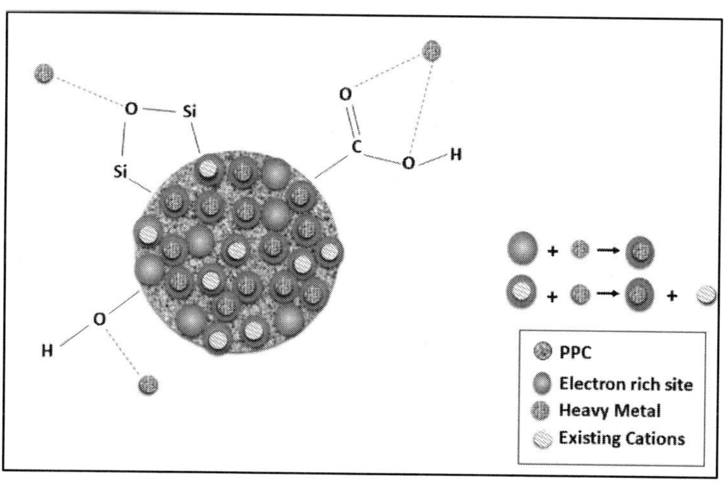

Figure 7. Proposed adsorption mechanism of heavy metals by PPC.

In contrast, a significant portion of adsorption occurs at the electron-dense sites of PPC. A substantial portion of heavy metals is retained in the pores by a strong electrostatic force of attraction. Additionally, adsorption can also occur due to PPC's cation exchange capacity, which was already confirmed through the sodium acetate method. The proposed mechanism of adsorption of heavy metals by PPC is illustrated in Figure 7.

Conclusion

Effective removal technologies are essential for managing landfill leachates and preventing their contamination consequences. The results of this study show that PPC performs better than other adsorbents already investigated for the simultaneous elimination of heavy metals, such as lead, mercury, cadmium, and chromium. A percentage removal from 93% to ≥99% was obtained for all metals. This can be attributed to its surface area, porous structure, and functional groups, which enable its effective adsorption of heavy metals.

In addition to its adsorptive effectiveness, PPC promotes plastic waste valorisation, providing a dual benefit of mitigating plastic pollution while addressing environmental risks associated with leachate contaminants. PPC's scalability and no cost make it a promising candidate for broader applications in wastewater and leachate treatment systems.

Future research should concentrate on optimising pyrolysis conditions for customised PPC characteristics, assessing long-term adsorption stability, and exploring potential regeneration techniques to improve PPC's reusability. This study highlights the viability of PPC as an efficient and sustainable adsorbent, reinforcing a circular economy framework in waste management and environmental remediation initiatives.

References

Adelopo, A. O., Haris, P. I., Alo, B. I., Huddersman, K., & Jenkins, R. O. (2018). Multivariate analysis of the effects of age, particle size and landfill depth on heavy metals pollution content of closed and active landfill precursors. *Waste Management,* 78, 227-237.

Al Raisi, S. A. H., Sulaiman, H., Suliman, F. E., & Abdallah, O. (2014). Assessment of heavy metals in leachate of an unlined landfill in the Sultanate of Oman. *International Journal of Environmental Science and Development*, 5(1), 60.

Balali-Mood, M., Naseri, K., Tahergorabi, Z., Khazdair, M. R., & Sadeghi, M. (2021). Toxic mechanisms of five heavy metals: mercury, lead, chromium, cadmium, and arsenic. *Frontiers in Pharmacology*, 12, 643972.

Barracco, F., Parisi, E., Pipitone, G., Simone, E., Bensaid, S., & Fino, D. (2024). Valorisation of pyrolytic plastic-derived char for adsorption of wastewater contaminants: A kinetic and thermodynamic investigation. *International Journal of Environmental Science and Technology*, 21(9), 6513-6530.

Chapman, H. D. (1965). Cation-exchange capacity. *Methods of soil analysis: Part 2 Chemical and microbiological properties*, 9, 891-901.

Dagwar, P. P., & Dutta, D. (2024). Landfill leachate a potential challenge towards sustainable environmental management. *Science of The Total Environment*, 171668.

Duwiejuah, A. B., Payne, J., & Yahaya, D. (2024). Adsorption of toxic metals from landfill leachate using orange and banana peel powders as adsorbents. *Sustainable Environment,* 10(1), 2405295.

Esfahani, A. R., Zhai, L., & Sadmani, A. A. (2021). Removing heavy metals from landfill leachate using electrospun polyelectrolyte fiber mat-laminated ultrafiltration membrane. *Journal of Environmental Chemical Engineering*, 9(4), 105355.

Essien, J. P., Ikpe, D. I., Inam, E. D., Okon, A. O., Ebong, G. A., & Benson, N. U. (2022). Occurrence and spatial distribution of heavy metals in landfill leachates and impacted freshwater ecosystem: An environmental and human health threat. *Plos one,* 17(2), e0263279.

Henmi, T. (1987). Increase in cation exchange capacity of coal fly ash by alkali treatment. *Clay science,* 6(6), 277-282.

Hredoy, R. H., Siddique, M. A. B., Akbor, M. A., Shaikh, M. A. A., & Rahman, M. M. (2022). Impacts of landfill leachate on the surrounding environment: A case study on Amin Bazar Landfill, Dhaka (Bangladesh). *Soil Systems*, 6(4), 90.

Hussein, M., Yoneda, K., Mohd-Zaki, Z., Amir, A., & Othman, N. (2021). Heavy metals in leachate, impacted soils and natural soils of different landfills in Malaysia: An alarming threat. *Chemosphere,* 267, 128874.

Jaradat, A. Q., Telfah, D. A. B., & Ismail, R. (2021). Heavy metals removal from landfill leachate by coagulation/flocculation process combined with continuous adsorption using eggshell waste materials. *Water Science and Technology*, 84(12), 3817-3832.

Kjeldsen, P., Barlaz, M. A., Rooker, A. P., Baun, A., Ledin, A., & Christensen, T. H. (2002). Present and long-term composition of MSW landfill leachate: a review. *Critical reviews in environmental science and technology*, 32(4), 297-336.

Kocaman, A., Savaş, B. F., & Ceyhan, D. İ. (2023). Removal efficacy of toxic metals in leachate through micro-organisms isolated from the natural environment. *BioResources,* 18(3), 5476.

Kumari, P., Kaur, A., Gupta, N. C., Tanwar, R., & Asthana, S. (2023). Removal of Hazardous Lead and Copper Ions from Landfill Leachate Using Kaolinite for Preventing Groundwater Contamination. *Environmental Claims Journal*, 35(2), 184-204.

Lim, C. K., Seow, T. W., Neoh, C. H., Md Nor, M. H., Ibrahim, Z., Ware, I., & Mat Sarip, S. H. (2016). Treatment of landfill leachate using ASBR combined with zeolite adsorption technology. *3 Biotech,* 6, 1-6.

Mamun, A. A., Alam, Z., Mohd, N. A. N., & Rashid, S. S. (2011). Adsorption of heavy metal from landfill leachate by wasted biosolids. *African Journal of Biotechnology*, 10(81), 18869-18881.

Martín-Lara, M. A., Piñar, A., Ligero, A., Blázquez, G., & Calero, M. (2021). Characterisation and use of char produced from pyrolysis of post-consumer mixed plastic waste. *Water,* 13(9), 1188.

Mitra, S., Chakraborty, A. J., Tareq, A. M., Emran, T. B., Nainu, F., Khusro, A., ... & Simal-Gandara, J. (2022). Impact of heavy metals on the environment and human health: Novel therapeutic insights to counter the toxicity. *Journal of King Saud University-Science,* 34(3), 101865.

Mohan, S., & Gandhimathi, R. (2009). Removal of heavy metal ions from municipal solid waste leachate using coal fly ash as an adsorbent. *Journal of hazardous materials,* 169(1-3), 351-359.

Parvin, F., & Tareq, S. M. (2021). Impact of landfill leachate contamination on surface and groundwater of Bangladesh: a systematic review and possible public health risks assessment. *Applied water science*, 11(6), 100.

Pathan, S. M., Aylmore, L. A. G., & Colmer, T. D. (2003). Properties of several fly ash materials in relation to use as soil amendments. *Journal of Environmental Quality,* 32(2), 687-693.

Pereira, L., Castillo, V., Calero, M., Blázquez, G., Solís, R. R., & Martín-Lara, M. Á. (2024). Conversion of char from pyrolysis of plastic wastes into alternative activated carbons for heavy metal removal. *Environmental Research,* 250, 118558.

Rabiee Abyaneh, M., Nabi Bidhendi, G., & Daryabeigi Zand, A. (2024). Pb (II), Cd (II), and Mn (II) adsorption onto pruning-derived biochar: physicochemical characterisation, modeling and application in real landfill leachate. *Scientific Reports*, 14(1), 3426.

Sabbas, T., Polettini, A., Pomi, R., Astrup, T., Hjelmar, O., Mostbauer, P., ... & pHOENIX Working Group on Management of MSWI Residues. (2003). Management of Municipal Solid Waste Incineration Residues. *Waste management*, 23(1), 61-88.

Schneider, J., Vehlow, J., & Vogg, H. (1994). Improving the MSWI Bottom Ash Quality by Simple in–Plant Measures. In *Studies in Environmental Science* (Vol. 60, pp. 605-620). Elsevier.

Singh, E., Kumar, A., Khapre, A., Saikia, P., Shukla, S.K., & Kumar, S. (2020). Efficient removal of arsenic using plastic waste char: Prevailing mechanism and sorption performance. *Journal of Water Process Engineering*, 33, 101095.

Singh, E., Kumar, A., Mishra, R., You, S., Singh, L., Kumar, S., & Kumar, R. (2021). Pyrolysis of waste biomass and plastics for production of biochar and its use for removal of heavy metals from aqueous solution. *Bioresource Technology*, 320, 124278.

Soudani, A., Youcef, L., Chebbi, M., Bulgariu, L., & Patel, N. (2024). Agricultural waste–based biochars for sustainable removal of heavy metals from stabilised landfill leachate. *Environmental Science and Pollution Research*, 31(47), 57733-57747.

Tchounwou, P. B., Yedjou, C. G., Patlolla, A. K., & Sutton, D. J. (2012). Heavy metal toxicity and the environment. *Molecular, clinical and environmental toxicology: volume 3: environmental toxicology,* 133-164.

Travar, I., Lidelöw, S., Andreas, L., Tham, G., & Lagerkvist, A. (2009). Assessing the environmental impact of ashes used in a landfill cover construction. *Waste Management,* 29(4), 1336-1346.

Türkmen, D., Bakhshpour, M., Akgönüllü, S., Aşır, S., & Denizli, A. (2022). Heavy metal ions removal from wastewater using cryogels: A review. *Frontiers in Sustainability,* 3, 765592.

UNEP 2008, Interim review of scientific information on lead, *Taken* 2008-05-18 fromhttp://www.chem.unep.ch/Pb_and_Cd/SR/Files/2008/UNEP_Lead_review_Interim-mar102008.doc.

UNEP, 2002, Global Mercury Assessment, Geneva, Switzerland, Taken 2008-05-10 from http://www.chem.unep.ch/mercury/Report/Chapter6.htm.

Wijesekara, D. A., Sargent, P., Ennis, C. J., & Hughes, D. (2021). Prospects of using chars derived from mixed post waste plastic pyrolysis in civil engineering applications. *Journal of Cleaner Production*, 317, 128212.

Xue, Q., Li, J. S., Wang, P., Liu, L., & Li, Z. Z. (2014). Removal of heavy metals from landfill leachate using municipal solid waste incineration fly ash as adsorbent. *Clean–soil, air, water,* 42(11), 1626-1631.

Yusof, N., Haraguchi, A., Hassan, M. A., Othman, M. R., Wakisaka, M., & Shirai, Y. (2009). Measuring organic carbon, nutrients and heavy metals in rivers receiving leachate from controlled and uncontrolled municipal solid waste (MSW) landfills. *Waste Management*, 29(10), 2666-2680.

Zand, A. D., & Abyaneh, M. R. (2020). Adsorption of Lead, manganese, and copper onto biochar in landfill leachate: implication of non-linear regression analysis. *Sustainable Environment Research*, 30, 1-16.

Chapter 18

Investigation of Unconventional Magnetism in Iridates

Shreemoyee Ganguly*
Centre of Excellence in Multifunctional Materials and Devices,
Department of Basic Science and Humanities, Institute of Engineering & Management,
University of Engineering and Management, Kolkata, India

Abstract

In this communication, we present a comparative study of various iridates, including the archetypal Sr_2IrO_4, the spin-orbit-driven insulator $Ba_3YIr_2O_9$, the 6H-perovskite-based $Ba_3IrTi_2O_9$ and $Ba_3TiIr_2O_9$, and the 3d-5d transition metal-based La_2CoIrO_6 and La_2ZnIrO_6. Our objective is to investigate the origin of magnetism in these novel compounds through first-principles calculations using multipolar analysis. It is essential to understand the origin of magnetism to manoeuvre magnetic crossovers in systems crucial for quantum computation and quantum information-based applications. Interestingly, the order parameter responsible for breaking time-reversal symmetry in these compounds is not necessarily the spin moment. In most of these compounds, the spin-orbit entangled higher-order multipole, namely the triakontadipole, is responsible for driving the unusual magnetic states in these systems. Remarkably, our study reveals that in strongly correlated La_2CoIrO_6,the magnetism of the 3d and 5d ions has entirely different origins. While the usual spin moment drives magnetism in Co, the higher-order magnetic multipole, the triakontadipole, is responsible for breaking the time-reversal symmetry in Ir. Finally, we have extended our investigation to an osmium-based [$5d^2$] compound, Ba_2ZnOsO_6, to

* Corresponding Author's Email: ganguly.shreemoyee@gmail.com

In: Breakthrough Research and Technologies in Physics …
Editors: Koyel Ganguly, Soumyadipta Pal and Kamakhya Prasad Ghatak
ISBN: 979-8-90134-111-7

determine whether our developed formalism aligns with the recent neutron diffraction study on the compound.

Keywords: spin-orbit coupling, iridates, multipolar analysis

Introduction

Since time immemorial, magnetism in 5d oxides has attracted condensed matter physicists (Witczak-Krempa W., Chen G., Kim Y. B. and Balents L., 2014) firstly due to the intriguingphysics arising due to correlated electron behavior and secondly due to the non trivial physics arising due to the presence of strong spin-orbit coupling (SOC) in these materials. Since SOC is a predominant phenomenon in heavy elements having a high value of atomic number Z, the magnetism arising due to spin-orbit entanglement was initially investigated in f-electron systems (Santini P., Carretta P., Amorreti R., Caciuffo R., Magnani N., and Lander G. H., 2009). However, currently the focus has shifted to 5d based compounds, especially the iridateswhere both electronic correlation and spin-orbit entanglement compete to determine the electronic structure and corresponding magnetic nature of the system (Kim B. J., Sin H., Moon S. J., Kim J. Y., Park B. J., Leem C. S., Yu J., Noh T. W., Kim C., Oh S. J., Park J. H., Durairaj V., Cao G. and Rotenberg E.., 2008) Although the correlation effect seems to be weaker in the 5d systems compared to their corresponding 3d counterparts, the predominant role of SOC in these compounds reinforces the impact of correlation physics in these compounds. Due to the widening of the d orbitals in the 5d elements as compared to their 3d counterparts and thus delocalisation of the d electrons, the effect of electronic repulsion and therefore their correlated behaviour seems to be diminished in such systems (Witczak-Krempa W., Chen G., Kim Y. B. and Balents L., 2014). However, due to SOC, the degenerate d orbitals split up into narrow levels, thus again reincarnating the features of correlated physics. Therefore, we observe SOC-driven Mott insulators in 5d compounds, such as iridates (Jackeli, G., and Khaliullin, G., 2009). Being able to control the conducting properties of a system by tuning its magnetic nature through doping or by applying mechanical stress makes these materials potential candidates for spintronics and quantum computer-based applications (Kitaev, A., 2006; Jackeli, G., and Khaliullin, G., 2009).

In this prospective SOC-driven correlated behaviour regime, experimental studies have shown interesting characteristics of the 5d compounds known as iridates, where iridium is generally present in the d^{4+} state (Ye F., Chi S., Chakoumakos B. C., Fernandez-Baca J. A., Qi T., and Cao G., 2013; Nichols J., Bray-Ali N., Ansary A., Cao A., and Ng K. W., 2014). As the Ir atom is in a d^5 state and located in an octahedral crystal field of the oxygen atoms in these compounds, the 6-fold degenerate t_{2g} levels (including the spin degree of freedom) have a single hole. Thus, the system is expected to be in a low-spin state (Jackeli, G., and Khaliullin, G., 2009). In this respect, the SOC-driven correlated insulator Sr_2IrO_4 has generated considerable interest (Nichols J., Bray-Ali N., Ansary A., Cao A., and Ng K. W., 2014; Dai J., Calleja E., Cao G., and McElroy K., 2014). This system is considered a weak ferromagnet having a substantial magnetic moment (Jackeli G. and Khaliullin G., 2009) ($0.14\mu_B$). Interest in this system is piqued by the proposal of an engineering "spin flop" transition in this compound. Sr_2IrO_4has a similar crystal structure to its 3d counterpart- the undoped high T_c cuprate La_2CuO_4 (Parschke E. M., Wohlfeld K., Foyevtsova K. and Brink J. V., 2017). In this compound, the Ir^{4+} ions form a square-shaped lattice structure. This happens because the IrO6 octahedra in the system are corner-sharing. These octahedra are also slightly distorted, having an elongation along the c-axis and a rotation of nearly 11 degrees about it. The corresponding spin canting angle of the Ir ions ($\varphi = 8$ degrees) is very close to the octahedral rotation angle, indicating a coupling between the geometric and spin structure of the system (Ganguly Shreemoyee, Granas Oscar and Nordstrom Lars, 2015). Thus, by distorting the system's geometric structure, the magnetic structure of the system can also be manipulated. The canted spin structure of the system can be converted to a collinear Néel antiferromagnetic structure (the said spin flop transition) by tuning the tetragonal distortion parameter θ of the system, as shown in Figure 1. Being able to manipulate such spin crossovers can be significant in quantum computation and quantum information theory. However, to handle such magnetic transitions, we should be able to understand and identify the primary magnetic order parameter (POP), responsible for driving magnetism in the system. We have studied a host of iridates with the primary objective of identifying the phase of the various systems and thus attempting to determine the root cause of magnetism in these compounds. For this purpose, we have developed a methodology (Ganguly Shreemoyee, Granas Oscar and Nordstrom Lars, 2015; Panda S. K., Bhowal S., Li Ying, Ganguly Shreemoyee, Valentí Rosa, Nordström Lars, and Dasgupta Indra, 2015; Bhowal Sayantika, Ganguly Shreemoyee, and Dasgupta Indra, 2019; Ganguly Shreemoyee and Bhowal Sayantika, 2020) based on magnetic multipolar analysis combined with constrained moment variation.

Next, after investigating and understanding magnetism in the iridates satisfactorily, we extended our study to analyse and identify the POP in an osmate ($5d^2$ electronic configuration), namely the compound Ba_2ZnOsO_6. This is because in recent times, there has been a remarkable claim through neutron and x-ray diffraction studies (Maharaj D. D., Sala G., Stone M. B., Kermarrec, Ritter C., Fauth F., Marjerisson C. A., Greedan J. A. and Paremakanti A., 2020) that in this compound, a strange octupolar magnetic order parameter seems to be predominant. We aim to confirm this claim through our study, as presented in the current communication.

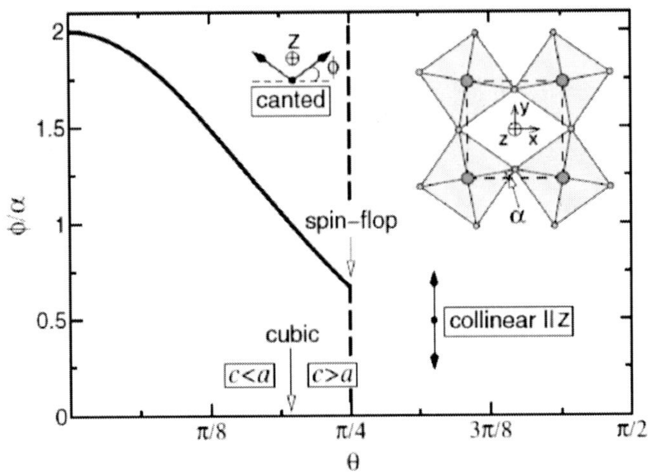

Figure 1. A study [Jackeli G. and Khaliullin G., 2009] of the angle of canting of spinφ (in terms of the angle of octahedral rotation α) as a function of θ, the parameter indicating the tetragonal distortion of the system, is demonstrated in the figure. A representation of the planar structure of the IrO_2 is shown in the inset. When $\theta = \pi/5$, then the system exhibits a perfect cubic geometry. In this situation, the spin canting angle is equal to the tetragonal distortion angle. However, the most significant feature observed here is the magnetic crossover from the canted spin structure to a collinear antiferromagnetic Néel order alongthe z-axis. This crossover is observed at $\theta = \frac{\pi}{4}$. Such spin crossovers are significant for quantum computation and quantum information theory.

Theoretical Understanding of the Iridates and Our Analysis

In literature, the Hubbard Hamiltonian (N. F. Mott, 1990; Hubbard, J., 1963) is widely used and highly successful in explaining the magnetic and conducting properties of 3d oxides. In 5d materials, it was previously believed that the widening of the d bands would significantly reduce the effective value of the Hubbard correlation U. Thus, correlation effects would be extremely weak. However, surprisingly, the 5d compound Sr_2IrO_4 turned out to be a correlation-driven insulator. To explain this SOC-driven correlation effect, a new model, popularly known as the J_{eff} = 1/2 model, was proposed (Kim B. J., Jin H., Moon S. J., Kim J. Y., Park B. G., Leem C. S., Yu J., Noh T. W., Kim C., Oh S. J., Park J. H., Durairaj V., Cao G., and Rotenberg E., 2008). This model introduced the concept of forming a new quantum state in these 5d compounds. The 10-fold degenerate (including the spin degree of freedom) d orbitals split into the t_{2g} and $e_g$levels in an octahedral environment. The six-fold degenerate t_{2g}levels are divided into the four-fold degenerate j_{eff}=3/2 and the doubly degenerate j_{eff}= 1/2 states due to strong spin-orbit coupling. The name "j_{eff}" model and "j_{eff}" state thus originates from the fact that the total angular momentum **J** = **L** + **S**is formed through the coupling of orbital angular momentum **L**and spin angular momentum **S**, andit is the interplay of orbital and spin angular momentum which is responsible for the formation of the j_{eff}= 1/2 state. The j_{eff}= 1/2 band splits into the upper and lower Hubbard bands under the influence of the Coulomb repulsion represented through the interaction parameter "U." The Ir atom in Sr_2IrO_4 is in a d^5 state, having 5 electrons. Thus, the four-fold degenerate j_{eff}=3/2 state is filled, and the higher j_{eff}=1/2 level is half-filled. While the lower Hubbard band is filled, the upper Hubbard band is empty. The gap between the upper and lower Hubbard bands makes the compound Sr_2IrO_4 an insulator.

While the J_{eff} model was extremely successful in explaining various features of this archetypal SOC-driven insulator Sr_2IrO_4, when we started studying the system, we realised that a few features of the system remained unexplained. Firstly, the J_{eff} model predicted the magnetic moment at the Ir site to be 1 μ_B, which is significantly higher than the experimentally obtained magnetic moment of 0.3 $\mu_B$per atom. Secondly, the experimentally determined band gap for the system is 240 meV, which is significantly larger than what would be expected if the spin moment and the corresponding magnetic exchange energy associated with it were solely responsible. To seek answers to these discrepancies, we conducted a magnetic multipolar-based study (Ganguly, Shreemoyee, Granas, Oscar, and Nordstrom, 2015). In our study, we employed the standard relativistic density functional theory approach, supplemented with an additional Hubbard parameter U that represents the enhanced Coulomb interaction. However, the additional analysis we conducted was a multipolar expansion of the corresponding enhanced

Coulomb interaction term in the mean-field limit. This expansion translates the information carried by the density matrix corresponding to the d orbital into various multipoles, including spin, orbital, dipole, quadrupole, octupole, hexadecapole, and triakontadipole. The exchange energy associated with the various multipoles can be estimated (Ganguly, Shreemoyee, Granas, Oscar, and Nordstrom, Lars, 2015), and by doing so, the hierarchy of the multipoles can be established. Through this analysis, we identified the magnetic multipole primarily responsible for breaking time-reversal symmetry and driving magnetism in our magnetic system. This dominant magnetic multipole was termed the primary order parameter (POP) of our system, and the other magnetic multipoles, which had a lesser exchange energy associated with them than the POP, were referred to as secondary order parameters (SOPs) in our study. In the above-discussed system, Sr_2IrO_4, the spin moment was identified as being driven solely by the triakontadipole moment (magnetic multipolar moment of order 5), which is the actual POP (Ganguly, Shreemoyee, Granas, Oscar, and Nordstrom, Lars, 2015). This is clear from Table 1. This moment is formed through the entanglement of spin and magnetic moments. This highlights the importance of spin-orbit coupling interaction (SOC) in the system. Next, we studied (Panda S. K., Bhowal S., Li Ying, Ganguly Shreemoyee, Valentí Rosa, Nordström Lars, and Dasgupta Indra, 2015) a $d^{4.5}$ system, namely the 6-H perovskite $Ba_3YIr_2O_9$. Here, as seen from Table 1, unlike the previously studied system, the spin moment is the POP. This is because an additional interaction, the intradimer Ir-Ir interaction, becomes relevant in this system. Due to the effect of this interaction, the d orbitals get widened, and thus the effect of SOC gets diminished. After this, we investigated [Bhowal Sayantika, Ganguly Shreemoyee, and Dasgupta Indra, 2019] a couple of hexagonal d^5 iridates, namely Ba3IrTi2O9and $Ba_3TiIr_2O_9$. While in one of the iridates ($Ba_3TiIr_2O_9$) the Ir-Ir dimer interaction plays a role, in another iridate ($Ba_3IrTi_2O_9$) such an interaction is not relevant. Despite this, in both sister compounds, the higher-order magnetic multipole, namely the triakontadipole formed through spin-orbit entanglement, is found to be the POP (as seen from Table 1). The reason for this was found to be that the $_{Jeff}$ model is preserved in the ^{D5} Ir atom, even in the dimeric state, to stabilise the intra-dimer interaction through hopping between the two $_{Jeff}$=1/2 states of the two Ir atoms forming the dimer. The next investigation was a fascinating and challenging study (Ganguly Shreemoyee and Bhowal Sayantika, 2020) because the compound studied comprised both a 5d and a 3d element. We studied the iridate La_2CoIrO_6 next. Here, while Coulomb localisation (introduced through Hubbard U) plays the primary role in driving magnetism in the 3d Co atom, we have already seen that SOC plays an important role in d^5 Ir atom. However, we wanted to understand how the magnetism of one Ir and Co atom is affected by the other when they are connected through two corner-sharing octahedra. We also conducted a comparative study of this compound with its isomorphous compound La_2ZnIrO_6. The Zn atom, though a 3d element, is nonmagnetic. So, in this isostructural compound, the magnetic nature of the Ir atom is certainly not influenced by the magnetism of any 3d transition metal species. Surprisingly, Table 1 reveals that in La_2CoIrO_6, two primary order parameters exist. In contrast, one of the order parameters (spin moment) is responsible for magnetism in the 3d Co atom, the other order parameter (triakontadipole moment) serves as the POP in the 5d Ir atom. The magnetic nature of the Ir species in La_2CoIrO_6 is similar to that in the isostructural La_2ZnIrO_6 species, indicating that the strongly correlated 3d species do not dominate the magnetism of the 5d Ir atom.

Table 1. A study of magnetic multipoles and the exchange energy associated with them in various iridates

Material		Multipole Moment	Value of Moment μ_B/atom		Exchange Energy (meV/atom)	
Ir in Sr_2IrO_4		Spin	0.08		-2.4	
		Orbital	0.24		-4.2	
		Octupolar	0.56		-41.3	
		Triakontadipole	3.18		-71.8	
Ir in $Ba_3YIr_2O_9$		Spin	0.55		-120.0	
		Orbital	0.34		-10.0	
		Octupolar	0.73		-35.0	
		Triakontadipole	1.96		-36.0	
Ir in $Ba_3IrTi_2O_9$	Ir in $Ba_3TiIr_2O_9$	Spin	0.14	0.13	-6.5	-5.0
		Orbital	0.29	0.28	-5.8	-8.3
		Octupolar	0.62	0.52	-50.7	-46.6
		Triakontadipole	3.68	2.94	-96.1	-74.5
Co in $La_2C_oIrO_6$	Ir in $La_2C_oIrO_6$	Spin	2.37	0.065	-1423.84	-0.71
		Orbital	0.224	0.149	-14.90	-3.11
		Octupolar	4.12	0.397	-89.92	-0.45
		Triakontadipole	8.80	4.07	-512.69	-58.75
Ir in La_2ZnIr_O6		Spin	0.091		-2.77	
		Orbital	0.325		-7.45	
		Octupolar	0.36		-0.74	
		Triakontadipole	4.11		-120.0	

Current Extension to Os in Ba_2ZnOsO_6

In this communication, we have extended our study to an osmium-based compound, Ba_2ZnOsO_6. Os^{6+} is in a $5d^2$ state in this compound. The motivation behind this extension is an experimental study on osmates (Maharaj D. D., Sala G., Stone M. B., Kermarrec, Ritter C., Fauth F., Marjerisson C. A., Greedan J. A., and Paremakanti A., 2020), conducted using neutron powder diffraction and X-ray diffraction techniques. This study indicates that a very interesting magnetic multipolar order parameter, namely the magnetic octupolar moment, is the primary order parameter. We wanted to test our developed methodology on one of the osmates studied in the mentioned work. Remarkably, in the osmate that we have studied, namely Ba_2ZnOsO_6, we found that the POP driving magnetism in the system is the magnetic octupole as seen from Table 2. This correspondence of our study with experimental results further establishes the validity and applicability of our developed methodology for various 5d-based oxides.

Table 2. A study of magnetic multipoles and the exchange energy associated with them in an osmate

Multipole Moment	Value of Moment (meV/atom)	Exchange Energy (meV/atom)
Spin	0.472	-67.33
Orbital	0.163	-6.024
Octupolar	4.05	-78.86
Triakontadipole	1.49	-13.43

Discussion

In the present study, we have initially discussed the magnetic nature of 5d oxides and how it differs from that of 3d oxides. While the Coulomb interaction is the dominant interaction in localised 3d oxides, 5d oxides are generally SOC-driven. It is essential to comprehend the magnetic order parameter, which is responsible for breaking time-reversal symmetry and driving magnetism in 5d oxides. Understanding this parameter would provide a means to control the magnetic order and conducting properties of these compounds, which could be beneficial for quantum computation and spintronics-based applications. Next, through a comparative analysis, we identified the POP driving magnetism in various iridates, as previously studied by us using a magnetic multipolar-based methodology developed through multiple studies. Our comparative analysis demonstrates (Table 1) that in systems where SOC is the dominant interaction, spin moment is merely a SOP. A higher-order moment formed through spin-orbit entanglement is the actual POP driving magnetism in such systems. To establish the validity and applicability of our developed methodology, in the present work, we tried to identify the POP for an osmate (Ba_2ZnOsO_6), which has been experimentally studied in recent times. The authenticity of our methodology is further established through this work, as our identified POP in this $5d^2$ osmate, namely the magnetic octupole, exactly corresponds to the experimentally indicated primary magnetic order parameter.

References

Bhowal Sayantika, Ganguly Shreemoyee, and Dasgupta Indra, *Journal of Physics: Condensed Matter*, (2019) (31):185802

Dai J., Calleja E., Cao G., and McElroy K., *Phys. Rev. B*, (2014) (90): 041102.

Ganguly Shreemoyee and Bhowal Sayantika, *Phys. Rev. B*, (2020) (101):085104.

Ganguly Shreemoyee, Granas Oscar and Nordstrom Lars, *Phys. Rev. B*, (2015) (91):020404®.

Hubbard J., *Proc. Roy. Soc. London*, (1963) (A276): 238.

Jackeli G. and Khaliullin G. *Phys. Rev. Lett.* (2009) (102):017209.

Kim B. J., Jin H., Moon S. J., Kim J. Y., Park B. G., Leem C. S., Yu J., Noh T. W., Kim C., Oh S. J., Park J. H., Durairaj V., Cao G., and Rotenberg E., *Phys. Rev. Lett.* (2008) (**101**):076402.

Kim B. J., Sin H., Moon S. J., Kim J. Y., Park B. J., Leem C. S., Yu J., Noh T. W., Kim C., Oh S. J., Park J. H., Durairaj V., Cao G. and Rotenberg E. *Phys. Rev. Lett.* (2008) (101): 076402.

Kitaev A. *Ann. Phys.* (2006) (321):2.

Maharaj D. D., Sala G., Stone M. B., Kermarrec, Ritter C., Fauth F., Marjerisson C. A., Greedan J. A. and Paremakanti A., *Phys. Rev. Lett.* (2020) (124): 087206.

N. F. Mott, Metal-Insulator Transitions (Taylor and Francis, London/Philadelphia, 1990).

Nichols J., Bray-Ali N., Ansary A., Cao A., and Ng K. W., *Phys. Rev. B*, (2014) (89): 085125.

Panda S. K., Bhowal S., Li Ying, Ganguly Shreemoyee, Valentí Rosa, Nordström Lars, and Dasgupta Indra, *Phys. Rev. B*, (2015) (92):180403®.

Parschke E. M., Wohlfeld K., Foyevtsova K. and Brink J. V., Nat Commun., (2017) (8):686.

Santini P., Carretta P., Amorreti R., Caciuffo R., Magnani N., and Lander G. H. *Rev. Mod. Phys.* (2009) (81): 807.

Witczak-Krempa W., Chen G., Kim Y. B. and Balents L. Correlated Quantum Phenomena in the Strong Spin-Orbit Regime. *Annual Review of Condensed Matter Physics* (2014) (5):57-82.

Ye F., Chi S., Chakoumakos B. C., Fernandez-Baca J. A., Qi T., and Cao G., *Phys. Rev. B* (2013) (87): 140406®.

Chapter 19

Assessing Seasonal Vegetation Dynamics for Sustainable Ecosystem Management: A Comparative Analysis of NDVI, NDMI, NDWI, and AMMI Indices on Henry Island

Shreyashi Das*
and Anindita Das Bhattacharjee
Department of Computer Science Engineering, Institute of Engineering & Management, University of Engineering and Management, Kolkata, West Bengal, India

Abstract

Mangroves have long been monitored using satellite photography; however, in remote areas, challenges such as deep mud, signal loss, and the complex root systems of mangroves present significant obstacles to this process. For climate resilience and sustainable ecosystem management, a precise and automated mangrove map that graphically represents canopy density using satellite imagery is desperately needed. To support conservation efforts, this project aims to develop a reliable, automated method for locating, mapping, and tracking mangrove forests. This method reduces expenses and improves efficiency by eliminating the need for time-consuming human digitisation, utilising publicly available satellite data. For experimental validation, Sentinel-2 footage from several mangrove forests is used to generate the canopy density index. The maps successfully depict the spatial extent of mangroves and visually match satellite pictures, indicating good accuracy. The study highlights the methodology's potential for extensive ecological

* Corresponding Author's Email: Shreyashidas197@gmail.com

In: Breakthrough Research and Technologies in Physics …
Editors: Koyel Ganguly, Soumyadipta Pal and Kamakhya Prasad Ghatak
ISBN: 979-8-90134-111-7

monitoring by applying it to the mangrove complex at Henry Island in the Sundarbans, West Bengal. The suggested technique is practical and easy to use. It enables the quick creation of mangrove maps, the updating of sustainability indices, and the monitoring of how mangrove conditions evolve. This study supports long-term conservation plans by offering a practical and affordable means of preserving coastal habitats and lessening the effects of climate change.

Keywords: vegetation indices, remote sensing for vegetation monitoring, Seasonal vegetation dynamics, Satellite-based vegetation assessment, Remote sensing

Introduction

Woody plants known as mangroves are found both inland and along the shore, particularly in tropical and subtropical regions [1, 2]. High environmental dynamics, including temperature, sedimentation, and tidal currents, are characteristic of mangrove ecosystems [3]. Due to their contributions to the coastal zone, productivity, and ecologically important habitats, mangroves are highly beneficial to shallow seas and coastal ecosystems [4]. Freshwater from the land via rivers and saltwater from the sea often influence the environment's unique features [5, 6]. The mangrove ecosystem offers several benefits, including nutrient cycling and serving as a nursery for various marine fisheries [7], providing habitat for diverse animal species, serving as landing grounds for thousands of migrating birds [3], and supporting biodiversity [8]. Mangroves are crucial for sequestering carbon in the context of climate change because they can do so through photosynthetic processes and because the majority of the carbon they store is also stored in the soil [9–14]. Mangroves are susceptible to both natural and artificial disturbances. According to the Food and Agriculture Organisation of the United Nations (FAO) [15], since the 1980s, deforestation has resulted in the destruction of 20% of the world's mangroves. Significant anthropogenic impacts include urbanisation [16–20]. Relevant data to monitor mangrove conditions include Sentinel-2 images, which have a spatial resolution of between 10 and 20 meters, as well as Landsat 5 Thematic Mapper, Landsat 7 Enhanced Thematic Mapper, and Landsat 8 Operational Land Imager, all of which have a spatial resolution of 15 meters or less. Mangroves were apparent when viewed through Red, Green, Blue (RGB) composite satellite imagery, which blended recognisable Near

Infrared (NIR) and Shortwave Infrared (SWIR) bands. For instance, mangroves appear more visually distinct when viewed through Landsat 8's RGB combination of NIR-SWIR1-Red.

Nevertheless, no translation technique has yet been developed to satisfactorily mechanically depict the canopy density and trace and record the presence of mangroves visible in the composite photos. Tracing the mangroves using the traditional method involves manually matching the appearance of the RGB composite image. Using the current techniques, the canopy thickness indices are then mapped [24]. The case study is located in Henry Island, West Bengal, India (Figure 1). Geographically, Henry Island is situated approximately at 21.5481° N latitude and 88.2227° E longitude. The island is part of the South 24 Parganas district, a region renowned for its rich biodiversity and serving as a gateway to the Sundarbans. The native populations and ecosystems of Henry Island are seriously threatened by the effects of climate change, which include increasing sea levels and more frequent storms. The authors thus suggested the following algorithm,

$$(\rho NIR - \rho Red)/ (\rho Red + \rho SWIR1) * (\rho NIR - \rho SWIR1)/ (\rho SWIR1 - 0.65\rho Red) \quad (1)$$

Figure 1. The case study is located on Henry Island, West Bengal, India.

Literature Review in Sustainable Ecosystem Management

The novel method of Automatic Mangrove Map and Index (AMMI) for effectively monitoring changes in mangrove ecosystems is presented in the paper "AMMI Automatic Mangrove Map and Index: Novelty for Efficiently Monitoring Mangrove Changes with the Case Study in Musi Delta, South Sumatra, Indonesia" by Suyarso and Praditya Avianto. The study focuses on the South Sumatra, Indonesia, Musi Delta region, where mangrove trees are essential for supporting livelihoods, conserving biodiversity, and protecting the coastline. The AMMI technique generates precise maps and indicators of mangrove distribution and health by integrating remote sensing technology with Geographic Information Systems (GIS). The authors efficiently evaluate changes in mangrove density, moisture content, and water availability over time by combining satellite imagery with vegetation indices, such as the Normalized Difference Vegetation Index (NDVI), Normalized Difference Moisture Index (NDMI), and Normalized Difference Water Index (NDWI). Policymakers, researchers, and local communities can monitor mangrove ecosystems extensively with the help of the AMMI technique, enabling them to make informed decisions about conservation and sustainable management projects. The Musi Delta case study highlights the potential for broader deployment of AMMI in other coastal regions facing similar issues by showcasing its usefulness and effectiveness in providing insightful information about mangrove dynamics and environmental changes [21]. A thorough analysis of the various spectral indices employed in remote sensing for mangrove ecosystems is presented in the work "A Review of Spectral Indices for Mangrove Remote Sensing" by T. Tran. R. Reef and Xuan Zhu, which was published in Remote Sensing in 2022. It emphasises the significance of mangroves for supporting biodiversity, protecting coastlines, and sequestering Carbon, while highlighting the necessity of efficient monitoring and management. Based on their uses in species recognition, stress identification, and evaluation of vegetation health, spectral indices are categorized and evaluated in this review. It examines the benefits and drawbacks of each indicator, taking into account ambient considerations and sensor compatibility. The writers also discuss the potential of remote sensing technology to improve mangrove tracking. To increase the precision and dependability of mangrove remote sensing, they advocate for uniform techniques and the integration of data from several sources. Researchers and practitioners seeking to create and utilise spectral indices for efficient mangrove management and preservation may find this publication to be a valuable resource [22].

Research Gap Analysis

Despite improvements in remote sensing and vegetation monitoring using indices, gaps remain in our understanding of seasonal vegetation changes. Although NDMI and NDWI offer more accurate moisture information, NDVI struggles to distinguish between soil moisture and vegetation water content. To investigate vegetation dynamics during the wet and dry seasons—a crucial aspect of sustainable ecosystem management—very few studies have integrated these indices. Particularly for complex ecosystems like mangroves, which are essential for biodiversity conservation, carbon sequestration, and coastal protection, a more comprehensive strategy is needed. Nevertheless, despite recent advancements, charting these ecosystems remains inconsistent and prone to error. The majority of mangrove mapping systems currently in use employ manual or semi-automated classification techniques, which restrict scalability and increase the risk of human error. Long-term ecological monitoring and conservation planning require the integration of multi-temporal and multispectral data for increased accuracy; however, there are currently no fully automated research frameworks that accomplish this. Although the AMMI index attempts to address these problems, additional verification is necessary to ensure regional accuracy. Mangrove mapping can be significantly enhanced by integrating data with machine learning and cutting-edge remote sensing methods, thereby supporting efforts to manage resources sustainably and build climate resilience.

Methodology

- *Materials:* The materials used for this research include high-resolution optical imagery from satellites with open-access data, specifically Sentinel 2.
- *Satellite Import:* Gather pertinent satellite imagery.
- *Visualisation Parameters:* Adjust the visualisation settings to improve the sharpness and detail of the images.
- *Date Range:* Establish a lengthy range of dates for your investigation.
- *Definition of Dry and Wet Seasons:* Determine and categorise the dataset's dry and wet seasons.
- *Index Calculation:* Utilise the designated calculation techniques to determine the vegetation indices (NDVI, NDWI, NDMI, and AMMI).

- *Season-wise Reporting*: Create reports broken down into the dry and rainy seasons for each index.
- *Dot Graph:* Construct dot graphs to show data points for every index across time.
- *DOY Charts:* To show seasonal and annual trends, create Day-of-Year (DOY) series charts and Day of Year-by-Year series charts.
- *Line Chart:* Create line charts that illustrate how the indices have changed throughout time.
- *Annual and Seasonal Images*: Create GIFs to illustrate annual and seasonal variance.

Result Analysis

Figure 2 shows the mean NDVI readings for each date throughout the wet season from 2018 to 2023. Because monsoon rains promote enhanced vegetation growth, the wet season usually has higher NDVI values. Variations within a single year can reveal the distribution and intensity of rainfall. Peaks in the NDVI represent times when the health and density of the vegetation are at their highest, indicating how the monsoon rains affect the growth of the vegetation.

Figure 3 shows the mean NDVI values for each date during the dry season. Lower NDVI readings may be observed during the dry season due to reduced moisture and slower plant growth. Peaks may still happen during this season, particularly if there is still rainwater from the previous wet season in the early dry season (October to December). The decrease in vegetative health resulting from the absence of rainfall is evident in the progressive decline of NDVI values from January to May.

The NDVI for the wet season is plotted against the day of the year in this graphic. As the monsoon intensifies, the Day of Year Series should exhibit an increasing tendency, which should either steady or decline as the season concludes. This pattern illustrates how the first upsurge in vegetation growth occurs as the rains begin, and then it stabilizes as the vegetation reaches its maximum.

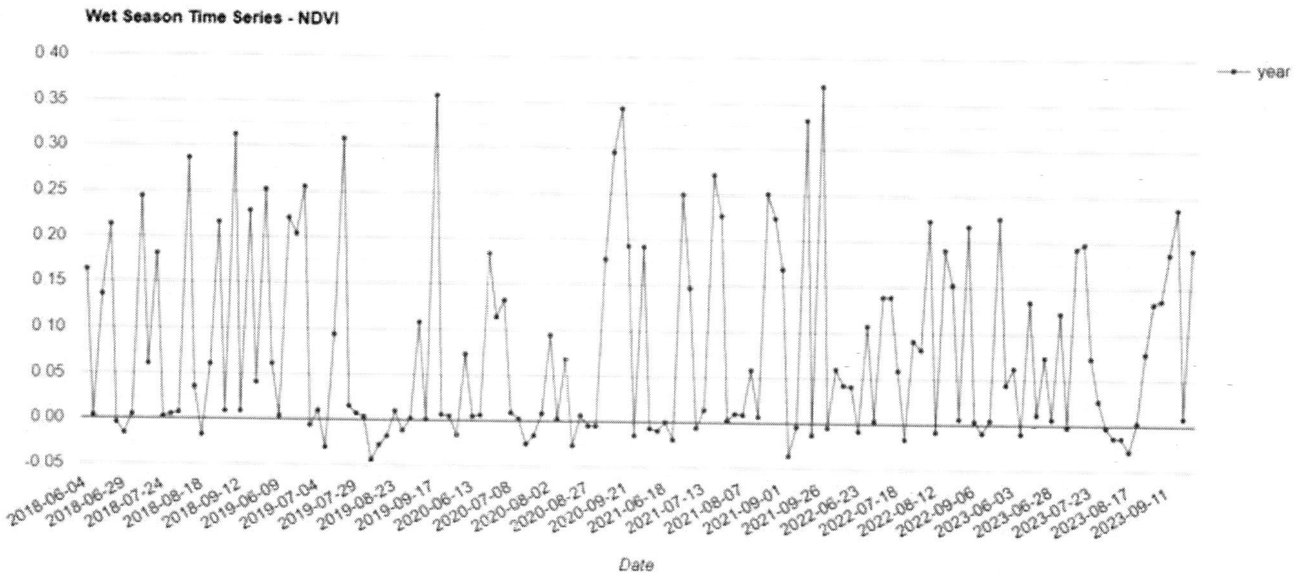

Figure 2. Wet Season Time Series – NDVI.

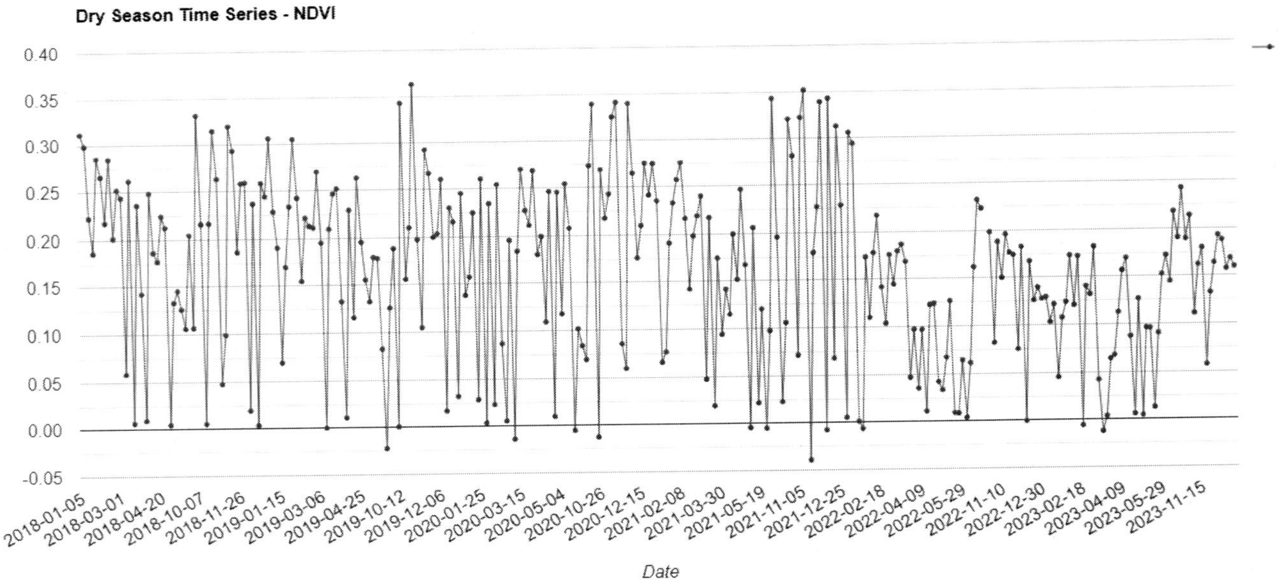

Figure 3. Dry Season Time Series – NDVI.

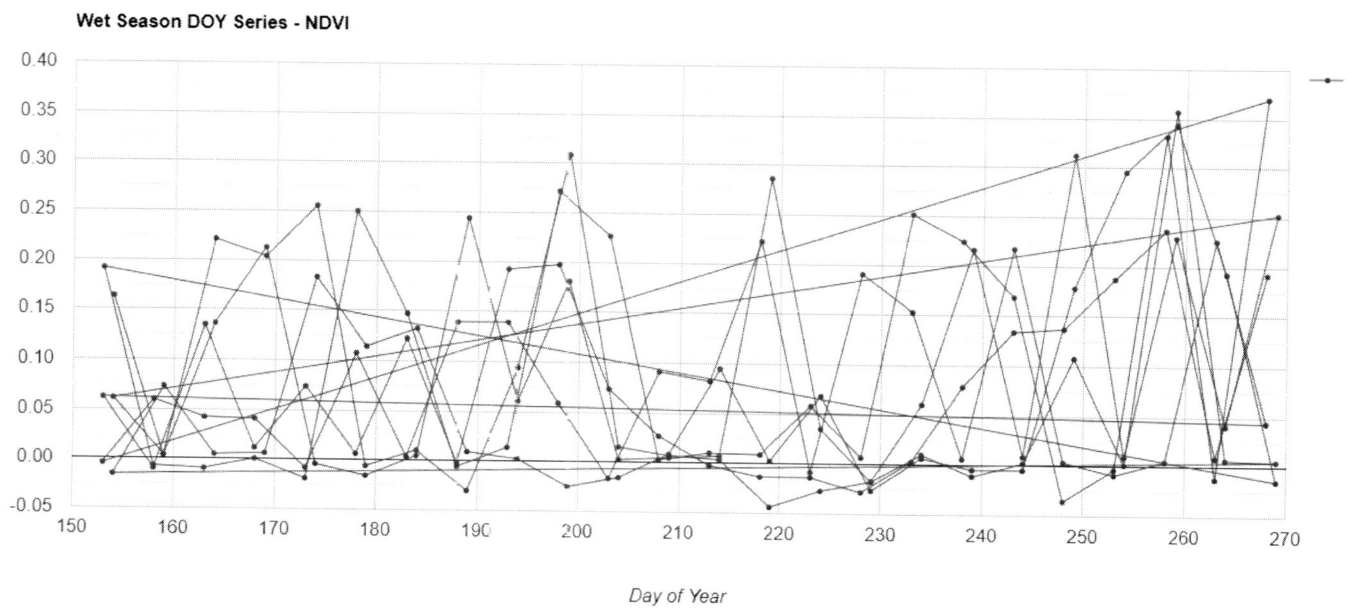

Figure 4. Wet Season Day of Year Series – NDVI.

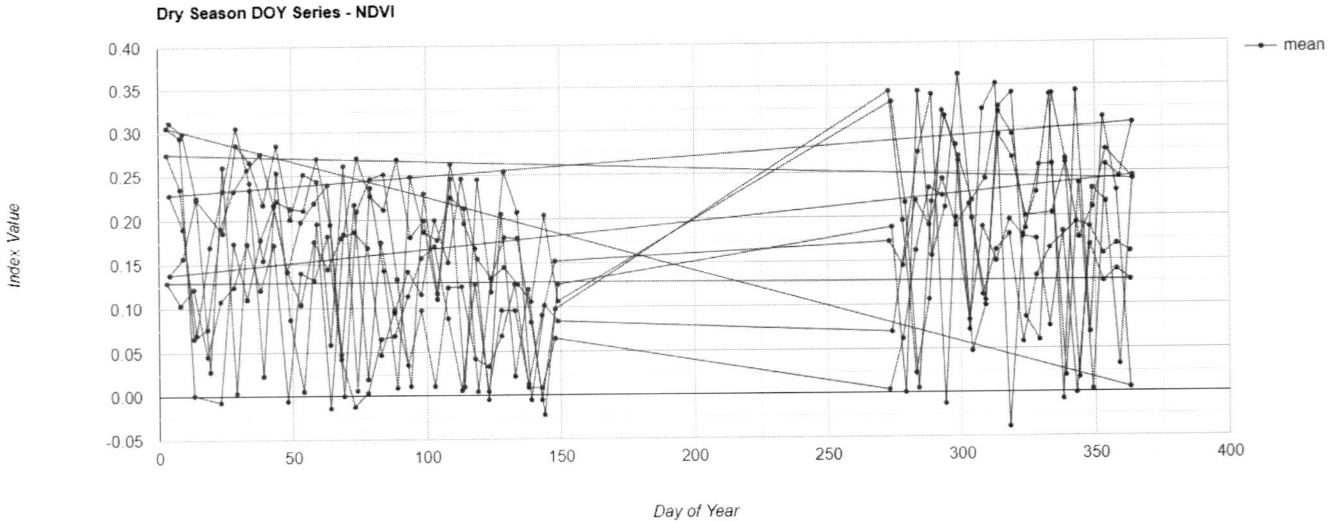

Figure 5. Dry Season Day of Year Series -NDVI.

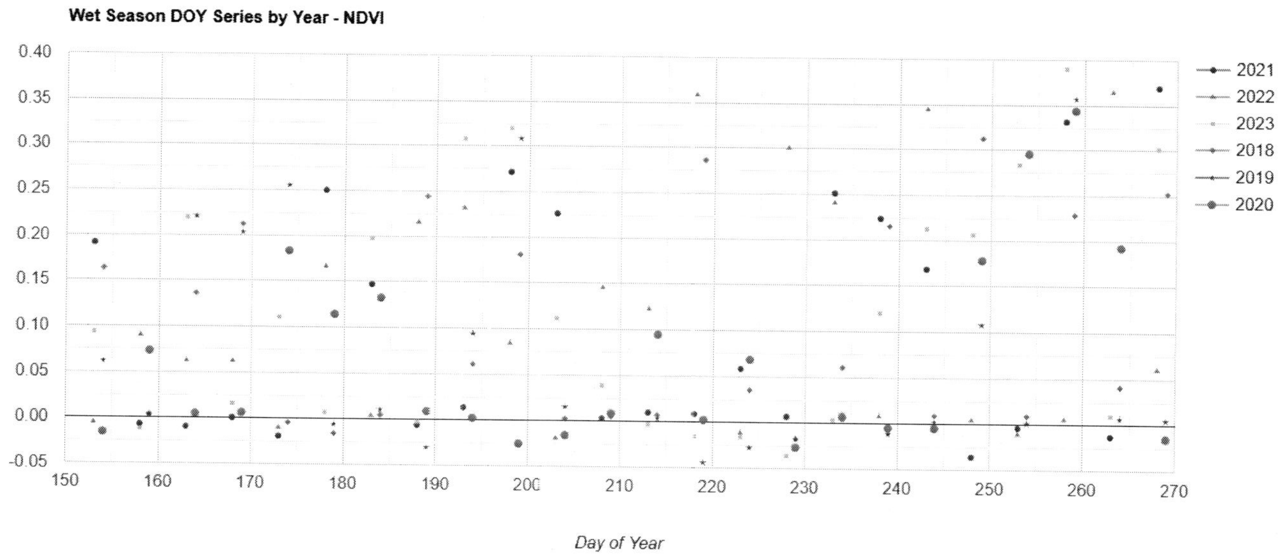

Figure 6. Wet Season Day of Year Series by Year – NDVI.

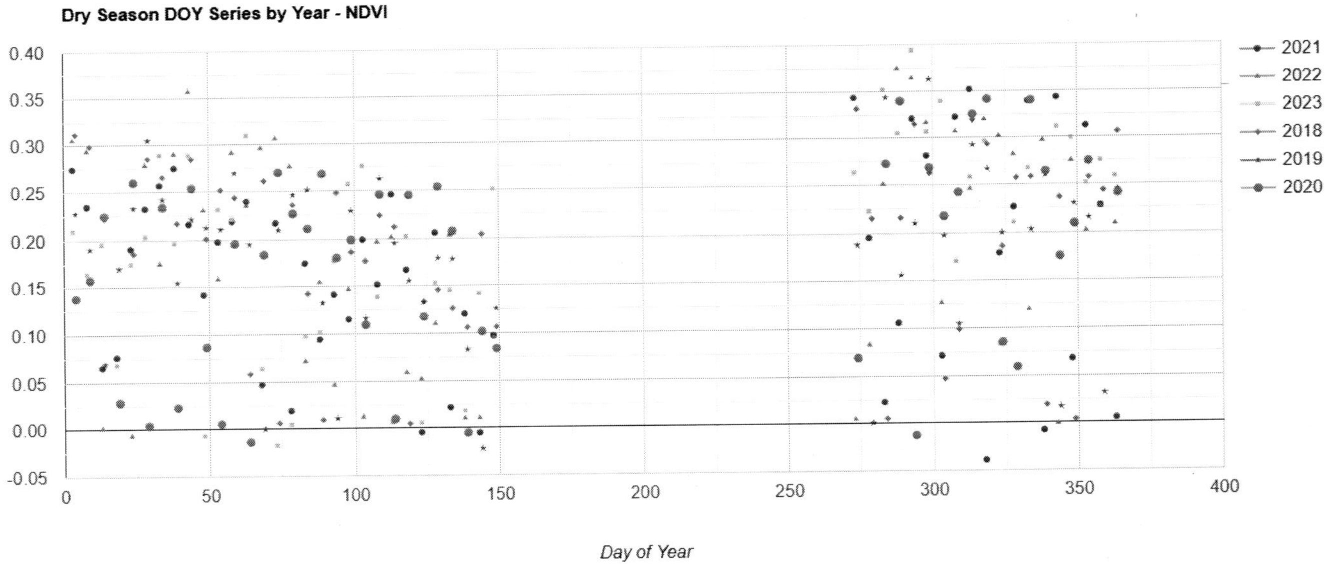

Figure 7. Dry Season Day of Year Series by Year – NDVI.

The NDVI for the dry season is plotted against the day of the year in this graphic. As moisture levels drop throughout the dry season, the NDVI may exhibit a declining trend, with sporadic maxima occurring in the months following the monsoon. The downward trend indicates the stress the plants are experiencing due to water deficiency.

This graph displays the NDVI readings for the wet season for several years. It enables the comparison of vegetation between years, demonstrating how fluctuations in rainfall and other climatic factors affect the response of plants each year. More favourable monsoon conditions were probably seen in years with higher NDVI levels.

This graph displays the NDVI values for the dry season for several years. The comparison highlights interannual variability, as well as the impact of dry season conditions on plant health. It's possible that years with higher NDVI readings during the dry season benefited from less extreme dry conditions or residual moisture. The mean NDMI values for each date throughout the wet season are displayed on the chart. Higher NDMI values indicate vegetation with a higher moisture content. As the monsoon intensifies, the NDMI tends to increase during the wet season, suggesting that vegetation is absorbing more precipitation. The NDMI peaks represent times when the vegetation moisture content is at its highest, signifying the effective monsoon rains. The mean NDMI values for each date throughout the dry season are displayed.

Lower NDMI values indicate the reduced moisture content in vegetation during the dry season. It is anticipated that the general trend will be downward, with sporadic peaks resulting from isolated rainstorms or rainwater leftover from the rainy season. The drying conditions typical of the dry season are reflected in the reduction in NDMI levels. The NDMI for the wet season is plotted against the day of the year. As the monsoon advances, the DOY series ought to exhibit a rising trend, indicating the vegetation's absorption of moisture. As the season closes and the amount of moisture available declines, the tendency either stabilises or reverses. The NDMI for the dry season is plotted against the day of the year. There may be a declining trend in the dry season series, which would suggest a lower moisture content. Remaining moisture or isolated rain events may cause occasional peaks in precipitation. The general trend is consistent with the drying conditions that occur during the dry season. The NDMI levels for the wet season in several years are compared, and the interannual comparison highlights changes in moisture content resulting from variations in the intensity and distribution of the monsoon. More effective monsoon rains were probably experienced in years with higher NDMI levels.

The NDMI levels have been stable for several years during the dry season. The comparison demonstrates how annual variations in the dry season circumstances affected the moisture content of the Vegetation. Higher NDMI readings during the dry season in a given year may indicate less extreme dry conditions or the effectiveness of efficient water management techniques. The mean NDWI values for each date during the wet season are displayed. Because there is more water in the soil and vegetation during the wet season, higher NDWI values are anticipated. The presence of surface water and vegetation moisture can be reflected in this indicator. Peaks in the NDWI occur when the most water is available, demonstrating how effectively the monsoon rains replenish water supplies. The mean NDWI values for each date throughout the dry season are displayed as the amount of water in the soil and vegetation decreases during the dry season, NDWI levels usually decline. However, because of the residual moisture from the wet season, some peaks may still occur, particularly in the early dry season. Generally, NDWI levels are expected to be lower within this time frame than they are throughout the rainy season. The NDWI for the wet season is plotted against the day of the year, and as the monsoon intensifies, the DOY series should exhibit an upward trend that corresponds with the increase in water content. As the season finishes and the amount of water available diminishes, this pattern either stabilises or reverses. The NDWI for the dry season is plotted against the day of the year. There may be a declining trend in the dry season series, which would suggest a lower moisture content. Remaining moisture or isolated rain events may cause occasional peaks. The general trend is consistent with the drying conditions that occur during the dry season. The NDWI values for the wet season in several years are compared, and the interannual comparison highlights changes in water content resulting from variations in the intensity and distribution of the monsoon. More effective monsoon rains were probably experienced in years with higher NDWI levels. The NDWI for the dry season is plotted against the day of the year. There may be a declining trend in the dry season series, which would suggest a lower moisture content. Remaining moisture or isolated rain events may cause occasional peaks. The general trend is consistent with the drying conditions that occur during the dry season.

The mean AMMI readings for each date in the wet season from 2018 to 2023 are displayed. AMMI values represent the distribution and condition of the mangrove vegetation during the rainy season. A healthy mangrove cover is characterised by higher AMMI values, which are essential for carbon sequestration, biodiversity, and coastal protection. AMMI peaks represent periods when mangrove growth and density are at their optimal levels. The mean AMMI values for each date in the dry season are displayed. The ability of mangrove ecosystems to tolerate environmental stressors, such as salinity and water scarcity, is demonstrated by AMMI levels during the dry season. Sustained AMMI values indicate the presence of robust mangrove stands that can withstand challenging conditions, even though they are often lower than during the wet season.

The AMMI for the wet season is plotted against the day of the year in this graphic. The temporal fluctuation in mangrove health and productivity over the wet season is depicted by the AMMI Day of Year (DOY) series. Peaks in AMMI, which are influenced by factors such as tidal inundation, sediment deposition, and freshwater intake, may correspond with periods of peak mangrove growth. The AMMI for the dry season is plotted against the day of the year in this graphic. The DOY series for AMMI in the dry season sheds light on the resilience of mangrove ecosystems to changing weather patterns throughout the year. While continuous or variable levels exhibit adaptive responses to changing environmental conditions, lower AMMI values may indicate periods of stress. The AMMI levels for the wet season in several years are compared in this chart. During the wet season, the graphic enables the comparison of mangrove productivity and health across several years by displaying AMMI values for each year. Changes in AMMI from year to year could be caused by variations in the surrounding environment, in management techniques, or by human influences on mangrove ecosystems. This research displays the AMMI readings for the dry season spanning several years. This figure enables the comparison of mangrove health and resilience throughout the dry season across several years, much as the wet season DOY series by year. An understanding of the long-term dynamics of mangrove ecosystems and how they adapt to shifting external conditions can be gained from variations in AMMI values throughout time.

Discussion and Conclusion

By evaluating the moisture content, structure, and health of the plants, indices such as NDVI, NDMI, NDWI, and AMMI can be used to track changes in the vegetation during the dry and wet seasons. While NDMI compares near-infrared and shortwave infrared reflectance to provide information on plant water content and moisture changes, NDVI evaluates vegetation density and greenness. Sentinel-2 satellite images covering Henry Island, West Bengal, from 2018 to 2023 are used in the study to assess vegetation and water indicators, including AMMI, NDVI, NDMI, and NDWI. With an emphasis on mangrove environments, it offers insights into vegetation health and moisture content by mapping seasonal and temporal patterns. To track differences between the rainy and dry seasons, the research also includes time series and day-of-year charts. The study highlights how a deeper understanding of vegetation responses to seasonal climate variations can be achieved by combining NDVI, NDMI, and NDWI. This comprehensive method enhances the monitoring of hydrological conditions, moisture content, and vegetation health, facilitating the identification of stress and recovery periods that are crucial for efficient ecosystem management. The study demonstrates that the combination of NDVI, NDMI, and NDWI is beneficial for environmental management, conservation, and policymaking, as it offers a more comprehensive understanding of vegetation dynamics. It also emphasises how effective computerised mangrove mapping is at removing human error and facilitating quicker, more precise monitoring for scientists and community members. Scientifically, the initial data for the forest condition in the upcoming research is the geographical distribution of mangroves across a larger region and the characterization of the surrounding areas. Other supporting data, including geochemical data, biota content statistics, and the density and age of trees per unit area, will contribute to determining the mangrove health index. Adopting advanced, automated, and scalable remote sensing methods that improve vegetation mapping accuracy and minimise human error is essential for sustainable ecosystem management. Better conservation and climate adaptation methods can be achieved by enhancing the identification of vegetation changes through the use of machine learning and multi-temporal data fusion. To develop more reliable, automated frameworks that support long-term monitoring and informed decision-making for biodiversity conservation, carbon sequestration, and ecosystem resilience, future research should focus on integrating these technologies. To make sure that scientific discoveries are translated into practical sustainability measures, interdisciplinary cooperation and policy support must be strengthened.

References

[1] Field C. D., "Impact of expected climate change on mangroves," *Hydrobiologia*, vol. 295, no. 1–3, pp. 75–81, 1995.

[2] K. Kathiresan and B. L. Bingham, "Biology of mangroves and mangrove ecosystems," *Advances in Marine Biology*, vol. 40, pp. 81–251, 2001.

[3] I. Nagelkerken, S. J. M. Blaber, S. Bouillon et al., "The habitat function of mangroves for terrestrial and marine fauna: a review," *Aquatic Botany*, vol. 89, no. 2, pp. 155–185, 2008.

[4] R. Costanza, R. d'Arge, R. de Groot et al., "The value of the world's ecosystem services and natural capital," Nature, vol. 387, no. 6630, pp. 253–260, 1997. M. Hutomo and M. K. Moosa, "Indonesian marine and coastal biodiversity: present status," *Indian Journal of Marine Sciences*, vol. 34, no. 1, pp. 88–97, 2005.

[5] S. Srikanth, S. K. Y. Lum, and Z. Chen, "Mangrove root: adaptations and ecological importance," *Trees*, vol. 30, no. 2, pp. 451–465, 2016.

[6] D. C. Donato, J. B. Kaufman, D. Murdiyarso, S. Kurnianto, M. Stidham, and M. Kanninen, "Mangroves among the most carbon-rich forests in the tropics," *Nature Geoscience*, vol. 4, no. 5, pp. 293–297, 2011.

[7] L. Carugati, B. Gatto, E. Rastelli et al., "Impact of mangrove forests degradation on biodiversity and ecosystem functioning," *Scientific Reports*, vol. 8, no. 1, Article ID 13298, 2018.

[8] T. E. Fatoyinbo, M. Simard, R. A. Washington-Allen, and H. H. Shugart, "Landscape-scale extent, height, biomass, and carbon estimation of Mozambique's mangrove forests with Landsat ETM+, and shuttle radar topography mission elevation data," *Journal of Geophysical Research: Biogeosciences,* vol. 113, no. G2, 2008.

[9] P. Wicaksono, P. Danoedoro, Hartono, and U. Nehren, "Mangrove biomass carbon stock mapping of the Karimunjawa Islands using multispectral remote sensing." *International Journal of Remote Sensing*, vol. 37, no. 1, pp. 26–52, 2016.

[10] A. W. Hastuti, K. I. Suniada, and F. Islamy, "Carbon stock estimation of mangrove vegetation using remote sensing in Perancak Estuary, Jembrana District, Bali," *International Journal of Remote Sensing and Earth Sciences (IJReSES),* vol. 14, no. 2, pp. 137–150, 2018.

[11] J. Sanderman, T. Hengl, G. Fiske et al., "A global map of mangrove forest soil carbon at 30 m spatial resolution," *Environmental Research Letters*, vol. 13, no. 5, Article ID 055002, 2018.

[12] A. R. Jones, R. Raja Segaran, K. D. Clarke, M. Waycott, W. S. H. Goh, and B. M. Gillanders, "Estimating mangrove tree biomass and carbon content: a comparison of forest inventory techniques and drone imagery," *Frontiers in Marine Science*, vol. 6, p. 784, 2020.

[13] M. Wang, W. Cao, Q. Guan, G. Wu, and F. Wang, "Assessing changes of mangrove forest in a coastal region of southeast China using multi-temporal satellite images," *Estuarine, Coastal and Shelf Science*, vol. 207, pp. 283–292, 2018.

[14] FAO (Food and Agriculture Organisation) of the United Nations 2005, The world's mangroves 1980– 2005, FAO Forestry Paper, 153, http://www.fao.org/3/a1427e/a1427e00.pdf.

[15] C. Giri, E. Ochieng, L. L. Tieszen et al., "Observation and monitoring of mangrove forests using remote sensing: opportunities and challenges," *Remote Sensing*, vol. 8, no. 9, pp. 783–159, 2011.

[16] N. C. Duke, J.-O. Meynecke, S. Dittmann et al., "A world without mangroves?" *Science,* vol. 317, no. 5834, pp. 41-42, 2007.

[17] K. D. Kanniah, A. Sheikhi, A. P. Cracknell et al., "Satellite images for monitoring mangrove cover changes in a fast-growing economic region in Southern Peninsular Malaysia," *Remote Sensing*, vol. 7, no. 11, pp. 14360–14385, 2015.

[18] D. M. Alongi, "Present state and future of the world's mangrove forests," Environmental Conservation, vol. 29, no. 3, pp. 331–349, 2002.

[19] I. Valiela, J. L. Bowen, and K. York, "Mangrove forests: one of the world's threatened major tropical environments," *BioScienc*e, vol. 51, no. 10, pp. 807–815, 2001.

[20] Suyarso, Praditya Avianto, "AMMI Automatic Mangrove Map and Index: Novelty for Efficiently Monitoring Mangrove Changes with the Case Study in Musi Delta, South Sumatra, Indonesia," *International Journal of Forestry Research*, vol. 2022, Article ID 8103242, 14 pages, 2022. https://doi.org/10.1155/2022/8103242 T. Tran, R. Reef, Xuan Zhu, "A Review of Spectral Indices for Mangrove Remote Sensing," *Remote Sens*. 2022, 14, 4868.

[21] Kamal M, Hartono, Wicaksono P, Adi NS, Arjasakusuma S, "*Assessment of mangrove forest degradation through canopy fractional cover in Karimunjawa, Central Java, Indonesia, Geoplanning.*" 2016;3(2):107116. doi:10.3390/rs70404753.

[22] S. (2022). AMMI Automatic Mangrove Map and Index: An Analytical Study on Satellite Imageries at Aru Islands, Maluku, Indonesia. *Emerging Challenges in Environment and Earth Science* Vol. 2, 106– 130. https://doi.org/10.9734/bpi/ecees/v2/3423E

[23] Jesús Delegido, Shari Van Wittenberghe, Jochem Verrelst, et al. Chlorophyll content mapping of urban vegetation in the city of Valencia based on the hyperspectral NAOC index. [J]. *Ecological Indicators*, 2014, 40(5):34-42.

[24] Gitelson A A, Peng Y, Arkebauer T J, et al. Relationships between gross primary production, green LAI, and canopy chlorophyll content in maise: Implications for remote sensing of primary production[J]. *Remote Sensing of Environment*, 2014, 144(4):65-72.

[25] C. J. Tucker, "Remote sensing of leaf water content in the near-infrared," *Remote Sensing of Environment*, vol. 10, no. 1, pp. 23–32, 1980.

Chapter 20

A Comprehensive Review of Nanocomposites as a Potential Candidate for Biosensors

Subarna Datta, PhD
and Soumyadipta Pal*, PhD
Institute of Engineering & Management, Salt Lake Electronics Complex, Kolkata, West Bengal, India
University of Engineering and Management, Kolkata, West Bengal, India

Abstract

Biosensors have emerged as invaluable tools in various fields, including healthcare, environmental monitoring, and food safety. Their ability to detect and quantify specific biological molecules or analytes has revolutionised diagnostics and monitoring processes. Recent advancements in nanotechnology have led to the development of nanocomposite-based biosensors, which offer enhanced Sensitivity, selectivity, and stability. This comprehensive review examines the application of nanocomposites in biosensors, providing a detailed analysis of their design, fabrication, and performance.

Keywords: nanomaterials, sensors, fabrication and characterisation techniques

---

* Corresponding Author's Email: soumya.phys88@gmail.com

In: Breakthrough Research and Technologies in Physics …
Editors: Koyel Ganguly, Soumyadipta Pal and Kamakhya Prasad Ghatak
ISBN: 979-8-90134-111-7

Introduction

Introduction to Biosensors and the Importance of Nanocomposites

Biosensors, at the intersection of biology, chemistry, and materials science, are powerful analytical devices designed to detect specific biological molecules or analytes and convert this information into measurable signals. These devices have found widespread application in healthcare [1], environmental monitoring [2], food safety [3], and biotechnology [4] due to their ability to provide rapid and accurate information about biological processes and the presence of specific substances. In recent years, the Integration of nanocomposites has significantly advanced the capabilities and versatility of biosensors, opening up new horizons for sensitive and selective detection. Biosensors have become indispensable tools in various fields for several compelling reasons.

- *Disease Diagnosis and Monitoring [5-7]:* In healthcare, biosensors facilitate the early detection of biomarkers associated with diseases, including cancer, diabetes, and infectious diseases. This enables timely diagnosis and more effective treatment.
- *Environmental Monitoring [8-10]:* Biosensors play a crucial role in monitoring environmental parameters, including pollutants, pathogens, and toxins. They help ensure the safety of air and water resources, contributing to ecological sustainability.
- *Food Safety [11, 12]:* In the food industry, biosensors are used to detect contaminants, allergens, and spoilage markers. This ensures the quality and safety of food products throughout the supply chain.
- *Biotechnology and Drug Development [13, 14]:* Biosensors play a crucial role in drug discovery and development, enabling researchers to investigate molecular interactions, screen for potential drug candidates, and monitor cellular responses to treatments.
- *Point-of-Care Testing [15]:* Miniaturized biosensors are utilized for point-of-care diagnostics, enabling rapid testing in remote or resource-limited settings. This is particularly valuable in global healthcare initiatives.

Furthermore, Nanocomposites, which consist of nanomaterials integrated into a matrix material, have revolutionised the field of biosensors by addressing key challenges and enhancing their performance. Here's why nanocomposites are crucial in this context.

- *Enhanced Sensitivity [16, 17]:* Nanocomposites offer a high surface area-to-volume ratio, enabling the immobilisation of a larger number of recognition elements (e.g., antibodies, aptamers, DNA strands). This results in higher Sensitivity, allowing the detection of trace amounts of analytes.
- *Selectivity [18, 19]:* The functionalisation of nanocomposite surfaces with specific receptors ensures selectivity, allowing the biosensor to interact only with the target analyte. This minimises false positives and enhances the accuracy of detection.
- *Rapid Response [20, 21]:* Nanocomposites, with their nanoscale dimensions, facilitate rapid analyte diffusion to the sensor surface. This leads to fast response times, which are crucial in real-time monitoring and diagnostics.
- *Versatility [22, 23]:* Nanocomposites can be tailored for specific applications by choosing different types of nanoparticles and matrix materials. This versatility allows researchers to design biosensors for a wide range of targets, including proteins, nucleic acids, and small molecules.

In summary, biosensors play a pivotal role in various domains, from healthcare to environmental protection, due to their ability to detect and monitor specific biological molecules and analytes. Nanocomposites have significantly improved the performance and versatility of biosensors, making them highly sensitive, selective, and rapid devices. As the field of biosensors continues to advance, nanocomposite technology is set to play a critical role in shaping the future of analytical and diagnostic tools.

Scope and Purpose of the Present Review

The scope and purpose of the review on nanocomposites as biosensors are to provide a comprehensive and in-depth exploration of the current state of research, advancements, challenges, and prospects in this exciting and rapidly evolving field. Here is a breakdown of the scope and purpose.

Scope

- *Nanocomposite Types:* The review will cover a wide range of nanocomposite materials commonly used in biosensors, including carbon-based nanocomposites (such as graphene and carbon nanotubes), metal nanoparticle-embedded nanocomposites, polymer-nanoparticle composites, and quantum dot nanocomposites.
- *Functionalisation and Surface Modification*: It will delve into the strategies and methods used for surface functionalisation and modification of nanocomposites to impart selectivity and specificity to biosensors. This will include discussions on the immobilisation of recognition elements, such as antibodies, aptamers, and DNA strands.
- *Advantages:* The review will highlight the benefits of using nanocomposites in biosensors, including their high Sensitivity, selectivity, rapid response times, and miniaturisation potential. These advantages will be illustrated with examples from various applications.
- *Applications:* This will encompass a broad spectrum of nanocomposite-based biosensor applications, including healthcare (disease diagnosis and monitoring), environmental monitoring (pollutant and pathogen detection), food safety (contaminant detection), and biotechnology (drug development and cellular response monitoring). The review will emphasise how nanocomposites are contributing to advancements in these areas.
- *Challenges and Limitations*: The challenges and limitations associated with nanocomposite-based biosensors, including issues related to reproducibility, scalability, and cost-effectiveness, will be discussed to provide a balanced perspective on the technology.

Purpose

- *Inform and Educate:* The primary purpose of the review is to inform and educate readers about the significance of nanocomposites in the context of biosensors. It aims to provide a clear understanding of how nanocomposites enhance the performance of biosensors, enabling sensitive and selective detection of biological molecules and analytes.
- *Highlight Recent Advances*: The review will showcase recent developments and breakthroughs in nanocomposite-based

biosensors, emphasising the most cutting-edge research in the field. This serves to keep researchers, scientists, and practitioners up-to-date with the latest advancements.

- *Identify Challenges and Future Directions*: By discussing challenges and limitations, the review aims to identify areas where further research is needed. It will suggest potential avenues for overcoming current limitations and improving the technology.
- *Inspire Future Research*: The comprehensive exploration of applications and potential uses of nanocomposite-based biosensors aims to inspire future research and innovation in this field. It highlights the potential for biosensors to address pressing challenges in healthcare, the environment, and biotechnology.
- *Facilitate Decision-Making*: For professionals and decision-makers in fields where biosensors are relevant, the review will provide valuable insights into the advantages, limitations, and areas of application for nanocomposite-based biosensors, helping them make informed choices about adopting and developing this technology.

In summary, the review of nanocomposites as biosensors aims to provide a comprehensive understanding of the current landscape, opportunities, and challenges in this field. Its purpose is to inform, inspire, and guide researchers, practitioners, and stakeholders interested in leveraging nanocomposites for the development of advanced biosensor technologies.

Nanocomposites in Biosensors

Defining Nanocomposites and Their Relevance in Biosensor Applications

Nanocomposites are materials composed of a matrix or host material that incorporates nanoparticles or nanoscale filler materials. These nanoparticles are typically 1 to 100 nanometers in size and are dispersed throughout the matrix material. Nanocomposites can be engineered to possess a wide range of desirable properties by adjusting the type, size, shape, and concentration of the nanoparticles, as well as the properties of the matrix material. These materials have found numerous applications across various industries due to their unique and tunable characteristics.

Nanocomposites have gained significant relevance in biosensor applications due to their ability to enhance the performance and Sensitivity of biosensors. Here are some key ways in which nanocomposites are utilised in biosensor technology.

- *Improved Sensitivity*: The incorporation of nanoparticles into the biosensor design can significantly enhance its sensitivity [24]. Nanoparticles offer a high surface area-to-volume ratio, which enables increased surface interactions with target molecules (e.g., biomarkers or analytes) [16, 25, 26]. This results in improved detection limits and the ability to detect low concentrations of analytes.
- *Enhanced Selectivity*: Functionalised nanoparticles can be designed to bind to specific biomolecules selectively. By using nanocomposites with tailored surface chemistry, biosensors can achieve high specificity for the target analyte while minimizing interference from other substances [27-29].
- *Faster Response Times*: Nanocomposite materials can facilitate faster signal transduction in biosensors. For instance, carbon nanotubes or graphene can enhance electron transfer rates, resulting in speedier sensor responses [30-33].
- *Stability and Longevity*: Nanocomposites can provide improved mechanical stability and resistance to environmental factors, making biosensors more robust and suitable for long-term use in various conditions [34, 35].
- *Miniaturization:* Nanocomposite-based biosensors can be designed on a nanoscale, enabling the development of miniaturized and portable devices. This is especially advantageous for point-of-care diagnostics and wearable biosensors [36-38].
- *Multiplexing:* Nanocomposites can be engineered to enable multiplexed detection, allowing for the simultaneous measurement of multiple analytes on a single biosensor platform. This is valuable for comprehensive analysis in medical diagnostics and research applications [39-42].
- *Functional Diversity:* Nanocomposites can be customised to incorporate various types of nanoparticles, such as metallic, semiconductor, or magnetic nanoparticles, depending on the specific

requirements of the biosensor application. This versatility allows for a wide range of functionalities [43-46].

- *Bioimaging:* Some nanocomposite materials, such as quantum dots and up-conversion nanoparticles, are not only used in biosensors but also serve as imaging agents for visualizing cellular and molecular processes in biological systems [47-49].

In summary, nanocomposites have revolutionised biosensor technology by improving Sensitivity, selectivity, speed, and overall performance. Their versatility and ability to be tailored for specific applications make them valuable tools in the fields of medical diagnostics, environmental monitoring, and life sciences research.

Various Types of Nanocomposites Used in Biosensors (e.g., Polymer-Based, Metal-Based, Carbon-Based)

Nanocomposites used in biosensors can be categorised into various types based on the nanomaterials incorporated into them. Each type offers distinct properties and advantages for specific biosensing applications. Here are some common types of nanocomposites used in biosensors.

Polymer-Based Nanocomposites

- *Polymer Nanocomposites [50-52]:* These are composed of a polymer matrix with nanoscale fillers, such as nanoparticles, nanofibers, or nanoclays. Polymer-based nanocomposites are flexible, biocompatible, and can be easily functionalized for biomolecule immobilization.
- *Conducting Polymer Nanocomposites [53-56]:* Incorporating conductive nanoparticles, such as carbon nanotubes or graphene, into conducting polymers enhances their electrical conductivity, making them suitable for electrochemical biosensors.
- *Hydrogel Nanocomposites [57-61]:* Hydrogel-based nanocomposites can absorb and retain a large volume of water, making them suitable for immobilising biomolecules. Nanoparticles or nanofibers can be added to enhance their mechanical and sensing properties.

Metal-Based Nanocomposites

- *Nanoparticle-Metal Nanocomposites [62-64]:* Metal nanoparticles (e.g., gold, silver) are often used due to their unique optical and electrochemical properties. These nanocomposites are employed in various biosensors, including colourimetric and surface-enhanced Raman spectroscopy (SERS) sensors.
- *Metal Oxide Nanocomposites [65-67]:* Metal oxide nanoparticles (e.g., zinc oxide, tin oxide) can be incorporated into sensors for gas sensing or pH monitoring. They provide high surface area and Sensitivity.
- *Magnetic Nanoparticle Nanocomposites [68-70]:* Magnetic nanoparticles, often coated with biocompatible polymers, enable magnetic manipulation of biomolecules in biosensors. They are used in magnetic biosensors and separation techniques.

Carbon-Based Nanocomposites

- *Carbon Nanotube (CNT) Nanocomposites [71-74]:* Single-walled or multi-walled carbon nanotubes can enhance the electrical conductivity and surface area of biosensors. CNT-based biosensors are utilised in both electrochemical and field-effect transistor (FET) biosensing applications.
- *Graphene-Based Nanocomposites [75-77]:* Graphene and its derivatives (e.g., graphene oxide) have exceptional electrical conductivity and a large surface area. They are used in various biosensing platforms, including electrochemical, field-effect transistor, and optical biosensors.
- *Carbon Nanofiber (CNF) Nanocomposites [78-81]:* CNF-based biosensors offer a combination of mechanical strength and electrical conductivity, making them suitable for various applications.

Hybrid Nanocomposites

- *Polymer-Metal Nanocomposites [82, 83]:* Combining polymers with metal nanoparticles or metal oxides can yield versatile biosensors that benefit from both the flexibility of polymers and the sensing capabilities of metals.

- *Polymer-Carbon Nanocomposites [84, 85]:* Hybrid nanocomposites of polymers and carbon-based materials, such as graphene-polymer composites, offer improved electrical conductivity and biocompatibility.
- *Multi-component Nanocomposites [86-88]:* Some biosensors incorporate multiple nanomaterials, such as metal nanoparticles, carbon nanotubes, and polymers, to achieve synergistic effects and enhanced sensing performance.

Quantum Dot Nanocomposites

Quantum Dots (QDs) are semiconductor nanocrystals with unique optical properties. QD-based nanocomposites are employed in fluorescence-based biosensors due to their tunable emission spectra and high quantum yields [89-91].

Lipid-Based Nanocomposites

- *Lipid Nanoparticles [92-94]:* Lipid-based nanocomposites, such as liposomes or lipid bilayers, are used for the encapsulation and controlled release of bioactive molecules in biosensors, especially in drug delivery and therapeutic monitoring applications.

The choice of nanocomposite type depends on the specific biosensing requirements, including the target analyte, desired Sensitivity, and environmental conditions. Researchers continue to explore new nanomaterials and innovative combinations to develop biosensors with improved performance and broader application potential.

Fabrication and Characterisation

Detailed Methods Used to Fabricate Nanocomposite Biosensors

Fabricating nanocomposite biosensors involves a series of steps to incorporate nanomaterials into a sensor platform for the detection of specific biomolecules or analytes. The following are some standard methods used to fabricate nanocomposite biosensors.

Chemical Synthesis

- *Solution-Based Synthesis [95-97]:* Nanocomposites can be prepared by mixing and reacting precursor materials in a solution. This method allows for precise control over the size, shape, and composition of nanoparticles. For example, metal nanoparticles (e.g., gold, silver) can be synthesised in solution and subsequently incorporated into a biosensor platform.
- *Sol-Gel Method [98, 99]:* This technique involves the hydrolysis and condensation of metal alkoxides or organosilanes to form a nanoscale network within a gel matrix. It is commonly used to produce metal oxide nanocomposites for biosensing applications.

Physical Deposition Techniques

- *Physical Vapour Deposition (PVD) [100-102]:* PVD methods, such as sputtering or thermal evaporation, can be used to deposit thin films of nanomaterials onto a substrate. This is often employed to create thin-film biosensors with specific surface properties.
- *Chemical Vapour Deposition (CVD) [101-106]:* CVD can be used to grow nanomaterials, such as carbon nanotubes or graphene, directly onto a substrate, allowing for precise control over the material's alignment and thickness.

Layer-by-Layer (LbL) Assembly

LbL assembly involves the sequential deposition of alternating layers of oppositely charged materials, including nanomaterials, polymers, and biomolecules. This technique is used to create nanocomposite thin films with precise control over layer thickness and composition. LbL assembly is often employed in electrochemical and optical biosensors [107-109].

Electrospinning

Electrospinning is used to create nanofibers or nanofiber mats by applying an electric field to a polymer solution or melt. Nanomaterials, such as nanoparticles or carbon nanotubes, can be incorporated into the polymer matrix during the electrospinning process. The resulting nanofibers can serve as a platform for biosensor construction [110-112].

Template-Assisted Synthesis

Templates, such as porous membranes or sacrificial materials, can be used to create nanoscale structures with specific shapes and sizes. Nanomaterials are deposited within these templates and subsequently released to form nanocomposites. This method is often employed to fabricate nanowire-based biosensors [113, 114].

Chemical Functionalisation

Nanocomposite surfaces can be functionalised with biomolecules (e.g., antibodies, DNA probes) through chemical reactions. For example, nanoparticles can be modified with thiolated molecules for attachment to a gold surface, enabling the immobilisation of biomolecules for biosensing [115-118].

Printing and Patterning Techniques

Printing techniques, such as inkjet printing or screen printing, can be used to create patterned nanocomposite layers on substrates. These techniques are suitable for the mass production of biosensors with well-defined sensor regions [119, 120].

Self-Assembly

Some nanocomposites can self-assemble into ordered structures due to their inherent properties. For instance, amphiphilic molecules can self-assemble into lipid bilayers or micelles, serving as platforms for the development of biosensors [121, 122].

3D Printing

3D printing technology allows for the fabrication of three-dimensional nanocomposite structures with precise control over geometry and material distribution. This is particularly useful for creating custom-designed biosensors [123, 124].

The choice of fabrication method depends on the specific nanomaterials used, the desired sensor configuration, and the intended application of the biosensor. Researchers often select the most appropriate method to optimise Sensitivity, specificity, and reliability for the detection of target biomolecules.

Techniques Employed for Characterising These Biosensors

Characterising nanocomposite biosensors is essential to assess their structure, properties, and performance. Various analytical techniques are employed to gain insights into the composition, morphology, surface properties, and functionality of these biosensors. Here are some commonly used characterisation techniques.

Scanning Electron Microscopy (SEM) [125]

- *Principle:* SEM uses a focused beam of electrons to scan the surface of a sample. The interaction between the electrons and the sample's surface generates high-resolution images, providing information on the sample's topography and morphology.
- *Application*: SEM is valuable for visualising the surface structure of nanocomposite biosensors, including the distribution and arrangement of nanoparticles or nanofibers on the substrate.

Transmission Electron Microscopy (TEM) [126]

- *Principle:* TEM involves transmitting an electron beam through a thin specimen. The interaction of electrons with the sample produces high-resolution images, allowing for detailed examination of the internal structure and nanoscale features.
- *Application:* TEM is used to study the morphology, size, and distribution of nanomaterials within the biosensor, providing insights into the nanocomposite's internal structure.

X-ray Diffraction (XRD) [127]

- *Principle:* XRD measures the diffraction of X-ray photons by the crystal lattice of a material. The resulting diffraction pattern provides information about the crystal structure, phase composition, and crystallinity of nanomaterials.
- *Application:* XRD is useful for characterising the crystalline nature of nanoparticles or nanocrystalline components within the biosensor, helping to identify the types of nanomaterials present.

Atomic Force Microscopy (AFM) [128, 129]

- *Principle:* AFM employs a sharp tip to scan the sample surface. It measures the interaction forces between the tip and the sample, creating high-resolution topographic images.
- *Application:* AFM is used to investigate the surface roughness, thickness, and mechanical properties of nanocomposite layers on biosensors. It can also help assess the uniformity of nanomaterial distribution.

Fourier-Transform Infrared Spectroscopy (FTIR) [130]

- *Principle:* FTIR measures the absorption and emission of infrared light by the chemical bonds in a material. It provides information about functional groups, chemical composition, and molecular interactions.
- *Application:* FTIR is employed to confirm the presence of specific functional groups, coatings, or chemical modifications on the biosensor's surface or within its nanocomposite components.

Raman Spectroscopy [131-133]

- *Principle:* Raman spectroscopy measures the scattering of laser light by molecules in a sample. It provides information about molecular vibrations and can identify specific chemical compounds.
- *Application:* Raman spectroscopy is used to analyse the chemical composition and structural properties of nanomaterials integrated into biosensors. It is particularly valuable for studying carbon-based nanomaterials, such as graphene or carbon nanotubes.

X-ray Photoelectron Spectroscopy (XPS) [134, 135]

- *Principle:* XPS analyses the energy distribution of photoelectrons emitted from the surface of a sample when exposed to X-rays. It provides information on elemental composition and chemical states.
- *Application:* XPS is employed to determine the chemical composition of nanomaterials, identify surface functional groups, and investigate the chemical environment of atoms within the biosensor.

Electrochemical Characterisation [136]

- *Principle:* Electrochemical techniques, including cyclic voltammetry and impedance spectroscopy, are used to assess the electrochemical behaviour of biosensors. These methods provide information about the sensor's Sensitivity, response time, and electrochemical performance.

Spectroscopy (UV-Vis, Fluorescence) [137, 138]

- *Principle:* UV-Vis and fluorescence spectroscopy measure the absorption and emission of light by materials. They are used to assess the optical properties, stability, and Sensitivity of biosensors, especially those incorporating nanoparticles with unique optical properties.

Surface Plasmon Resonance (SPR) [139-141]

- *Principle: SPR* measures changes in the refractive index at the biosensor surface due to biomolecular binding events. It provides real-time information on binding kinetics, affinity, and specificity.
- *Application:* SPR is a powerful tool for studying the interaction between biomolecules and nanocomposite surfaces in biosensors, offering insights into their binding behaviour.

These characterisation techniques are essential for ensuring the quality and performance of nanocomposite biosensors, aiding in the optimisation of their design and functionality for specific applications in areas such as medical diagnostics, environmental monitoring, and biotechnology.

Functionalisation and Surface Modification

How Nanocomposite Surfaces Are Functionalised to Enhance Biosensor Performance

Functionalising nanocomposite surfaces is a critical step in enhancing the performance of biosensors. Functionalisation involves modifying the surface of the nanocomposite with specific molecules or chemical groups to achieve desired properties or functionalities. Here are several ways nanocomposite surfaces can be functionalised to enhance biosensor performance.

Biomolecule Immobilisation

- *Antibodies and Antigens [142]:* Nanocomposite surfaces can be functionalised with antibodies or antigens to create immunosensors. These biomolecules selectively bind to target antigens, enhancing the biosensor's specificity for particular analytes.
- *DNA Probes [143-145]:* DNA-functionalized nanocomposites are used in DNA biosensors. DNA probes with complementary sequences can hybridise with target DNA, allowing for the specific detection of genetic sequences.

Enzyme Immobilisation

- *Enzymes [146,147]:* Enzymes can be immobilised on nanocomposite surfaces to create enzyme-based biosensors (enzymatic biosensors). These enzymes catalyse specific reactions with target analytes, leading to a measurable signal change.
- Nanomaterial-Enzyme Conjugates [148]: Nanocomposites can be functionalised with enzyme-nanomaterial conjugates to enhance enzyme stability and activity, allowing for more robust enzymatic biosensors.

Chemical Functionalisation

- *Amine, Thiol, or Carboxyl Groups [149]:* Nanocomposite surfaces can be modified with chemical functional groups, such as amino (-NH2), thiol (-SH), or carboxyl (-COOH) groups. These groups can be used to attach biomolecules or linkers for further functionalisation covalently.
- *PEGylation [150]:* Polyethylene glycol (PEG) can be attached to nanocomposite surfaces to improve biocompatibility, reduce non-specific adsorption, and extend sensor stability.

Affinity Ligands and Aptamers

- *Affinity Ligands [151]:* Nanocomposite surfaces can be functionalised with affinity ligands, such as peptides or small molecules, that have high binding affinity for specific target molecules. These ligands enhance the selectivity of biosensors.
- *Aptamers [152, 153]:* Aptamers are synthetic, single-stranded DNA or RNA molecules that bind to target analytes with high specificity. Functionalising nanocomposite surfaces with aptamers creates aptamer-based biosensors with excellent target recognition capabilities.

Molecularly Imprinted Polymers (MIPs) [154, 155, 156]

Molecularly imprinted polymers are synthetic polymers that are designed to recognise and bind to specific target molecules. Nanocomposite surfaces can be functionalised with MIPs to create highly selective biosensors for various analytes.

Conductive Nanomaterials

- *Carbon Nanotubes (CNTs) and Graphene [157, 158]:* Nanocomposite surfaces incorporating CNTs or graphene can be functionalised with redox-active molecules or conductive polymers to enhance the electrochemical properties of biosensors.
- *Metal Nanoparticles [96]:* Functionalising nanocomposites with metal nanoparticles, such as gold or silver, can enhance the electrical conductivity and surface area of electrochemical biosensors.

Nanostructures for Signal Amplification

- *Nanozymes [159, 160]:* Nanocomposite surfaces can be functionalised with nanozymes, which are nanomaterials with enzyme-like catalytic activity. Nanozymes can amplify the signal generated by biochemical reactions, increasing biosensor sensitivity.
- *Quantum Dots [161, 162]:* Functionalising nanocomposite surfaces with quantum dots can enhance the fluorescence signal in optical biosensors, improving the detection limits and Sensitivity.

pH and Ionic Strength Adjustment

- *Buffer Solutions [43, 163]:* Adjusting the pH and ionic strength of the biosensor's environment can optimise the binding kinetics and stability of biomolecules on the nanocomposite surface, enhancing biosensor performance.

Functionalisation of nanocomposite surfaces is a highly tailored process, and the choice of functionalisation method and molecules depends on the specific biosensing application and the desired sensor performance parameters, including Sensitivity, selectivity, and stability. Proper functionalisation is crucial for the successful development of effective biosensors for a wide range of applications in healthcare, environmental monitoring, and biotechnology.

Importance of Surface Modification for Specific Biomolecule Detection

Surface modification plays a crucial role in detecting specific biomolecules and is essential for various applications in biotechnology, diagnostics, and medical research. Here are several key points highlighting the importance of surface modification for the specific detection of biomolecules.

- *Enhanced Selectivity [164]:* Surface modification allows for the design of highly selective detection platforms. By functionalising surfaces with specific ligands, such as antibodies, aptamers, or molecular imprints, researchers can ensure that only the target biomolecule interacts with the surface. This reduces false positives and increases the accuracy of detection.
- *Increased Sensitivity [165, 166]:* Surface modification techniques can improve the Sensitivity of detection methods. Functionalised surfaces can concentrate biomolecules at the sensing interface, enhancing the signal-to-noise ratio. This is particularly important when detecting low-abundance biomarkers or when working with complex samples, such as blood or tissue.
- *Reduced Non-Specific Binding [167]:* Unmodified surfaces often exhibit non-specific binding, where molecules other than the target

biomolecule adhere to the surface. Surface modification can reduce this non-specific binding, leading to cleaner and more reliable detection results.

- *Multiplexed Detection [168, 169]:* Surface modification allows for the creation of multiplexed detection platforms, where multiple biomolecules can be detected simultaneously on the same surface. Different regions of the surface can be functionalised with other ligands, enabling the detection of various targets in a single experiment.
- *Tailored Surface Chemistry [170, 171]:* Surface modification techniques can be tailored to the specific properties of the target biomolecule. For example, hydrophobic or hydrophilic surfaces, as well as surfaces with different charge characteristics, can be created to optimise interactions with other types of biomolecules.
- *Long-Term Stability [172, 173]:* Surface modification can improve the stability of the detection platform over time. Coatings and changes can protect the sensing surface from degradation or fouling, ensuring consistent and reliable performance.
- *Miniaturisation and Integration [174-177]:* Surface modification techniques are compatible with miniaturisation and Integration into microfluidic devices and lab-on-a-chip systems. This enables point-of-care diagnostics, where rapid and specific detection of biomolecules can be performed outside traditional laboratory settings.
- *Biomedical Applications [178-180]:* Surface modification is crucial in various biomedical applications, including disease diagnosis, drug development, and personalised medicine. It enables the detection of specific biomarkers associated with diseases, facilitating early diagnosis and treatment monitoring.
- *Environmental Monitoring [181, 182]:* Surface modification is also necessary in environmental monitoring, where it can be used to detect and quantify specific biomolecules, pollutants, or pathogens in water, air, or soil samples.
- *Research Advancements [183, 184]:* Surface modification techniques continue to advance, providing researchers with more options and tools for biomolecule detection. This contributes to the progress of scientific knowledge and the development of new diagnostic methods.

In summary, surface modification is a fundamental aspect of biomolecule detection, enabling increased selectivity, Sensitivity, and reliability in various applications. It is a critical component of advancing research, diagnostics, and technologies in the fields of biology, medicine, and environmental science.

Biosensing Applications Enabled by Nanocomposites

Nanocomposites, which are materials composed of nanoparticles dispersed within a matrix, have enabled a wide range of biosensing applications due to their unique properties and versatility. These applications span various fields, including healthcare, environmental monitoring, and food safety. Here's an overview of diverse biosensing applications enabled by nanocomposites.

Healthcare Applications

- *Disease Diagnosis [185-188]:* Nanocomposites are used in biosensors for the early diagnosis of diseases. Functionalised nanoparticles can bind to specific biomarkers, enabling the detection of diseases like cancer, diabetes, and infectious diseases.
- *Point-of-Care Testing (POCT) [189, 190]:* Nanocomposite-based POCT devices are portable and can provide rapid results for various medical tests, such as glucose monitoring, cardiac biomarker detection, and pregnancy testing.
- *Drug Delivery [191, 192]:* Nanocomposite materials can be used to create drug delivery systems with enhanced precision and efficiency. They can be designed to release drugs in response to specific biological triggers or at controlled rates.

Environmental Monitoring

- *Water Quality Monitoring [193-195]:* Nanocomposites are utilised in sensors to detect contaminants in water, including heavy metals, organic pollutants, and pathogens. They enable real-time monitoring and help ensure the safety of drinking water.
- *Air Quality Monitoring [196, 197]:* Nanocomposite-based sensors can detect gases and particulate matter, contributing to air pollution monitoring and the assessment of indoor air quality.

- *Soil Analysis [198, 199]:* Nanocomposites can be integrated into soil sensors to measure factors such as pH, moisture content, and the presence of pollutants, thereby aiding in agricultural and environmental studies.

Food Safety and Quality Control

- Foodborne Pathogen Detection [200,201]: Nanocomposite-based biosensors are utilised to detect pathogens such as Salmonella and E. coli in food products, thereby ensuring food safety.
- Food Spoilage Detection [202,203]: Nanocomposite sensors can detect spoilage markers, such as amines and volatile organic compounds (VOCs), thereby helping to maintain food quality and reduce food waste.
- Authentication and Traceability [204-207]: Nanocomposites are utilised in anti-counterfeiting measures for food products by creating unique, traceable labels or tags that alter their properties in response to environmental factors.

Environmental Remediation

- *Wastewater Treatment [208-211]:* Nanocomposites are used in water treatment processes to remove pollutants and heavy metals from wastewater more efficiently.
- *Oil Spill Cleanup [212, 213]:* Nanocomposites with high adsorption capacities can be applied to absorb and remove oil and hydrophobic pollutants from water surfaces during oil spills.

Biotechnology

- *Protein and DNA Detection [214, 215]:* Nanocomposite-based biosensors are used for the detection of proteins and nucleic acids, supporting research in genomics, proteomics, and molecular diagnostics.
- *Cell-Based Assays [216, 217]:* Nanocomposites are incorporated into cell culture substrates to support cell growth, differentiation, and monitoring in tissue engineering and drug testing applications.

Energy Conversion and Storage

- *Energy Harvesting [218, 219]:* Nanocomposites are utilised in energy-harvesting devices, such as piezoelectric or thermoelectric nanogenerators, for biosensors that can harness energy from biological processes or body movements.
- *Biosolar Cells [220, 221]:* Nanocomposites can enhance the efficiency of biosolar cells that use biological molecules like chlorophyll for energy conversion.

Nanocomposites have revolutionised biosensing by offering improved Sensitivity, selectivity, and versatility in detecting a wide range of biological and environmental analytes. Their unique properties, tunability, and compatibility with various sensor platforms make them valuable tools for addressing critical challenges in healthcare, environmental monitoring, and food safety.

Performance Metrics Used to Evaluate Nanocomposite Biosensors

Evaluating the performance of nanocomposite biosensors is crucial for determining their effectiveness in detecting specific analytes or biomolecules. Several key performance metrics are commonly used to assess nanocomposite biosensors, including Sensitivity, selectivity, response time, and sometimes also factors like stability and reproducibility [222, 223]. Here's an explanation of these key metrics.

Sensitivity

- *Definition:* Sensitivity measures the ability of a biosensor to detect small changes in analyte concentration. It quantifies how effectively the sensor can convert an analyte's concentration into a measurable signal.
- *Measurement:* Sensitivity is typically expressed as the slope of the sensor's calibration curve, which illustrates the relationship between signal output and analyte concentration. Higher Sensitivity means a steeper slope.

- *Importance:* High Sensitivity is crucial when detecting low concentrations of analytes, such as in medical diagnostics or environmental monitoring. It ensures that even minute amounts of the target analyte can be accurately detected and quantified.

Selectivity

- *Definition:* Selectivity assesses the ability of the biosensor to distinguish the target analyte from interfering substances or other molecules present in the sample.
- *Measurement:* Selectivity is often expressed as a selectivity coefficient, which compares the sensor's response to the target analyte with its response to potential interferents. A higher selectivity coefficient indicates better selectivity.
- *Importance:* Selectivity is crucial to avoid false positives or false negatives. A highly selective biosensor ensures that it specifically detects the target analyte even in the presence of other substances, enhancing its reliability.

Response Time

- *Definition:* Response time measures the speed at which a biosensor produces a signal in response to changes in analyte concentration. It quantifies the sensor's speed in reporting changes.
- *Measurement:* Response time is typically measured as the time it takes for the biosensor signal to reach a steady-state value (e.g., 90% of the final signal) after a step change in analyte concentration is introduced.
- *Importance:* A shorter response time is critical in applications where rapid detection or monitoring is required, such as point-of-care diagnostics or real-time environmental monitoring.

Stability

- *Definition:* Stability refers to the biosensor's ability to maintain its performance over time and under various conditions, including changes in temperature, humidity, and storage duration.

- *Measurement*: Stability can be assessed by monitoring the sensor's signal over an extended period, exposing it to various environmental conditions, or analyzing its performance after repeated use.
- *Importance:* A stable biosensor ensures consistent and reliable results throughout its operational lifespan, reducing the need for frequent recalibration or replacement.

Reproducibility

- *Definition:* Reproducibility measures the ability of the biosensor to produce consistent results when tested multiple times with the same analyte concentration.
- *Measurement:* Reproducibility is typically assessed by conducting multiple tests under identical conditions and calculating the variation in sensor responses.
- *Importance:* High reproducibility is essential for ensuring the reliability and accuracy of biosensor measurements, particularly in research, quality control, and clinical applications.

These key performance metrics collectively provide a comprehensive assessment of a nanocomposite biosensor's capabilities. Researchers and developers often optimise these metrics to meet specific application requirements, ensuring that the biosensor performs effectively in real-world scenarios.

Challenges and Limitations Associated with Nanocomposite Biosensors

Nanocomposite biosensors offer numerous advantages, but they also face several challenges and limitations that must be considered during their development and application. Here are some of the key challenges and constraints associated with nanocomposite biosensors.

Stability

- Chemical and Environmental Stability: Nanocomposites can be sensitive to changes in temperature, humidity, and pH, which can affect their stability and performance over time. Proper encapsulation and protective coatings are often required to mitigate these effects.

- Long-Term Stability: Ensuring that nanocomposite biosensors maintain their performance over extended periods can be challenging, particularly in applications that require continuous monitoring or use in harsh environments.

Reproducibility

- *Batch-to-Batch Variability*: The Synthesis of nanocomposites can result in variations from one batch to another, impacting sensor performance and reproducibility. Quality control measures and standardised protocols are needed to minimise batch-to-batch differences.
- *Sensor-to-Sensor Variability*: Variability in the fabrication process can lead to differences in sensor responses even within the same batch, requiring careful calibration and quality assurance procedures.

Cost

- *Material Costs:* Some nanocomposite materials can be expensive to produce or source, which can increase the overall cost of biosensor development. Finding cost-effective materials and fabrication methods is a challenge.
- *Manufacturing Complexity:* The fabrication of nanocomposite biosensors can be a complex and labour-intensive process, resulting in higher production costs. Scaling up production while maintaining quality can be a challenging task.

Selectivity

- *Interference from Matrix Components*: In complex sample matrices, such as blood or soil, nanocomposite biosensors may encounter interference from other substances that can affect their selectivity and specificity. Pre-treatment or sample preparation steps may be necessary to address this limitation.

Sensitivity

- *Detection Limits:* While nanocomposite biosensors can offer high Sensitivity, they may still have limitations in detecting extremely low concentrations of analytes, especially in applications where ultratrace detection is required.
- *Signal-to-Noise Ratio:* Achieving a high signal-to-noise ratio can be challenging, particularly when dealing with weak signals or when there are non-specific interactions at the sensor surface.

Biocompatibility

- *Toxicity Concerns:* Certain nanomaterials employed in nanocomposite biosensors may raise toxicity concerns, particularly when intended for use in biological systems. Ensuring the biocompatibility of these materials is essential.

Integration and Miniaturisation

- *Integration Complexity*: Integrating nanocomposite biosensors into miniaturised or portable devices can be technically challenging and may require advanced microfabrication techniques.
- *Power Requirements:* Some nanocomposite biosensors may have power consumption requirements that limit their suitability for specific applications, particularly in remote or resource-constrained environments.

Regulatory and Safety Considerations

- *Regulatory Approval:* Meeting regulatory requirements for biosensors used in healthcare or other critical applications can be a lengthy and expensive process.
- *Safety and Environmental Impact*: The disposal of nanomaterial-containing biosensors and their potential environmental impact raise concerns that need to be addressed responsibly.

Sample Complexity

- *Complex Samples:* Biosensors often face challenges when dealing with complex samples, such as bodily fluids or environmental samples, where the presence of multiple analytes or interfering substances can complicate detection.

Cross-Contamination

In specific applications, particularly those involving point-of-care testing or diagnostics, the risk of cross-contamination between samples can be a significant concern; therefore, proper sensor design and handling protocols are essential.

Despite these challenges and limitations, nanocomposite biosensors continue to advance and find applications in various fields due to their unique capabilities. Addressing these limitations through research, innovation, and technology development is crucial for unlocking their full potential and expanding their practical use.

Future Directions

The future of nanocomposite biosensors holds great promise, as ongoing research and technological advancements are likely to overcome current limitations and open up new possibilities for their applications. Here are some speculative advancements and trends in the field.

- *Enhanced Sensitivity and Detection Limits*: Future nanocomposite biosensors may achieve even higher levels of Sensitivity, allowing for the detection of analytes at ultra-low concentrations. This will be particularly valuable in early disease diagnosis and environmental monitoring.
- *Multi-Analyte Detection*: Advances in nanocomposite design may enable the simultaneous detection of multiple analytes within a single sample. This capability will be crucial in complex diagnostic scenarios and multiplexed environmental monitoring.
- *Real-time Monitoring*: Nanocomposite biosensors with rapid response times may enable real-time monitoring of dynamic

processes, such as continuous glucose monitoring for diabetes management or tracking the progression of diseases.

- *Wearable and Implantable Sensors*: Nanocomposite biosensors integrated into wearable devices or implanted in the body could provide continuous health monitoring and facilitate early intervention in medical conditions.
- *Point-of-Care Testing (POCT):* The development of portable, user-friendly nanocomposite-based POCT devices may revolutionise healthcare by enabling rapid and accurate diagnostics at the patient's bedside, in clinics, or resource-limited settings.
- *Nanomaterial Innovation:* The exploration of novel nanomaterials with unique properties, such as 2D materials (e.g., graphene), quantum dots, and metal-organic frameworks, could lead to improved sensor performance and new functionalities.
- *Biocompatibility and Safety*: Researchers will continue to address concerns regarding the biocompatibility and potential toxicity of nanomaterials used in biosensors, ensuring their safe use in medical applications.
- *AI and Data Analytics Integration*: Integrating artificial intelligence (AI) and machine learning algorithms will enable more effective data analysis and interpretation, making biosensors more intelligent and capable of providing actionable insights.
- *Environmental Applications*: Nanocomposite biosensors will play a pivotal role in monitoring and mitigating environmental pollution, thereby contributing to improved resource management and enhanced sustainability efforts.
- *Regulatory Approval and Commercialisation*: As the field matures, more nanocomposite biosensors are likely to receive regulatory approval and enter the market for various applications, from healthcare to environmental monitoring.
- *Self-Powered Sensors*: Advances in energy-harvesting technologies may lead to self-powered nanocomposite biosensors that can generate energy from their environment, reducing the need for external power sources.
- *Advanced Manufacturing Techniques*: Innovative fabrication methods, such as 3D printing and microfluidics, will continue to evolve, enabling the cost-effective and high-throughput production of nanocomposite biosensors.

- *Environmental Impact Considerations*: Researchers and manufacturers will focus on developing biosensors with minimal ecological impact, including biodegradable components and sustainable materials.
- *Global Accessibility:* Efforts to make nanocomposite biosensors more affordable and accessible to an international population, particularly in resource-limited regions, will drive innovation in manufacturing and distribution.
- *Customization and Personalization*: Advances in nanocomposite biosensors may enable tailored solutions for individual healthcare needs, offering personalized diagnostics and treatment recommendations.

The future of nanocomposite biosensors is exciting, with the potential to revolutionise healthcare, environmental monitoring, and various industries. Continued collaboration between researchers, engineers, and healthcare professionals, along with investment in research and development, will be essential to unlock these advancements and bring nanocomposite biosensors closer to their full potential.

Summary and Conclusion

The review of nanocomposites in advancing biosensor technology has revealed several key findings and insights that underscore the significant role of nanocomposites in this field.

- *Enhanced Sensitivity and Selectivity:* Nanocomposites, with their large surface area and unique properties, improve the Sensitivity and selectivity of biosensors. They enable the detection of specific biomolecules at low concentrations, reducing the likelihood of false positives and false negatives.
- *Multifunctionality:* Nanocomposites can be engineered to serve multiple functions within a biosensor, such as signal amplification, immobilisation of recognition elements (e.g., antibodies), and transduction of biological interactions into measurable signals. This multifunctionality streamlines biosensor design and improves overall performance.

- *Miniaturisation and Portability*: The use of nanocomposites facilitates the miniaturisation of biosensors, making them suitable for portable and point-of-care applications. This trend aligns with the growing demand for on-the-spot diagnostics and monitoring.
- *Real-Time Monitoring*: Nanocomposite-based biosensors enable real-time monitoring of analyte concentrations, which is crucial in fields like healthcare and environmental monitoring. This capability supports timely decision-making and intervention.
- *Diverse Applications*: Nanocomposite biosensors have applications across various domains, including healthcare (disease diagnosis and drug monitoring), environmental monitoring (assessment of water and air quality), food safety (pathogen detection), and beyond. Their versatility enables them to address various societal challenges.
- *Emerging Nanomaterials*: This review highlights the emergence of novel nanomaterials, including graphene, quantum dots, and carbon nanotubes, in the development of biosensors. These materials possess unique properties, including high electrical conductivity and biocompatibility, which expand the possibilities for biosensor innovation.
- *Challenges and Considerations*: The review also identifies challenges in nanocomposite biosensor technology, such as stability, reproducibility, and potential toxicity of nanomaterials. Researchers must address these issues to ensure the safe and reliable deployment of biosensors.
- *Interdisciplinary Collaboration*: The field of nanocomposite biosensors relies heavily on multidisciplinary collaboration among materials scientists, chemists, biologists, and engineers. This collaboration fosters innovation by combining expertise from various domains.
- *Sustainable and Green Approaches*: An emerging trend involves the development of sustainable and environmentally friendly nanocomposites. The review emphasises the importance of considering the life cycle of nanocomposite materials and their environmental impact.

In conclusion, the review underscores the significant impact of nanocomposites in advancing biosensor technology. Their unique properties, versatility, and multifunctionality have revolutionised the field by enabling highly sensitive, selective, and portable biosensors for a wide range of applications. Nanocomposite-based biosensors are poised to play a crucial role in improving healthcare, environmental monitoring, and food safety, among other areas, as they continue to evolve and address the challenges associated with their implementation.

References

[1] P. Mohankumar, J. Ajayan, T. Mohanraj, R. Yasodharan, *Measurement*167, 108293 (2021).

[2] S. R. Mozaz, M. J. L. de Alda, M. -P. Marco, D. Barceló, *Talanta* 65, 291 (2005).

[3] J. Yasmin, M. R. Ahmed, B.-K. Cho, *J. of Biosystems Eng.* 41, 240 (2016).

[4] Frank Lammers & Thomas Scheper, *Thermal Biosensors in Biotechnology*. In : Bhatia, P. K., et al. Thermal Biosensors, Bioactivity, Bioaffinitty. *Advances in Biochemical Engineering/Biotechnology,* vol 64. Springer, Berlin, Heidelberg (1999).

[5] S. A. El-Safty, M. A. Shenashen, *Mater. Today Bio* 5, 100044 (2020).

[6] R. G. Kerry, K. E. Ukhurebor, S. Kumari, G. K. Maurya, S. Patra, B. Panigrahi, S. Majhi, J. R. Rout, M. del P. R. Torres, G. Das, H.-S. Shinji, J. K. Patra, *Biomater. Sci.* 9, 3576 (2021).

[7] Z. Altintas, I. E. Tothill, *Nanobiosens. Dis. Diagn.* 4, 1-10(2016).

[8] X. Lu, Y. Ye, Y. Zhang, X. Sun, *Crit. Rev. Food Sci. Nutr.* 61, 3819(2021).

[9] I. Palchettia, M. Mascini, *Analyst* 133, 846 (2008).

[10] X. Xu, Y. Ying, *Food Rev. Int.* 27, 300(2011).

[11] A. Vasilescu, C. Polonschii, A. M. Titoiu, R. Mishra, S. Peteu, J. -L. Marty, Chapter Nine - Bioassays and biosensors for food analysis : focus on allergens and food packaging, Editor(s): M. K. Sezgintürk, *Commercial Biosensors and Their Applications*, Elsevier, Pages 217-258 (2020).

[12] M. Weston, S. Geng, R. Chandrawati, *Adv. Mater. Technol.* 6, 2001242 (2021).

[13] D. Yu, B. Blankert, J. -C. Viré, J. -M. Kauffmann, *Anal. Lett.* 38, 1687(2005).

[14] K. A. Giuliano, D. L. Taylor, *Trends. Biotechnol.* 16, 135 (1998).

[15] D. Liu, J. Wang, L. Wu, Y. Huang, Y. Zhang, M. Zhu, Y. Wang, Z. Zhu, C. Yang, *TrAC Trends Anal. Chem.* 122, 115701 (2020).

[16] Y. Li, H. J. Schluesener, S. Xu, *Gold Bull.* 43, 29 (2010).

[17] X. Li, T. Zhao, L. Sun, K. E. Aifantis, Y. Fan, Q. Feng, F. Cui, F. Watari, *J. Biomed. Mater. Res. – A* 104, 322 (2016).

[18] R. Pinalli, A. Pedrini, E. Dalcanale, *Chem. Soc. Rev.* 47, 7006 (2018).

[19] J. Sengupta, C. M. Hussain, *Biomolecules* 13, 1024 (2023).

[20] S. K. Krishnan, E. Singh, P. Singh, M. Meyyappan, H. S. Nalwa, *RSC Adv.* 9, 8778 (2019).

[21] T. Sen, S. Mishra, N. G. Shimpi, *RSC Adv.* 6, 42196 (2016).

[22] A. Kausar, *C* 9, 13 (2023).

[23] R. H. Althomali, K. A. Alamry, M. A. Hussein, G. S. Tay, *Chemistry Select* 7, e202200843 (2022).

[24] A. Shi, J. Wang, X. Han, X. Fang, Y. Zhang, *Sens. Actuators B: Chem.* 200, 206 (2014).

[25] W. A. Williams, A. J. Denslow, P. W. Radulovic, D. J. Denmark, S. S. Mohapatra, *A Computational Method Involving Surface Area to Volume Ratio to Estimate Inorganic Nanoparticle Efficacy.* Preprints 2021, 2021080138.

[26] M. Swierczewska, G. Liu, S. Lee, X. Chen, *Chem. Soc. Rev.* 41, 2641 (2012).

[27] M. A. Holgado, L. M. Banderas, J. A. Fuentes, M. F. Arevalo, J. L. Arias, *Curr. Med. Chem.* 19, 3188 (2012).

[28] N. J. Ronkainen, H. B. Halsall, W. R. Heineman, *Chem. Soc. Rev.* 39, 1747 (2010).

[29] S. Prakash, T. Chakrabarty, A. K. Singh, V. K. Shahi, *Biosens. Bioelectron.* 41, 43 (2013).

[30] M. Oves, M. Shahadat, S. A. Ansari, M. Aslam, I. I. M Ismail, Chapter 6 : Polyaniline Nanocomposite Materials for Biosensor Designing, Book Editor(s) : A. Khan, M. Jawaid, A. A. P. Khan, A. M. Asiri, *Electrically Conductive Polymer and Polymer Composites : From Synthesis to Biomedical Applications* (2018).

[31] S. Kumar, W. Ahlawat, R. Kumar, N. Dilbaghi, *Biosens. Bioelectron.* 70, 498 (2015).

[32] S. Bagyalakshmi, A. Sivakami, K. Pal, R. Sarankumar, C. Mahendran, *J. Nanopart. Res.* 24, 201 (2022).

[33] N. P. Shetti, D. S. Nayak, K. R. Reddy, T. M. Aminabhvi, Chapter 10 - Graphene–Clay-Based Hybrid Nanostructures for Electrochemical Sensors and Biosensors, Editor(s) : Alagarsamy Pandikumar, Perumal Rameshkumar, In Micro and Nano Technologies, *Graphene-Based Electrochemical Sensors for Biomolecules*, Elsevier, Pages 235-274 (2019).

[34] D. M. Nascimento, Y. L. Nunes, M. C. B. Figueirêdo, H. M. C. de Azeredo, F. A. Aouada, J. P. A. Feitosa, M. F. Rosa, A. Dufresne, *Green Chem.* 20, 2428 (2018).

[35] D. Chen, H. Zhu, S. Yang, N. Li, Q. Xu, H. Li, J. He, J. Lu, *Adv. Mater.* 28, 10443 (2016).

[36] G. Ebrahimi, P. S. Pakchin, A. Shamloo, A. Mota, M. de la Guardia, H. Omidian, Y. Omidi, *Microchim. Acta* 189, 252 (2022).

[37] D. -G. Macovei, M. -B. Irimes, O. Hosu, C. Cristea, M. Tertis, *Anal. Bioanal. Chem.* 415, 1033 (2023).

[38] A. Kaushal, A. Seth, D. Kala, S. Gupta, L. Krishnia, V. Verma, Chapter 22 - Miniaturised devices for point-of-care testing/miniaturisation and Integration with microfluidic systems, Editor(s) : R. Khan, A. Parihar, S. K. Sanghi, *Biosensor Based Advanced Cancer Diagnostics*, Academic Press, Pages 375-383 (2022).

[39] N. Liu, Z. Ma, *Biosens. Bioelectron.* 51, 184 (2014).

[40] S. S. Timilsina, P. Jolly, N. Durr, M. Yafia, D. E. Ingber, *Acc. Chem. Res.* 54, 3529 (2021).

[41] X. Chen, X. Jia, J. Han, J. Ma, Z. Ma, *Biosens. Bioelectron.* 50, 356 (2013).

[42] Y. Song, Y, Luo, C. Zhu, H. Li, D. Du, Y. Lin, *Biosens. Bioelectron.* 76, 195 (2016).

[43] Y. Jiang, J. Wu, *Electrophoresis* 40, 2084 (2019).

[44] I. H. Cho, D. H., Kim, S., Park, *Biomater. Res.* 24, 6 (2020).
[45] S. Shrivastava, N. Jadon, R. Jain, *TrAC Trends Anal. Chem.* 82, 55 (2016).
[46] X. Li, J. Wei, K. E. Aifantis, Y. Fan, Q. Feng, F. Z. Cui, F., Watari, *J. Biomed. Mater. Res. A* 104, 1285 (2016).
[47] F. T. Wang, L.-N. Wang, J. Xu, K. -J. Huang, X. Wu, *Analyst* 146, 4418 (2021).
[48] A. P. Demchenko, *Methods Appl. Fluoresc.* 1, 022001 (2013).
[49] B. Wang, H. Cai, G. I. N. Waterhouse, X. Qu, B. Yang, S. Lu, *Small Science* 2, 2200012 (2022).
[50] M. Naseri, L. Fotouhi, A. Ehsani, *Chem. Rec.* 18, 599 (2018).
[51] D. R. Paul, L. M. Robeson, *Polymer* 49, 3187 (2008).
[52] E.W. Gacitua, A. A. Ballerini, J. Zhang, *Maderas. Cienciaytecnologia* 7,159 (2005).
[53] A. John, L. Benny, A. R. Cherian, S. Y. Narahari, A. Varghese, G. Hegde, *J. Nanostruct. Chem.* 11, 1 (2021).
[54] M. E. Rahzi, S. Majid, M. Elbasri, F. E. Salih, L. Oularbi, K. Lafdi, *Int. Nano Lett.* 8, 79 (2018).
[55] E. Lahiff, C. Lynam, N. Gilmartin, R. O'Kennedy, D. Diamond, *Anal. Bioanal. Chem.* 398, 1575 (2010).
[56] M. H. Naveen, N. G. Gurudatt, Y. -B. Shim, *Appl. Mater. Today* 9, 419 (2017).
[57] S. Rani, T. Gupta, V. Garg, S. Bandyopadhyay-Ghosh, S. B. Ghosh, G. Liu, *Mater. Today : Proc.* 62, 638 (2022).
[58] N. S. Satarkar, D. Biswal, J. Z. Hilt, *Soft Matter* 6, 2364 (2010).
[59] E. B. Catton, M. L. Ross, P. Asuri, *Polymers* 13, 856 (2021).
[60] N. S. Satarkar, J. Z. Hilt, *Acta Biomater.* 4, 11 (2008).
[61] Z. Weian, L. Wei, F. Yue'e, *Mater. Lett.* 59, 2876 (2005).
[62] I. Khalil, N. M. Julkapli, W. A. Yehye, W. J. Basirun, S. K. Bhargava, *Materials* 9, 406 (2016).
[63] N. Yusoff, A. Pandikumar, R. Ramaraj, H. N. Lim, N. M. Huang, *Microchim Acta* 182, 2091 (2015).
[64] M. I. Kim, J. Shim, T. Li, J. Lee, H. G. Park, *Chem. Eur. J.* 17, 10700 (2011).
[65] S. G. Chatterjee, S. Chatterjee, A. K. Ray, A. K. Chakraborty, *Sens. Actuators B : Chem.* 221, 1170 (2015).
[66] C. Wang, L. Yin, L. Zhang, D. Xiang, R. Gao, *Sensors* 10, 2088 (2010).
[67] F. Chandra, P. Laha, F. Benyettou, T. Skorjanc, N. Saleh, *Sci. Rep.* 13, 11287 (2023).
[68] E. M. Materón, C. M. Miyazaki, O. Carr, N. Joshi, P. H. S. Picciani, C. J. Dalmaschio, F. Davis, F. M. Shimizu, *Appl. Surf. Sci. Adv.* 6, 100163 (2021).
[69] G. K. Kouassi, *Curr. Nanosci.*7, 510 (2011).
[70] S. Natarajan, K. Harini, G. P. Gajula, B. Sarmento, M. T. N. Petersen, V. Thiagarajan, *BMC Mat.* 1, 2 (2019).
[71] T. T. Tran, A. Mulchandani, *TrAC Trends Anal. Chem.* 79, 222 (2016).
[72] P. Ramnani, N. M. Saucedo, A. Mulchandani, *Chemosphere* 143, 85 (2016).
[73] S. Gupta, C. N. Murthy, C. R. Prabha, *Int. J. Biol. Macromol.* 108, 687 (2018).
[74] S. Alabsi, A. Y. Ahmed, J. O. Dennis, M. H. M. Khir and A. S. Algamili, *IEEE Access* 8,69509 (2020).

[75] E. M. Narváez, L. B. Pires, A. Z. Gálvez, A. Merkoçi, *Adv. Mater.* 29, 1604905 (2017).

[76] J. Sengupta, C. M. Hussain, *Carbon Trends* 2, 100011 (2021).

[77] F. Arshad, F. Nabi, S. Iqbal, R. H. Khan, *Colloids Surf. B* 212, 112356 (2022).

[78] J. Huang, Y. Liu, T. You, *Anal. Methods* 2, 202 (2010).

[79] L. Natrayan, A. Merneedi, G. Bharathiraja, S. Kaliappan, D. Veeman, P. Murugan, *J. Nanomater.* Volume 2021, Article ID 7323885, 7 pages.

[80] S. Mallakpour, E. Khadem, *J. Chem. Eng.* 302, 344 (2016).

[81] D. Yadav, F. Amini, A. Ehrmann, *Eur. Polym. J.* 138, 109963 (2020).

[82] F. Parnianchi, M. Nazari, J. Maleki, M. Mohebi, *Int. Nano Lett.* 8, 229 (2018).

[83] S. Sarkar, E. Guibal, F. Quignard, A. K. SenGupta, *J. Nanopart. Res.* 14, 715 (2012).

[84] G. Kaur, R. Adhikari, P. Cass, M. Bown, P. Gunatillake, *RSC Adv.* 5, 37553 (2015).

[85] A. M. D. Pascual, *Polymers* 12, 872 (2020).

[86] H. J. Salavagione, A. M. D. Pascual, E. Lázaro, S. Vera, M. A. G. Fatou, *J. Mater. Chem. A* 2, 14289 (2014).

[87] Z. Su, J. Ding, G. Wei, *RSC Adv.* 4, 52598 (2014).

[88] G. K. Nezhad, A. Sarkary, Z. Khorablou, P. S. Dorraji, *Iran. J. Pharm. Res.* 17, 52 (2018).

[89] A. M. Bagher, *Sens. Transducers.* 198, 37 (2016).

[90] M. Pourmadadi, E. Rahmani, M. R. Khosroshahi, A. Samadi, R. Behzadmehr, A. Rahdar, L. F. R. Ferreira, *J. Drug Deliv. Sci. Technol.* 80, 104156 (2023).

[91] F. Ma, C. C. Li, C. Y. Zhang, *J. Mater. Chem. B* 6, 6173 (2018).

[92] R. Tenchov, R. Bird, A. E. Curtze, Q. Zhou, *ACS Nano* 15, 16982 (2021).

[93] B. Kandregula, S. Narisepalli, D. Chitkara, A. Mittal, *Mol. Pharmaceutics* 19, 1977 (2022).

[94] V. D. Leo, A. M. Maurelli, L. Giotta, L. Catucci, *Colloids Surf. B : Biointerfaces* 218, 112737 (2022).

[95] L. Fei, M. Naeemi, G. Zou, H. Luo, *Chem. Rec.* 13, 85 (2013).

[96] J. Wang, *Microchim. Acta* 177, 245 (2012).

[97] H. Ma, B. Yin, S. Wang, Y. Jiao, W. Pan, S. Huang, S. Chen, F. Meng, *Chem Phys Chem* 5, 68 (2004).

[98] A. P. Chiriac, I. Neamtu, L. E. Nita, M. T. Nistor, *Mini-Rev. Med. Chem.* 10, 990 (2010).

[99] A. Zadehnazari, *Polym. -Plast. Technol. Mater.* 62, 655 (2023).

[100] S. M. Rossnagel, *J. Vac. Sci. Technol. A* 21, S74 (2003).

[101] M. Dominik, A. Leśniewski, M. Janczuk, J. N. Jönsson, M. Hołdyński, Ł. Wachnicki, M. Godlewski, W. J. Bock, M. Śmietana, *Biosens. Bioelectron.* 93, 102 (2017).

[102] I. V. Shishkovsky, P. N. Lebedev, 3 - Chemical and physical vapour deposition methods for nanocoatings, Editor(s) : A. S. H. Makhlouf, I. Tiginyanu, In Woodhead Publishing Series in Metals and Surface Engineering, *Nanocoatings and Ultra-Thin Films*, Woodhead Publishing, Pages 57-77 (2011).

[103] K. A. Shah, B. A. Tali, *Mater. Sci. Semicond. Process.* 41, 67 (2016).

[104] M. Kumar, Y. Ando, *J. Nanosci. Nanotechnol.* 10, 3739 (2010).

[105] Y. M. Manawi, Ihsanullah, A. Samara, T. A. Ansari, M. A. Atieh, *Materials* 11, 822 (2018).
[106] U. Chadha, S. Sinha, J. Jonna, M. Goswami, H. Ghani, K. Nair, N. Pandey, T. Kataray, S. K. Selvaraj, P. Bhardwaj, *ECS J. Solid State Sci. Technol.* 11, 041003 (2022).
[107] H. Ai, S. A. Jones, Y. M. Lvov, *Cell Biochem. Biophys.* 39, 23 (2003).
[108] W. Zhao, J. -J. Xu, H. -Y. Chen, *Electroanalysis* 18, 1737 (2006).
[109] M. M. Barsan, C. M. A. Brett, *TrAC Trends Anal. Chem.* 79, 286 (2016).
[110] C. Feng, K. C. Khulbe, T. Matsuura, *J. Appl. Polym. Sci.* 115, 756 (2010).
[111] A. Frenot, I. S. Chronakis, *Curr. Opin. Colloid. Interface. Sci.* 8, 64 (2003).
[112] E. Sapountzi, M. Braiek, J. -F. Chateaux, N. J. Renault, F. Lagarde, *Sensors* 17, 1887 (2017).
[113] D. Wang, *Template-assisted Synthesis of nanomaterials* (Doctoral dissertation, City University of Hong Kong) (2006).
[114] J. Wang, Z. Li, Z. Gu, *Sens. Actuators Rep.* 3, 100029 (2021).
[115] C. F. Wang, X. -Y. Sun, M. Su, Y. -P. Wang, Y. -K. Lv, *Analyst* 145, 1550 (2020).
[116] B. P. Simon, M. Campas, J.-L. Marty, *Protein Pept. Lett.* 15, 757 (2008).
[117] W. Putzbach, N. J. Ronkainen, *Sensors* 13, 4811 (2013).
[118] F. R. R. Teles, L. P. Fonseca, *Mater. Sci. Eng. C* 28, 1530 (2008).
[119] S. Khan, S. Ali, A. Khan, M. Ahmed, B. Wang, A. Bermak, *Sens. Actuator A Phys.* 323, 112638 (2021).
[120] R. A. S. Couto, J. L. F. C. Lima, M. B. Quinaz, *Talanta* 146, 801 (2016).
[121] H. Cui, M. J. Webber, S. I. Stupp, *Pept. Sci.* 94, 1 (2010).
[122] S. Misra, H. Wang, *Mater. Horiz.*8, 869 (2021).
[123] J. Ni, H. Ling, S. Zhang, Z. Wang, Z. Peng, C. Benyshek, R. Zan, A. K. Miri, Z. Li, X. Zhang, J. Lee, K.-J. Lee, H.-J. Kim, P. Tebon, T. Hoffman, M. R. Dokmeci, N. Ashammakhi, X. Li, A. Khademhosseini, *Mater. Today Bio* 3, 100024 (2019).
[124] M. Silva, I. S. Pinho, J. A. Covas, N. M. Alves, M. C. Paiva, *Funct. Compos. Mater.* 2, 8 (2021).
[125] K. Akhtar, S. A. Khan, S. B. Khan, A. M. Asiri, Scanning Electron Microscopy : Principle and Applications in Nanomaterials Characterisation. In : S. Sharma, (eds) *Handbook of Materials Characterisation*. Springer, Pages 113–145 (2018).
[126] C. Y. Tang, Z. Yang, Chapter 8 - Transmission Electron Microscopy (TEM), Editor(s) : N. Hilal, A. F. Ismail, T. Matsuura, D. O. Radcliffe, *Membrane Characterisation*, Elsevier, Pages 145-159 (2017).
[127] R. Das, E. Ali, A. Hamid, S. Bee, *Rev. Adv. Mater. Sci.* 38, 95 (2014).
[128] S. Vahabi, B. N. Salman, A. Javanmard, *Iran. J. Med. Sci.* 38, 76 (2013).
[129] N. Jalili, K. Laxminarayana, *Mechatronics* 14, 907 (2004).
[130] S. A. Khan, S. B. Khan, L. U. Khan, A. Farooq, K. Akhtar, A. M. Asiri, Fourier Transform Infrared Spectroscopy : Fundamentals and Applications in Functional Groups and Nanomaterials Characterisation. In : S. Sharma (eds) *Handbook of Materials Characterisation*. Springer, Pages 317–344, (2018).
[131] R. S. Das, Y. K. Agrawal, *Vib. Spectrosc.* 57, 163 (2011).
[132] I. Notingher, *Sensors* 7, 1343 (2007).

[133] X. Liang, N. Li, R. Zhang, P. Yin, C. Zhang, N. Yang, K. Liang, B. Kong, *NPG Asia. Mater.* 13, 8 (2021).
[134] E. Desimoni, B. Brunetti, *Chemosensors* 3, 70 (2015).
[135] N. Fairley, V. Fernandez, M. R. Plouet, C. G. Deudon, J. Walton, E. Smith, D. Flahaut, M. Greiner, M. Biesinger, S. Tougaard, D. Morgan, J. Baltrusaitis, *Appl. Surf. Sci. Adv.* 5, 100112 (2021).
[136] D. Grieshaber, R. MacKenzie, J. Vörös, E. Reimhult, *Sensors* 8, 1400 (2008).
[137] J. Yao, M. Yang, Y. Duan, *Chem. Rev.* 114, 6130 (2014).
[138] M. Wamsley, S. Nawalage, J. Hu, W. E. Collier, D. Zhang, *Anal. Chem* 94, 7123 (2022).
[139] P. Englebienne, A. V. Hoonacker, M. Verhas, *Spectroscopy* 17, 255 (2003).
[140] H. H. Nguyen, J. Park, S. Kang, M. Kim, *Sensors* 15, 10481 (2015).
[141] P. O. Patil, G. R. Pandey, A. G. Patil, V. B. Borse, P. K. Deshmukh, D. R. Patil, R. S. Tade, S. N. Nangare, Z. G. Khan, A. M. Patil, M. P. More, M. Veerapandian, S. B. Bari, *Biosens. Bioelectron.* 139, 111324 (2019).
[142] M. A. Sadique, S. Yadav, P. Ranjan, R. Khan, F. Khan, A. Kumar, D. Biswas, *ACS Appl. Bio Mater.* 5, 2421 (2022).
[143] K. M. Abu-Salah, M. M. Zourob, F. Mouffouk, S. A. Alrokayan, M. A. Alaamery, A. A. Ansari, *Sensors* 15, 14539 (2015).
[144] A. Kowalczyk, *Curr. Opin. Electrochem.* 23, 36 (2020).
[145] S. Daniel, T. P. Rao, K. S. Rao, S. U. Rani, G. R. K. Naidu, H. –Y. Lee, T. Kawai, *Sens. Actuators B : Chem.* 122, 672 (2007).
[146] G. Rocchitta, A. Spanu, S. Babudieri, G. Latte, G. Madeddu, G. Galleri, S. Nuvoli, P. Bagella, M. I. Demartis, V. Fiore, R. Manetti, P. A. Serra, *Sensors* 16, 780 (2016).
[147] H. H. Nguyen, S. H. Lee, U. J. Lee, C. D. Fermin, M. Kim, *Materials* 12, 121 (2019).
[148] S. Ding, A. A. Cargill, I. L. Medintz, J. C. Claussen, *Curr. Opin. Biotechnol.* 34, 242 (2015).
[149] S. Kroll, Chapter 3: Surface Modification of Ceramic Materials, *Surface-Functionalised Ceramics: For Biotechnological and Environmental Applications*, Book Editor(s): L. Treccani, F. Meder (2022).
[150] H. Otsuka, Y. Nagasaki, K. Kataoka, *Adv. Drug Deliv. Rev.* 55, 403 (2003).
[151] H. S. Mansur, *WIREs Nanomed Nanobiotechnol* 2, 113 (2010).
[152] A. V. Kulbachinskiy, *Biochemistry Moscow* 72, 1505 (2007).
[153] G. Mayer, Chemical Biology of Aptamers, *Angewandte Chemie International Edition*, 48, 2672 (2009).
[154] K. Haupt, A. V. Linares, M. Bompart, B. T. S. Bui, Molecularly Imprinted Polymers. In: K. Haupt (eds) Molecular Imprinting. *Topics in Current Chemistry*, vol 325, pp 1–28, Springer, Berlin, Heidelberg (2011).
[155] S. Ramanavicius, A. Ramanavicius, *Adv. Colloid Interface Sci.* 305, 102693 (2022).
[156] M. Bompart, K. Haupt, C. Ayela, Micro and Nanofabrication of Molecularly Imprinted Polymers. In: Haupt, K. (eds) *Molecular Imprinting. Topics in Current Chemistry*, vol 325, pp 83–110 Springer, Berlin, Heidelberg (2011).

[157] J. Xu, Y. Wang, S. Hu, *Microchim. Acta* 184, 1 (2017).
[158] A. T. Lawal, *Mater. Res. Bull.* 73, 308 (2016).
[159] Q. Wang, H. Wei, Z. Zhang, E. Wang, S. Dong, *TrAC Trends Anal. Chem* 105, 218 (2018).
[160] L. Liu, H. Jiang, X. Wang, *TrAC Trends Anal. Chem* 143, 116376 (2021).
[161] J. Belza, A. Opletalová, K. Poláková, *Microchim. Acta* 188, 430 (2021).
[162] R. Ghasemi, S. Z. Mirahmadi-zare, M. H. Nasr-Esfahani, A. Allafchian, M. Behmanesh, *Anal. Bioanal. Chem.* 411, 6733 (2019).
[163] V. Naresh, N. Lee, *Sensors* 21, 1109 (2021).
[164] H. Amani, H. Arzaghi, M. Bayandori, A. S. Dezfuli, H. P. Toroudi, A. Shafiee, L. Moradi, *Adv. Mater. Interfaces* 6, 1900572 (2019).
[165] S. Nishitani, T. Sakata, *ACS Appl. Mater. Interfaces* 12, 14761 (2020).
[166] N. Sandhyarani, Chapter 3 - Surface modification methods for electrochemical biosensors, Editor(s): A. A. Ensafi, *Electrochemical Biosensors*, Elsevier, Pages 45-75 (2019).
[167] J. M. Goddard, J. H. Hotchkiss, *Prog. Polym. Sci.* 32, 698 (2007).
[168] J. Smolsky, S. Kaur, C. Hayashi, S. K. Batra, A. V. Krasnoslobodtsev, *Biosensors* 7, 7 (2017).
[169] H. H. Gorris, O. S. Wolfbeis, *Angew. Chem. Int. Ed.* 52, 2 (2013).
[170] Z. Ma, Z. Mao, C. Gao, *Colloids Surf. B : Biointerfaces* 60, 137 (2007).
[171] C. Auría-Soro, T. Nesma, P. Juanes-Velasco, A. Landeira-Viñuela, H. Fidalgo-Gomez, V. Acebes-Fernandez, R. Gongora, M. J. A. Parra, R. Manzano-Roman, M. Fuentes, *Nanomaterials* 9, 1365 (2019).
[172] B. L. Hanssen, S. Siraj, D. K. Y. Wong, *Rev. Anal. Chem.* 35, 1 (2016).
[173] A. M. C. Maan, A. H. Hofman, W. M. de Vos, M. Kamperman, *Adv. Funct. Mater.* 30, 2000936 (2020).
[174] A. J. Tüdős, G. A. J. Besselink, R. B. M. Schasfoort, *Lab. Chip.* 1, 83 (2001).
[175] V. Vaishampayan, A. Kapoor, S. P. Gumfekar, *Can. J. Chem. Eng.* 101, 5208 (2023).
[176] L. Gervais, N. de Rooij, E. Delamarche, *Adv. Mater.* 23, H151 (2011).
[177] F. B. Myers, L. P. Lee, *Lab. Chip.* 8, 2015 (2008).
[178] J. Horne, L. McLoughlin, B. Bridgers, E. K. Wujcik, *Sens. Actuators Rep.* 2, 100005 (2020).
[179] A. Chen, S. Chatterjee, *Chem. Soc. Rev.*42, 5425 (2013).
[180] M. Liu, G. Zeng, K. Wang, Q. Wan, L. Tao, X. Zhang, Y. Wei, *Nanoscale* 8, 16819 (2016).
[181] H. Park, Y. Park, W. Kim, W. Choi, *J. Photochem. Photobiol. C : Photochem. Rev.* 15, 1 (2013).
[182] A. Arenillas, F. Rubiera, J. B. Parra, C. O. Ania, J. J. Pis, *Appl. Surf. Sci.* 252, 619 (2005).
[183] M. Mozetič, *Materials* 12, 441 (2019).
[184] A. Thakur, A. Kumar, S. Kaya, R. Marzouki, F. Zhang, L. Guo, *Coatings* 12, 1459 (2022).
[185] S. Eskandarinezhad, I. A. Wani, M. Nourollahileilan, A. Khosla, T. Ahmad, *J. Electrochem. Soc.* 169, 047504 (2022).

[186] N. Shoaie, M. Daneshpour, M. Azimzadeh, S. Mahshid, S. M. Khoshfetrat, F. Jahanpeyma, A. Gholaminejad, K. Omidfar, M. Foruzandeh, *Microchim. Acta* 186, 465 (2019).

[187] J. H. Kim, Y. J. Suh, D. Park, H. Yim, H., Kim, H. J., Kim, D. S., Yoon, K. S., Hwang, *Biomed. Eng. Lett.* 11, 309 (2021).

[188] K. K. Jain, *Clin. Chim. Acta* 358, 37 (2005).

[189] W. Zhang, S. Bu, J. Zhang, L. Ma, X. Liu, X. Wang, Z. Li, Z. Hao, Z. Li, J. Wan, *Microchem. J.* 175, 107142 (2022).

[190] B. Daurai, M. Gogoi, Point-of-Care Biosensors for Healthcare Applications. In : M. Gogoi, S. Patra, D. Kundu, (eds) *Nanobiosensors for point-of-care medical diagnostics*, Springer, Singapore, pp 1–23 (2022).

[191] W. Cui, J. Li, G. Decher, *Adv. Mater.* 28, 1302 (2016).

[192] M. C. Scicluna, L. Vella-Zarb, *ACS Appl. Nano Mater.* 3, 3097 (2020).

[193] T. Gupte, T. Pradeep, Chapter 3 - Nanosensors for water quality monitoring, Editor(s) : S. Ahuja, *Separation Science and Technology*, Academic Press, Volume 15, Pages 37-53 (2022).

[194] S. Mao, J. Chang, G. Zhou, J. Chen, *Small* 11, 5336 (2015).

[195] P. J. Vikesland, *Nature Nanotech.* 13, 651 (2018).

[196] S. Tajik, H. Beitollahi, F. G. Nejad, Z. Dourandish, M. A. Khalilzadeh, H. W. Jang, R. A. Venditti, R. S. Varma, M. Shokouhimehr, *Ind. Eng. Chem. Res.* 60, 1112 (2021).

[197] A. Kausar, I. Ahmad, T. Zhu, H. Shahzad, M. H. Eisa, *Pollutants* 3, 123 (2023).

[198] K. M. Aguilar-Pérez, M. S. Heya, R. Parra-Saldívar, H. M. N. Iqbal, *Case Stud. Chem. Environ. Eng.* 2, 100055 (2020).

[199] Md. A. Ali, X. Wang, Y. Chen, Y. Jiao, N. K. Mahal, S. Moru, M. J. Castellano, J. C. Schnable, P. S. Schnable, L. Dong, *ACS Appl. Mater. Interfaces* 11, 29195 (2019).

[200] Y. Wang, T. V. Duncan, *Curr. Opin. Biotechnol.* 44, 74 (2017).

[201] M. Lv, Y. Liu, J. Geng, X. Kou, Z. Xin, D. Yang, *Biosens. Bioelectron.* 106, 122 (2018).

[202] C. M. Cova, E. Rincón, E. Espinosa, L. Serrano, A. Zuliani, *Biosensors* 12, 51 (2022).

[203] T. R. Pavase, H. Lin, Q. Shaikh, S. Hussain, Z. Li, I. Ahmed, L. Lv, L. Sun, S. B. H. Shah, M. T. Kalhoro, *Sens. Actuators B: Chem.* 273, 1113 (2018).

[204] G. A. Saqqa, *J. Adv. Res. Food Sci. Nutr.* 3, 24 (2020).

[205] E. Osmólska, M. Stoma, A. S. -Wójcicka, *Sensors* 22, 9956 (2022).

[206] D. Aggarwal, R. Idrishi, Chapter 23 - Nanotechnology applications for food traceability, Editor(s): A. Sharma, P. S. Vijayakumar, P. K Prabhakar, R. Kumar, *Nanotechnology Applications for Food Safety and Quality Monitoring*, Academic Press, Pages 457-472 (2023).

[207] A. Sobhan, K. Muthukumarappan, L. Wei, *Food Packag. Shelf Life.* 30, 100745 (2021).

[208] O. A. Oyewo, E. E. Elemike, D. C. Onwudiwe, M. S. Onyango, *Int. J. Biol. Macromol.* 164, 2477 (2020).

[209] N. B. Singh, A. B. H. Susan,21 - Polymer nanocomposites for water treatments, Editor(s) : M. Jawaid, M. M. Khan, In Woodhead Publishing Series in Composites Science and Engineering, *Polymer-based Nanocomposites for Energy and Environmental Applications*, Woodhead Publishing, Pages 569-595 (2018).

[210] V. B. Yadav, R. Gadi, S. Kalra, *J. Environ. Manage.* 232, 803 (2019).

[211] H. D. Beyene, T. G. Ambaye, Application of Sustainable Nanocomposites for Water Purification Process. In : Inamuddin, S. Thomas, R. K. Mishra, A. M. Asiri, (eds) *Sustainable Polymer Composites and Nanocomposites*. Springer, pp 387–412 (2019).

[212] J. Gu, W. Jiang, F. Wang, M. Chen, J. Mao, T. Xie, *Appl. Surf. Sci.* 301, 492 (2014).

[213] M. Chen, W. Jiang, F. Wang, P. Shen, P. Ma, J. Gu, J. Mao, F. Li, *Appl. Surf. Sci.* 286, 249 (2013).

[214] Y. Luo, Metal-Functional Nucleic Acid Nanocomposite-Based Biosensor. In: *Functional Nucleic Acid-Based Biosensors for Food Safety Detection*. Springer, Singapore, pp 205–219 (2018).

[215] Y. Gao, Y. Wang, Y. Wang, P. Magaud, Y. Liu, F. Zeng, J. Yang, L. Baldas, Y. Song, *TrAC Trends Anal. Chem* 158, 116887 (2023).

[216] H. Nalvuran, A. E. Elçin, Y. M. Elçin, *Int. J. Biol. Macromol.* 114, 77 (2018).

[217] N. S. Kehr, S. Atay, B. Ergün, *Macromol. Biosci.* 15, 445 (2015).

[218] S. K. Si, S. K. Karan, S. Paria, A. Maitra, A. K. Das, R. Bera, A. Bera, L. Halder, B. B. Khatua, *Mater. Chem. Phys.* 213, 525 (2018).

[219] I. Sobianin, S. D. Psoma, A. Tourlidakis, *Energies* 15, 7959 (2022).

[220] C. Zhang, W. Zhao, S. Sasaki, H. Tamiaki, X. F. Wang, *Electrochim. Acta* 347, 136283 (2020).

[221] B. G. Krishna, S. Tiwari, Chapter 19 - Bioinspired solar cells : contribution of biology to light harvesting systems, Editor(s) : M. A. Fraga, D. Amos, S. Sonmezoglu, V. Subramaniam, In *Solar Cell Engineering, Sustainable Material Solutions for Solar Energy Technologies*, Elsevier, Pages 593-632, (2021).

[222] Y. V. M. Reddy, J. H. Shin, V. N. Palakollu, B. Sravani, C. -H. Choi, K. Park, S. -K. Kim, G. Madhavi, J. P. Park, N. P. Shetti, *Adv. Colloid Interface Sci.* 304, 102664 (2022).

[223] B. Lakard, *Appl. Sci.* 10, 6614 (2020).

Chapter 21

An Overview of Nuclear Energy, Along with Its Sustainability, Damage-Control Techniques, and Applications

Subhajit Samaddar*, **PhD**
Debiprasad Jana†
and Malay Purkait‡, **PhD**
Ramakrishna Mission College, Vivekananda Centre for Research
(Autonomous under University of Calcutta), India

Abstract

Stability and instability are two terms that correspond to the storage and emission or intake of energy. The latter one pertains to a change in energy meeting prime interest as an activated type. Nuclear energy falls in the order of MeV. The challenge is to store and use it properly. The energy of the nuclei is transformed through alpha, beta, and gamma decay. The nuclei with mass number above 210 and atomic number above 82 are mostly fit for the transformation via the production of daughter nuclei (Except gamma decay). Amongst the nuclear models, the liquid drop model and the shell model (mostly) assume a Spherical structure. However, the collective model assumes an ellipsoidal structure, due to which quadrupole terms appear because of the motion of loosely bound electrons outside the core. The spherically symmetric potential is best suited for the liquid drop model, but the experimental data support the collective model due to the peripheral nucleons. Again, from the

* Corresponding Author's Email: ssamaddar1988@gmail.com
† Corresponding Author's Email: djana.physics@gmail.com
‡ Corresponding Author's Email: mpurkait_2007@rediffmail.com

In: Breakthrough Research and Technologies in Physics …
Editors: Koyel Ganguly, Soumyadipta Pal and Kamakhya Prasad Ghatak
ISBN: 979-8-90134-111-7

perspective of better stability, the liquid drop model is the right one in ignoring pairing terms. Another crucial decider is the neutron-to-proton ratio.

Moreover, fusion is the source of energy that allows a star to combat the gravitational pull. When the nuclear reactions stop, the evolution proceeds to the end of the stellar evolution step by step. Radiation can be used in physical effects (R.B.E. is noteworthy) and dosimetry as a sustainable process. Nuclear Magnetic Resonance is a popular process that helps archaeology develop, finding missing links through chemical shift. Furthermore, cyclotrons and synchrotrons help increase the velocity of charged particles by a higher order of magnitude. The scope of study encompasses detection and counting in the atomic and subatomic domains. Finite volume effects on atomic spectra (hyperfine structure).

Keywords: radiation, dosimetry, damage-control, sustainability, stability, reaction

Introduction

The most potent force and energy of the universe lies in the nucleus of atoms. The challenge is to control, extract and use it properly. The immense amount of energy may manifest as both a curse and a blessing for society. It is said to be a necessary evil. The elements with atomic numbers greater than 82 are called radioactive elements, as they have decreased life spans with an increase in atomic and mass numbers. The ratio of neutrons to protons, Disintegration constant, arrangement of nucleons, Collision cross-sections in the case of bombardment, Q value, Binding energy, and medium are the controlling parameters. Like chemical reactions, nuclear reactions also obey specific conservation rules [1]. Along with these, nuclear energy may be harnessed sustainably with zero emissions [2]. It has, on average, the same capacity as fossil fuels (e.g. coal, petroleum) and its waste is minimal [3]. However, at the same time, precautionary measures should be taken to ensure that proper claddings are used, thereby avoiding leakage, which may be hazardous. Coolants with high specific heat and conductivity are to be used [4] so that the released thermal energy can be quickly removed from the core and carried to the electric generators [5]. Apart from this, there are many applications of this, ranging from medical science to various other areas.

Classical vs Quantum Context

The nucleus contains protons (+ve charge) and neutrons (neutral). The classical context recalls Coulomb's law in the context of electrostatics. The pairwise repulsion of protons is a logical possibility. So, the challenge is to counteract the question of how the protons are confined and bound together inside such a small radius of the nucleus. Rutherford's experiment [6] is another classical counterpart where the alpha particles are allowed to impinge on the nucleus as projectiles. At a considerable distance from the centre of repulsion, the perpendicular drawn to the asymptotic path of the projectile from the centre of attraction is called the impact parameter. The angle made between the impact parameter before collision and after collision is called the angle of scattering. The classical expression of cross-section occurs from the conservation of angular momentum and energy. Later, the concept was modified by quantum treatment. The alpha disintegration [44] was the first to allow the quantum idea to enter the paradigm. Inside the nuclear radius, the strong force lets the positively charged protons and neutrally charged neutrons remain assembled. It occurs due to the exchange of π-mesons amongst nuclei. When the atomic number of the nucleus increases above 82, the Coulombic repulsion becomes significant, which disturbs the stability of the nucleus. Actually, in the case of a stable nucleus, there is a substantial potential barrier at the centre of the nucleus. But at the instability above, the alpha decay occurs, but the alpha particle does not emit with the kinetic energy required to overcome the barrier. Upon considering quantum mechanics, it is found that although the barrier of a range of the order of $10^{\wedge -14}$ meters is classically impenetrable, quantum mechanically, there lies a finite probability of finding the alpha particle outside the barrier. This is said to be a tunnelling effect [8]. Beyond the potential barrier, the alpha particle encounters the Coulombic repulsion from the residual nucleus, which causes it to drift or increase its emissive velocity.

Stability of Nucleus

To study how a nucleus is, the help of some models has been considered. Among them, stability has been primarily described by the liquid drop model. The binding energy is analogous to the latent heat of vaporisation. The surface tension can maintain the spheroidal shape of the nucleus [9]. But, in case of a heavy nucleus, as the number of protons increases, repulsive forces come into play pairwise, which try to reduce stability. The law of nature also comes into play to implement stability by increasing the number of neutrons. As the number of neutrons outnumbers the number of protons, stability is attained. It goes in the opposite direction of volume energy, which occurs due to the sharing of potential energy among the nucleons in contact in the smallest volume. 'Bethe-Wiizacker's semi-empirical formula' gives a relation between atomic number and mass number [13] which almost accords with magic numbers [10]. The Collective model [11] deviates from the spheroidal structure of the nucleus. The strong coupling among the core nucleons and loosely bound nucleons outside the core causes a pressure resulting in deformation.

For the better reproducibility of stability, a graphical representation of the N (number of neutrons) vs. Z (number of protons) curve is essential. For Z = 20, it follows a linear graph. The nuclei are stable. But, as the number of neutrons increases, the number of neutrons starts to outnumber those given in the previous stanza to balance the Coulombic repulsion. The curve bends leftwards. Radioactive elements (α, β) start emerging from nucleons having Z>82.

To know the extent of nuclear energy, the relation between Packing Fraction and Binding energy [12] is to be known. Both of them can be derived with the help of the other. Binding energy is related to the Packing fraction with the help of a constant of proportionality c^2(c is the velocity of light in a vacuum). What happens in the case of the break-up of a mother nucleus into its constituent nuclei is that there comes about a difference between the mass of the mother nucleus and its constituents when summed separately. This difference of mass is known as the mass defect, which is converted into binding energies of the component nucleons. Packing fraction (f) is a scaling which is given as a ratio between the departure (subtraction) of the mass of a nucleus from its mass number (M-A) to its mass number (A). It's a semi-empirical formula [14]. The difference between M and A may seem unnatural, as its negative value cannot persist for the conversion of binding energy into a stable nucleus. But the normalisation of the difference by the corresponding mass number buffers the idea. Packing fraction may be both positive and negative. Up to the mass number 16, the (f) is non-negative. So, there is no question of stability. The same logic also holds for very massive nuclei (A = 180 to A = 200). In case of A>16, the (f) is negative, and the negative values exhaust the maximum number of nuclei. The f-A curve has a minimum which corresponds to Fe (A=56) as the most stable nucleus. All of the nuclei with mass numbers below 56 and above 56 tend to adopt the nuclear structure of Fe (A=56). Some of the lighter nuclei fuse, which is called fusion, and some heavier nuclei are broken up into two or many lighter products to accord with the condition of stability. The object of both types of nuclei is to release energy to procure a stable state.

Nuclear Reactions

Nuclear reaction is generally a mechanism where a projectile exists that hits a target nucleus, by which one or more particles are formed along with a new nucleus. Moreover, in nuclear reactions, some matter is converted into energy, and vice versa. This is known as the 'Mass-Energy' equivalence. The conservation law exists for neither the mass nor the energy separately; moreover, the conservation of mass-energy [15] occurs here.

There are two types of reactions to a collision. One is an elastic collision, and the other is an inelastic collision. In case of the first one, there is no change in the internal structure and energy state of the reactant particle. The sum of kinetic energies both before and after the collision is the same. But in the case of an inelastic collision, the situation is different. The internal structure or identity, as well as the energy state, may differ before and after the crash occurs. The equations where the 'Mass-Energy' equivalence takes place cannot be an elastic collision. The reaction is said to be exothermic or endothermic depending on the sign of the kinetic energy difference (after and before), which is called the Q value [16]. The energy equation reads as

$$M_x c^2 + (m_a c^2 + E_a) = (M_Y c^2 + E_Y) + (m_b c^2 + E_b) \ldots \qquad 1.(A)$$

The Q value is given by

$$Q = (E_b + E_Y) - E_a \ldots. \qquad 1.(B)$$

where E_a and E_b are kinetic energies of the projectile and emitted particles, whereas E_X is the rest mass energy of the initial nucleus, and E_Y is the recoil kinetic energy of the final nucleus. The equation mentioned above does not occur instantly; there exists some complication. There is an intermediate stage where a compound nucleus is formed, the period of which is of the order of 10^{-15} seconds [17]. The mass of the compound nucleus is-

$$M_c = M_x + m_a \ldots. \qquad 2.(A)$$

and the kinetic energy is-

$$E_c = E_Y + E_b \ldots.. \qquad 2.(B)$$

It is found that to enable the reaction to start, a minimum kinetic energy of the projectile is required, which is referred to as the threshold energy; this threshold energy varies for different responses. Exploiting equation (2), the threshold energy of the projectile is given by [18]-

$$E_{th} = -Q\,(1+\frac{m_a}{M_x}) \ldots. \qquad (3)$$

In case of a non-matter beam projectile (e.g., Gamma ray), the threshold energy is nothing but the negative of the Q value.

The reaction needs more discussion. In case of the bombardment by external particles, the projectiles (parallel beam) impinge on the target slab with the assumption that there is no overlap of nuclei, i.e., the depth of the target slab is minimal yet not negligible (The arrangement is shown in the figure (PPT)). Recapitulating Rutherford's experiment, the lion's share portion of the atom is empty. So, even if the beam falls perpendicular to the target slab, a proportion of the projectile beam does not interact with the incoming beam. They are collected by a detector placed behind the slab. It is found by simple calculus that the number of particles that do not interact shows an exponentially decreasing trend with the depth of the slab. The expression is given as

$$N = N_0\, e^{-\sigma n x} \ldots. \qquad (4)$$

where,

N→ Number of survived particles that did not take part in the reaction

N_0 → The total number of projectile particles in the beam initially

σ → Reaction of the Cross-section of each target nucleus

n → Number of nuclei per unit volume of the target slab

x → Depth of the slab

In the nuclear domain, it is almost impossible to know what is happening. So, the knowledge of probability appears to be indispensable. Reaction Cross-

Section is a probabilistic measure. It is defined as the probability of a reaction to occur if the projectile is shot perpendicular to the target, which consists of one particle per unit area.

Nuclear Magnetic Radiation (NMR) and Chemical Shift

First of all, it needs a classical explanation to discuss, which is "Larmor Precession." If there is a precessing charged particle moving around a centre of force, it produces a magnetic field. When an external magnetic field is applied to the precessing particle, the former magnetic field tries to oppose the latter one. Upon arrival at the steady state, the particle proceeds with a frequency different from the previous frequency. The new frequency is referred to as the Larmor frequency [20]. In the present discussion on nuclear physics, the angular momentum is provided by the spin of the nucleon. When the frequency of the external magnetic field matches that of the nucleon, resonance occurs, resulting in the exchange of energy [21]. The magnitude of the magnetic moment is proportional to the intensity of the applied magnetic field. The transition of a neutron from one energy state to another resembles a laser [22]. Another important word is relaxation [22]. If nucleons in a bulk obtain energies in excess, they may share them with the neighbours, and the time taken to dissipate 37% of energy is called the relaxation time [22]. The physical significance of the relaxation time is the time it takes for particles to transition from an upper energy state to a lower energy state.

Another essential concept is the chemical shift [23], which is also associated with Nuclear Magnetic Resonance (NMR). It pertains to nuclei in a bulk rather than isolated ones. It refers to a cooperative phenomenon [24] that describes how much the magnetic field is affected by the shielding of the bonds. The opposing magnetic field tends to diminish the effect of the exerted one. For example, the shielding constant of the CH bond is greater than that of the OH bond (for O being more electronegative than C); hence, resonance requires a greater magnetic field for the CH bond than the OH bond. Chemical shift is the measure of whether identical nuclei are present or not in the vicinity. The shift of the absorption peak determines the degree of the impurity peak. It also has an impact on medical dosimetry [25].

Nuclear Energy and Sustainability

Nuclear energy is a plentiful source of energy. It is said to be a clean energy [26]. The reason for this lies in the fact that-

- It does not cause air pollution
- It produces a small Land footprint / low installation area
- It generates a minimal quantity of waste

Nuclear energy is preferred due to its stable price and affordability. It is independent of weather, and one can provide a 24/7 supply. The calorific value is very high. One kilogram of Uranium has the calorific value of 2.7 kilograms of coal [27]. The sustainable development occurs when three goals overlap/meet in a common ground [28] –

- Social
- Envoronmetal
- Economic

The nuclear source does not emit carbon dioxide (CO_2) gas or any Greenhouse gas. The energy here can be converted to electric energy without leaving any harmful effects on the surroundings [29]. Uranium is in high abundance on Earth (average concentration being 2.8 parts per million) [30], which can contribute much to sustainable development.

Damage Control and Solutions

It has been stated earlier that nuclear energy can have a devastating effect if it is not adequately controlled. The Three Mile Island (U.S.A.), Fukushima (Japan), and Chernobyl (Ukraine) accidents [31] have taught the world how harmful the occurrence of leakage can be.

Some measures have been taken both nationally and internationally to serve the purpose. The hazards arise when an explosion occurs or transportation/shipping is conducted without taking any precautions. As per [32], the three ways out from already occurred nuclear hazards are-

- To get inside
- To stay inside
- To stay tuned

According to [33]. The U.S.A. has made some international contracts to prevent harmful effects, which are-

- "no 1st use"
- "sole authority"
- "de-altering"
- "smart spending"
- "International Agreement"

Moreover, from the perspective of physics, in the case of chain reactions, a proper moderator and coolant must be used. For shipping purposes, walls should have sufficient reflectance to be implemented. Proper cladding is to be designed so that it can prevent fission products from escaping [34]. Nowadays, some portable reactors (Small Modular Reactor) [29] are also being used for better accessibility with prevention.

Radiation and Dosimetry

The most used area of nuclear physics is Radiobiology. Because of the high energy of Nuclear radiation, it can damage targeted cells and tissues. The key principle of dosimetry lies in the fact that the maximum dose to be applied to the lesion is such that surrounding normal tissues are not affected in any way. The targeted part refers to Planned Targeted Volume (PVT) [36]. In the case of ion-beam therapy, ragg peaks [35, 36] occur, a group of maxima with high to intermediate kurtosis [37] that reverse forward with the beam [as shown in the figure]. As per [ref. My thesis] The projectile velocity is $v_p \sim Z_P^{\frac{3}{2}} v_0$, where Z_P is the charge of the projectile and v_0 is the Bohr velocity [37]. Another necessary quantity is Radio Biological Effect (R.B.E.) [36]. It is an amount of ionic dose equivalent to an X-ray that has the same biological effect as an X-ray. Another critical factor is Linear Energy Transfer (LET) [36], which measures the energy deposition at the point of practical interest.

For X-ray and Gamma-Ray Beams, the intensity of the projected beam falls exponentially with distance [38]. Although the primary energy of the beam is very high, it cannot withstand the stopping power over a considerable distance. Yet, in cancer treatment and cataract removal, the importance of gamma irradiation cannot be denied. These matterless beams mainly ionise a chemical structure. In the case of medical science, a compound structure is formed in the lesion that prevents the symptoms from growing. But, the projected ionic/non-ionic beam disrupts the harmful chemical bond and frees the electron(s). Thus, the cause of the disease may be removed if metals are exposed to high-energy irradiation, causing one or more atoms to be displaced from their equilibrium position, which can lead to defects in the lattice. If the displaced atom is unable to find an equilibrium vacant position, it becomes bound in some interstitial site, and the defect becomes permanent [39].

So, it is evident that the controlled dose of radiation is as good as a blessing, but an overdose may have destructive residues. A chart of dosage is given below [44] –

Dosage in Roentgen (R)	Effect
25-100	Change in blood composition
100-400	Miscellaneous disabilities
>400	50% Lethality
>600	Lethal

Peaceful Applications of Nuclear Energy

Nuclear energy is gigantic in order of magnitude. Although it can be extracted in large amounts with modern technologies, due to the dearth of scientific storage and control, its impact may be devastating. Some of the techniques are-

- Reduction of the size of the reactor
- Using a proper reflector to avoid leakage
- Proper moderation in a chain reactor for chain reactions
- Use coolants with high specific heat and conductivity
- Proper structural and cladding on packages to prevent the escape of fission arrangements
- Shielding of the reactor with an appropriate barrier, like barite concrete [40]

Nuclear energy has immense applications in the following fields-

- Power industries
- Chemical sciences
- Biological effects of radiation in cancer therapy
- Dosimetry
- Magnetic mirror (confinement of plasma) [41, 42]

Conclusion

Now, it's time to compile the above-mentioned ideas and facts. Symmetry and stability are essential factors. Nature loves both the symmetry to occur and the symmetry to break down. Nuclear energy is one of the primary sources of energy used to advance humanity and society. It is prevailing everywhere in the universe, and it is one of the prime factors which control different stages of stardom [43] in the interstitial space. This energy is neither renewable nor inexhaustible. Yet, due to its ability to meet present needs and provide future assurance, it has carved out a niche as a sustainable source of energy [31]. It is noteworthy that fusion is safer than fission reactions, and the waste resulting from fusion reactions is recyclable [29]. Nuclear energy has numerous applications, especially in biology and medical science, which cannot be ignored. That has been given thoroughly. Chemical shift allows one to determine whether similar types of cells are present in a tissue or if any abnormality has occurred.

Finally, the further scope of research will be presented. It is worth mentioning that there are many accelerators and controllers to speed up or slow down the ionic particles. In the case of scattering, counters are used to determine the number of scattered particles. The following fields of research are to be taken care of to develop the future-

- Nature of potential
- Moderators with low absorbance and high reflectance
- Cladding on packets to prevent fission residues from escaping
- Taking care of the behaviour of a particle when it is isolated and when it is in a bulk
- Effect of the nucleus in a molecule/compound
- Statistics and randomness (Radioactivity and Compound nucleus)

- Increasing the sustainability of nuclear energy and channelising the wastes properly

References

[1] *"Nuclear Physics,"* Ghoshal S. N., S. Chand & Company LTD. (2009); ISBN: 81-219-0413-7; pg-437.
[2] https://www.nationalgrid.com>stories>energy-explained.
[3] https://www.iaea.org>sustainability.
[4] *"Nuclear Physics,"* Ghoshal S. N., S. Chand & Company LTD. (2009); ISBN: 81-219-0413-7; pg-727.
[5] Nuclear Reactor Coolants; Suraya Omar, Physics 241; Stanford University, Winter 2011.
[6] *"Nuclear Physics,"* Ghoshal S.N.; S. Chand & Company LTD. (2009); ISBN: 81-219-0413-7; pg-2-7.
[7] *"Quantum Physics"* Gasiorowicz S.; Willey Student Edition (2009); ISBN: 978-81-265-1117-4; pg-302-305.
[8] http://hyperphysics phy-astr. gsu.edu>Nuclear>aiptun.
[9] https://en.wikipedia,org>wiki>Surface_tension.
[10] https://en.wikipedia.org/wiki/Magic_number_(physics).
[11] *"Nuclear Physics,"* Ghoshal S. N., S. Chand & Company LTD. (2009); ISBN: 81-219-0413-7; pg-413.
[12] *"Nuclear Physics,"* Ghoshal S. N., S. Chand & Company LTD. (2009); ISBN: 81-219-0413-7; pg-15-22.
[13] *"Nuclear Physics,"* Ghoshal S. N., S. Chand & Company LTD. (2009); ISBN: 81-219-0413-7; pg-372-379.
[14] https://www.quora.com/What-is-the-difference-between-empirical-and-semi-empirical-equations-formulas-in-physics-and-science-in-general.
[15] https://www.toppr.com>question.
[16] https://en.wikipedia.org/wiki/Q_value_(nuclear_science).
[17] *"Nuclear Physics,"* Ghoshal S. N., S. Chand & Company LTD. (2009); ISBN: 81-219-0413-7; pg-487-490.
[18] *"Nuclear Physics,"* Ghoshal S. N., S. Chand & Company LTD. (2009); ISBN: 81-219-0413-7; pg-445-448.
[19] *"Nuclear Physics,"* Ghoshal S. N., S. Chand & Company LTD. (2009); ISBN: 81-219-0413-7; pg-454-455.
[20] *"Solid State Physics";* Puri R. K., Babbar V.K. , S. Chand & Company LTD.; Third Edition 1997, ISBN: 81-219-1476-0.
[21] *"Fundamentals of Molecular Spectroscopy,"* Banwell C.N., McCash E.M.; Tata McGraw-Hill, Fourth Edition; ISBN-13:978-0-07-462025-0, ISBN-10: 0-07-462025-8; pg-206.

[22] *"Fundamentals of Molecular Spectroscopy,"* Banwell C. N., McCash E. M., Tata McGraw-Hill, Fourth Edition; ISBN-13:978-0-07-462025-0, ISBN-10: 0-07-462025-8; pg-207.

[23] *"Fundamentals of Molecular Spectroscopy,"* Banwell C. N., McCash E. M., Tata McGraw-Hill, Fourth Edition; ISBN-13:978-0-07-462025-0, ISBN-10: 0-07-462025-8; pp. 215- 220.

[24] https://www.sciencedirect.com/topics/chemistry/cooperative-phenomena.

[25] https://www.medicaldosimetry.org/about/medical-dosimetrist/.

[26] https://www.energy.gov/ne/articles/3-reasons-why-nuclear-clean-and-sustainable.

[27] National Grid Group (https://www.nationalgrid.com>stories>energy-explained).

[28] https://world-nuclear.org/information-library/energy-and-the-environment/nuclear-energy-and-sustainable-development.

[29] National Grid Group (https://www.nationalgrid.com>stories>energy-explained).

[30] Search Labs|AI Overviews.

[31] Barry, W. Brook et al. Sustainable Materials and Technologies 1-2 (2014) 8-16.

[32] Ready.Gov; https://www.ready.gov>radiation.

[33] Union of Concerned Scientists; www.ucsusa.org.

[34] *"Nuclear Physics,"* Ghoshal S. N., S. Chand & Company LTD. (2009); ISBN: 81-219-0413-7; pg-728.

[35] https://www.sciencedirect.com/topics/medicine-and-dentistry/bragg-peak.

[36] Schardt D., Elsasser T. (2010); Revs. Mod. Phys. 82 383-425.

[37] *"Basic Statistics,"* Gun A.M., Gupta M.K., Dasgupta B.; The World Press Private Limited Kolkata (2017); ISBN: 978-93-82878-04-9.

[38] https://www.nde-ed.org/Physics/X-Ray/radmatinteraction.xhtml.

[39] *"Nuclear Physics,"* Ghoshal S.N.; S. Chand & Company LTD. (2009); ISBN: 81-219-0413-7; pg-746-747.

[40] *"Nuclear Physics,"* Ghoshal S. N., S. Chand & Company LTD. (2009); ISBN: 81-219-0413-7; pg-717-730.

[41] *"Nuclear Physics,"* Ghoshal S. N., S. Chand & Company LTD. (2009); ISBN: 81-219-0413-7; pg-746-750.

[42] *"Nuclear Physics,"* Ghoshal S. N., S. Chand & Company LTD. (2009); ISBN: 81-219-0413-7; pg-765-766.

[43] *"Mahabishwer Katha"* (Bengali); Surjendrabikash KarMohapatra; Sahityabihar; 1st Edition (1989), ISBN: 978-93-5264-605-0.

[44] *"Nuclear Physics,"* Ghoshal S. N., S. Chand & Company LTD. (2009); ISBN: 81-219-0413-7; pg-100-105.

Chapter 22

Nanotechnology for Environmental Remediation

Sujeet Kumar Pandey*
Department of Chemical and Biochemical Engineering,
Rajiv Gandhi Institute of Petroleum Technology, Jais, Amethi, India

Abstract

Nanotechnology has become a revolutionary method for cleaning up the environment by utilising the distinct characteristics of nanostructures to tackle pollution efficiently. This emerging discipline offers advanced approaches for eliminating diverse pollutants from the atmosphere, aquatic bodies, and land, demonstrating notable benefits compared to conventional methods. The removal of contaminants is a crucial characteristic of this emerging discipline. Nanomaterials, such as zero-valent iron and titanium dioxide, exhibit remarkable efficacy in eliminating heavy metals, organic pollutants, and pathogens owing to their substantial surface area and reactivity. The presence of various structures, including nanoparticles, nanotubes, and polymer composites, enhances the ability to absorb and facilitate chemical reactions, thereby aiding in the breakdown of complex pollutants such as dyes and volatile organic chemicals. Nanomaterials have a dual role in sustainable development considerations.

Although nanomaterials offer advantages, their integration into ecosystems raises concerns about potential toxicity, particularly in aquatic environments, where they can harm marine organisms. The focus of ongoing research is to maintain a delicate equilibrium between the efficient utilisation of nanotechnology and its ecological consequences.

---

* Corresponding Author's Email: 20ce0012@rgipt.ac.in.

In: Breakthrough Research and Technologies in Physics …
Editors: Koyel Ganguly, Soumyadipta Pal and Kamakhya Prasad Ghatak
ISBN: 979-8-90134-111-7

For example, phytoremediation and biostimulation, when coupled with nanotechnology, provide environmentally acceptable alternatives for efficient cleanup. Although nanotechnology offers potential solutions for environmental cleanup, it is crucial to assess its ecological effects to ensure careful and sustainable use.

Keywords: nanotechnology, environmental remediation, nanoparticles

Introduction

Nanotechnology is a revolutionary discipline that alters matter at the nanoscale, often ranging from 1 to 100 nanometers, resulting in distinct physical, chemical, and mechanical characteristics. This subject has transformed several areas, including health, electronics, and materials science, by facilitating innovations such as medication delivery systems, energy-efficient gadgets, new materials, and environmental remediation. The synthesis of nanoparticles can be accomplished through several processes, each offering distinct advantages and applications. The subsequent sections provide an in-depth exploration of the scope, applications, and consequences of nanotechnology. The production of nanoparticles involves several processes, each offering unique benefits and uses. Conventional approaches encompass physical and chemical procedures; however, green synthesis has arisen as a sustainable alternative. Nanoparticles are influenced by several factors that determine their characteristics and applications in various domains. Understanding these aspects is crucial for improving their efficacy in nanotechnology, medicinal applications, and environmental sustainability.

The utilisation of nanotechnology in environmental remediation is significant. Ecological remediation involves several procedures designed to eliminate toxins from the environment, hence safeguarding ecological equilibrium and public health. Principal methodologies encompass phytoremediation, bioremediation, and green chemistry, each providing distinct advantages and uses in mitigating pollution challenges.

Nanotechnology offers advanced methods for environmental remediation, particularly through the use of nanoparticles (NPs) that enhance pollutant removal, providing novel options for addressing pollution in air, water, and soil. The processes encompass adsorption, catalysis, and redox reactions, which are augmented by the distinctive characteristics of nanomaterials, including their raised surface area and reactivity; hence, nanotechnology improves the efficiency and efficacy of conventional remediation techniques. This chapter will examine the fundamental elements of nanotechnology's contribution to environmental cleanup.

Although nanotechnology offers considerable benefits for environmental cleanup, it is essential to address the related hazards and regulatory challenges to ensure its sustainable implementation. Harmonising innovation with ecological safety is a crucial factor in the progression of nanotechnology in this domain.

Mechanisms in Nanotechnology

The subsequent sections delineate essential mechanisms in nanotechnology (Figure 1).

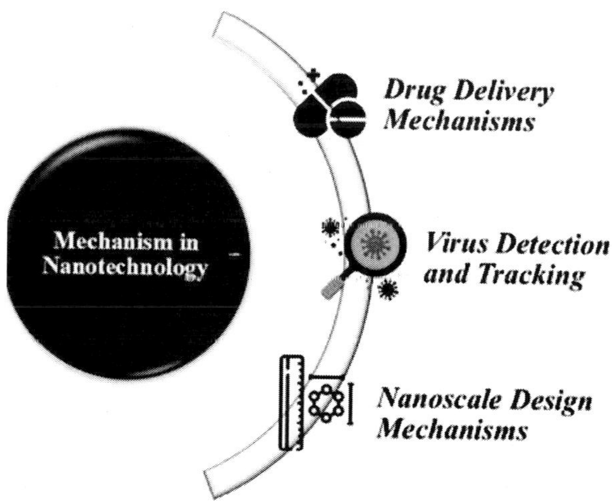

Figure 1. Different mechanisms in nanotechnology.

Nanoscale Mechanisms Design

The design of nanoscale mechanisms in nanotechnology involves the fabrication and manipulation of devices at the molecular level, utilising the distinctive features of nanomaterials. This discipline amalgamates concepts from physics and engineering to create operational nanomachines and nanorobots, which possess substantial applicability across diverse sectors, including healthcare and energy. The subsequent sections delineate essential elements of nanoscale mechanism design. Nanoscale devices exhibit quantum phenomena that influence their behaviour and characteristics. The self-assembly process enables the spontaneous arrangement of nanomaterials into structured shapes, which is crucial for the development of sophisticated methods. Employing proven kinematics, researchers can methodically construct nanomachines from molecular primitives, thereby facilitating regulated motion and function (Chorsi et al., 2021). The advancement of modular nanorobots enables numerous applications, particularly in nanomedicine and industrial operations (Kong et al., 2023).

Despite the considerable benefits offered by breakthroughs in nanotechnology, obstacles persist, including the need for enhanced detection precision and the complexities of nanoscale interactions that can influence functionality and reliability.

Drug Delivery Mechanisms

Nanotechnology has transformed medication delivery systems by improving the effectiveness, targeting, and bioavailability of medicines. This progress is primarily achieved through the development of diverse nanoparticle systems that enable precise medication delivery and controlled release. Diverse nanocarriers, including dendrimers, solid lipid nanoparticles, and polymeric micelles, are utilised for drug delivery, each offering distinct advantages in terms of stability and targeting efficacy (Patra et al., 2018). The interactions of nanoparticles with biological systems are profoundly influenced by their surface characteristics, which can be modified to enhance targeting and mitigate toxicity. Advanced computational models, such as molecular dynamics and pharmacokinetic modelling, are utilised to predict drug behaviour and to improve nanoparticle formulations (Mikayilov et al., 2024). Computational fluid dynamics facilitates the comprehension of nanoparticle mobility in biological systems, ensuring efficient distribution (Mikayilov et al., 2024). Nanocarriers enhance the solubility and stability of pharmaceuticals, leading to improved absorption and therapeutic efficacy (Wei et al., 2020). Functionalised nanoparticles can be engineered to selectively attach to sick cells, thereby reducing side effects and enhancing therapeutic efficacy (Elumalai et al., 2024).

Virus Detection and Tracking

Virus detection and tracking systems in nanotechnology utilise the distinctive characteristics of nanomaterials to improve diagnostic efficacy and facilitate real-time surveillance of viral infections. These developments are crucial for prompt responses to viral epidemics, as they provide enhanced sensitivity, specificity, and rapidity compared to conventional approaches. Frequently utilised nanomaterials encompass gold nanoparticles (AuNPs), magnetic nanoparticles (MNPs), quantum dots (QDs), and carbon-based substances such as graphene and carbon nanotubes (Pirzada & Altintas, 2022; Vitor & Parreira, 2022). Nanobiosensors utilise these materials to enhance signals in the presence of viral biomarkers, thereby improving detection capabilities in point-of-care environments (Vitor & Parreira, 2022). Single Virus Tracking (SVT) technology facilitates the study of individual viral life cycles within host cells, yielding insights into viral dynamics (Kang et al., 2021). Real-time tracking methodologies, such as nanoparticle tracking analysis (NTA) and nano-digital in-line holographic microscopy (Nano-DIHM), facilitate the real-time observation of virus particles, thereby enhancing comprehension of their behaviour in diverse contexts (Pal et al., 2023).

Synthesis of Nanoparticles

The production of nanoparticles involves several processes, each offering unique benefits and uses. Conventional procedures encompass physical and chemical processes, although green synthesis has arisen as a sustainable alternative (Figure 2):

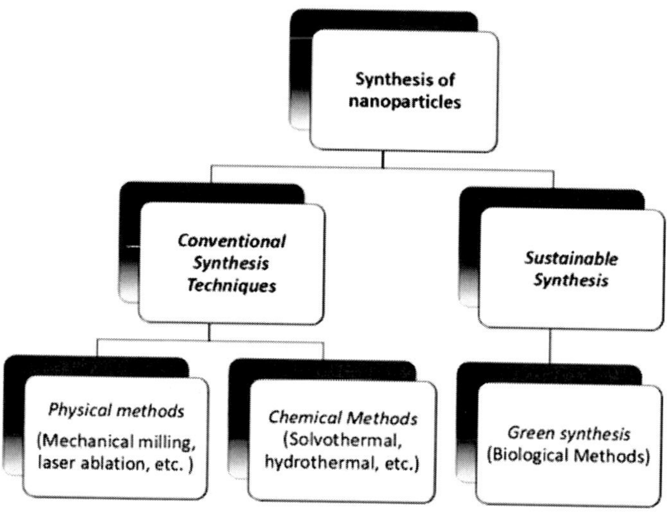

Figure 2. Different methods of nanomaterials synthesis.

Conventional Synthesis Techniques

The synthesis of nanoparticles can be accomplished using various physical and chemical techniques, each offering distinct benefits and applications. These technologies are crucial for synthesising nanoparticles with specific attributes necessary for several domains, including environmental remediation. The following are essential synthesis techniques:

Physical Methods

Mechanical milling is a top-down methodology that reduces bulk materials into nanoparticles by mechanical forces. It is economical and scalable, making it suitable for large-scale manufacturing. The laser ablation technology utilises high-energy lasers to evaporate material, generating nanoparticles in a regulated environment. It enables the exact regulation of particle size and shape, making it suitable for applications that necessitate ligand-free surfaces (Nyabadza et al., 2023). High-energy irradiation techniques utilise radiation to initiate chemical processes, resulting in the synthesis of nanoparticles. This technique can generate nanoparticles with distinctive features owing to the significant energy required. Spray pyrolysis involves atomising a precursor solution into a heated chamber, leading to nanoparticle synthesis through solvent evaporation. It is adaptable and capable of generating nanoparticles with a consistent size and content (Workie et al., 2023).

Chemical Methods

Solvothermal and hydrothermal methods utilise solvents at elevated temperatures and pressures to promote nanoparticle production, resulting in uniform particles (Li & Yu, 2024). The chemical vapour deposition (CVD) procedure involves gaseous reactants that generate solid nanoparticles on a substrate, allowing for meticulous control of particle size and composition (Li & Yu, 2024). Microemulsion and sonochemical methods utilise surfactants or ultrasonic waves to produce nanoscale particles, thereby enhancing the homogeneity and stability of nanoparticles (Nyabadza et al., 2023).

Sustainable Synthesis

The green synthesis of nanoparticles (NPs) is a novel method that utilises biological systems, particularly plant extracts, to produce eco-friendly nanomaterials. This process is becoming increasingly popular since it reduces harmful byproducts and waste, rendering it a sustainable alternative to conventional synthesis techniques. The subsequent sections outline the key aspects of green nanoparticle synthesis.

Biological Methods: Utilising plant extracts and microorganisms for nanoparticle manufacturing is an environmentally sustainable and economically efficient approach. This technique has demonstrated potential in generating metallic nanoparticles with considerable medicinal applications, such as drug delivery and antibacterial properties (Kazemi et al., 2023). It utilises cost-effective botanical resources, thereby reducing production expenses. It generates non-toxic nanoparticles appropriate for biological purposes.

Although conventional techniques provide accuracy, green synthesis offers a sustainable alternative that aligns with environmental objectives, underscoring the necessity for greater improvement in both approaches.

Approaches to Environmental Remediation

Environmental remediation includes several methods used to restore polluted areas. Bioremediation and phytoremediation (Figure 3) emerge as viable, environmentally sustainable techniques that employ biological processes to alleviate pollution.

Bioremediation

Bioremediation is a crucial technique that utilises biological agents, primarily bacteria, to detoxify and eliminate contaminants from the environment. This technology is increasingly preferred due to its sustainability, efficiency, and cost-effectiveness compared to conventional cleanup procedures.

Microorganisms, including bacteria and fungi, metabolise contaminants, transforming them into non-toxic metabolites. Bacteria can decompose hydrocarbons and heavy metals, promoting mineralisation (Dinakarkumar et al., 2024). Some microorganisms, such as cyanobacteria, generate extracellular polymeric compounds that facilitate the immobilisation and sequestration of contaminants, hence improving their detoxifying efficacy (Gu et al., 2023). Bioremediation methods are essential for addressing polluted water, especially in the treatment of pharmaceutical waste (Saeed et al., 2022). Methods such as bioaugmentation and biostimulation are utilised to enhance the breakdown of organic contaminants in soil (Kuppan et al., 2024).

Phytoremediation

Phytoremediation is a novel and environmentally sustainable approach to remediation, utilising the inherent capabilities of plants and algae to absorb, collect, and transform contaminants from soil and water. This technology is highly efficacious against heavy metals, organic pollutants, and radioactive materials, providing a sustainable alternative to traditional remediation procedures. The subsequent sections delineate the significant elements of phytoremediation.

Some plants, such as Noccaea caerulescens, can accumulate heavy metals to levels of 30,000 ppm, demonstrating their capacity as hyperaccumulators (Lavanya et al., 2024). Plants can enhance microbial populations in the rhizosphere, facilitating the breakdown of pollutants (Maurya et al., 2023). Phytoremediation encompasses several techniques, including phytodegradation, rhizodegradation, and phytostabilisation, each offering distinct advantages and applications (Lavanya et al., 2024). The integration of phytoremediation with stabilising approaches has proven to enhance effectiveness in remediating polluted soils, particularly for cadmium and arsenic (Cao et al., 2024).

Despite the acknowledged efficacy of bioremediation and phytoremediation, issues such as identifying suitable species and determining the necessary duration for remediation persist as substantial obstacles to their application.

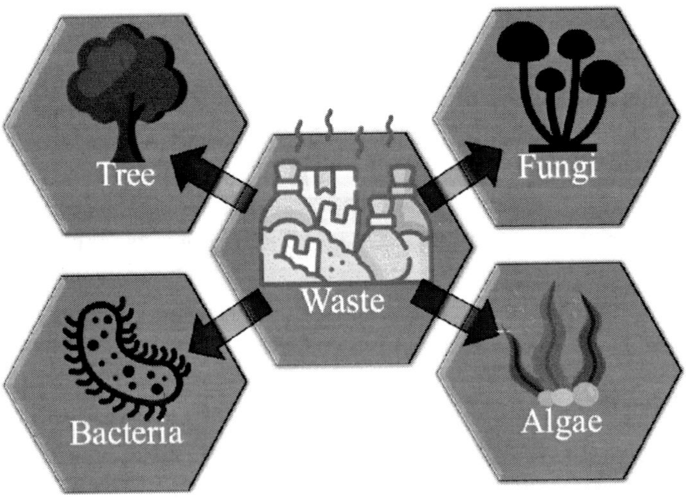

Figure 3. Schematic diagram of bioremediation (bacteria and fungi) and phytoremediation (tree and algae).

Case Studies

Environmental remediation case studies offer valuable insights into the practical application of various technologies and techniques for addressing pollution challenges. These studies highlight the challenges, approaches, and outcomes of remediation initiatives across multiple settings and pollutants. The following sections examine pivotal case examples from the published papers, each demonstrating unique methodologies for environmental cleanup.

Remediation of Total Petroleum Hydrocarbons (TPHs) in these case studies from many nations illustrates the use of technology to remediate areas polluted with total petroleum hydrocarbons (TPHs). These studies tackle practical challenges, including the formulation of remediation objectives and the selection of suitable treatment alternatives. They assess variations in pollutant concentrations before and after treatment, taking into account cost considerations and economic viability (Kuppusamy et al., 2020).

Soil Remediation Techniques, in which soil washing is identified as an efficient technique for remediating polluted soil, including organics, heavy metals, and hydrocarbons. Case studies from Long Island, Saudi Arabia, and Kuwait illustrate its efficacy in addressing various pollution challenges. The method is preferred for its rapidity, compactness, and sustainability attributes (Wilson & Conway, 2024).

Algal-Derived Biomaterials Algae-based systems are being investigated for their potential in carbon dioxide sequestration and wastewater remediation. These systems offer an environmentally sustainable and economical method for cleanup, although obstacles persist in scaling them up to demonstration and pilot systems (Nguyen et al., 2022).

A brownfield site in Michigan exemplifies the significance of comprehensive scientific investigation and efficient management in rehabilitation efforts. The project involved numerous investigative stages and advanced technologies, culminating in a cost-effective solution that achieved all remediation goals (Buehlman et al., 1998).

Nanotechnology's Effect on Environmental Remediation

Nanotechnology has evolved into a revolutionary method for environmental remediation, offering novel strategies to address pollution in various media. Nanotechnology enhances the efficiency and efficacy of conventional remediation techniques by leveraging the distinctive characteristics of nanoparticles, including their high surface area and reactivity.

Enhanced Efficiency

Nanomaterials have enhanced reactivity and adsorption properties, facilitating more efficient pollutant sequestration (Arabzadeh Nosratabad et al., 2024a). Methods like nano-phytoremediation employ plants alongside nanomaterials to collect and neutralise pollutants, enhancing overall remediation efficacy (Dhanapal et al., 2024).

Cost-Effectiveness

Methods utilising nanotechnology are often more cost-effective than traditional cleanup procedures, resulting in a substantial decrease in operational expenses (Rehman et al., 2024). The capacity to customise nanoparticles for particular pollutants increases their effectiveness, resulting in reduced resource expenditure during remediation operations (Ningthoujam et al., 2022).

Eco-Friendliness

Nanoremediation is an environmentally friendly strategy that reduces the ecological impact linked to conventional remediation techniques (Kumar et al., 2024). Utilising biodegradable nanoparticles advances the objective of rehabilitating polluted environments without introducing additional contaminants (Dhanapal et al., 2024).

Although nanotechnology offers considerable benefits for environmental remediation, it is essential to mitigate the associated hazards and promote sustainable practices in its implementation. Harmonising innovation with ecological safety will be crucial for the future of nanotechnology in this domain.

Environmental Applications

Environmental remediation encompasses various methods aimed at removing pollutants from soil, water, and air, thereby restoring and safeguarding ecosystems. The incorporation of new materials, particularly nanotechnology, has significantly enhanced the efficacy of these procedures. This response examines the various applications of environmental remediation, with a focus on nanomaterials, photocatalysis, and biochar modifications.

Nanomaterials, including inorganic, carbon-based, and polymeric types, are used for the remediation of heavy metals, dyes, and organic chemicals owing to their elevated reactivity and surface area-to-volume ratio (Guerra et al., 2018).

Graphene oxide is notably efficient in wastewater treatment, functioning as an adsorbent and membrane modulator to eliminate contaminants such as heavy metals, dyes, and pesticides (Sontakke et al., 2022).

In photocatalytic Applications, transition metal oxides, including TiO2, ZnO, and CuO, serve as photocatalysts for the degradation of organic contaminants in aquatic systems. These materials are economical, stable, and ecologically sustainable (Danish et al., 2021).

Heterogeneous Photocatalysis: This method utilises solar energy to decompose organic contaminants, reduce heavy metals, and eliminate microorganisms, making it a viable solution for air and water purification (Wang et al., 2024).

Modifications of Biochar: Derived from organic matter, biochar is utilised for its adsorptive properties. The incorporation of nanoparticles significantly improves their efficacy in eliminating heavy metals and organic pollutants (Arabzadeh Nosratabad et al., 2024b).

Challenges and Future Directions

Nanotechnology holds substantial promise for environmental cleanup, but it faces particular challenges that must be addressed for successful deployment. The incorporation of nanomaterials (NMs) into remediation systems has the potential to remove pollutants from water and soil; however, challenges such as scalability, cost, and potential ecological consequences persist as significant obstacles. The subsequent sections outline the primary difficulties and potential trajectories in this domain.

Challenges in Nanotechnology for Environmental Remediation

Scalability

The transition from laboratory-scale to field applications is often hindered by challenges in scaling up the production of nanoparticles without compromising their efficacy (Herdiana et al., 2022).

Cost-Effectiveness

The manufacture of nanoparticles often requires expensive ingredients and complex techniques, which can potentially limit their economic feasibility for large-scale applications. The substantial expenses associated with increasing output may deter enterprises from implementing nanotechnology for environmental cleanup.

Environmental Impact

Concerns about the long-term toxicity of nanoparticles on ecosystems are substantial. The potential detrimental impacts on soil organisms and terrestrial flora necessitate a comprehensive examination (Rathod et al., 2024). Regulatory frameworks are continually evolving, which may complicate the approval processes for nanotechnology applications in environmental settings.

Technical Limitations

The integration of nanomaterials with current remediation methods requires a comprehensive study to enhance their efficacy and safety (Mondal & Palit, 2022). The real-time monitoring and regulation of nanoparticle activity in diverse environmental media are little investigated, hindering their practical use (Guo et al., 2023).

Future Directions

Prospects in nanotechnology for environmental cleanup are encouraging, emphasising novel uses and tackling current obstacles. Researchers are increasingly investigating the integration of nanomaterials (NMs) with bioremediation methods, enhancing the efficacy of pollution extraction while minimising ecological impacts. The integration of nanotechnology with bioremediation is anticipated to provide more rapid and efficient remediation procedures (Dhanapal et al., 2024).

Diverse nanomaterials, encompassing carbon-based, metallic, and metal oxide variants, are being engineered for targeted pollutant categories (Asghar et al., 2024). Future study focuses on investigating ecologically friendly and sustainable green synthesis techniques (Osman et al., 2024). Nanomaterials are utilised to enhance wastewater treatment processes, efficiently removing pollutants such as heavy metals and microplastics (Tyagi et al., 2024). The creation of multifunctional nanocomposites is expected to address numerous contaminants concurrently, hence improving treatment efficacy (Tyagi et al., 2024). Subsequent research will focus on elucidating the ecological implications of nanomaterials, ensuring that their use does not compromise ecosystems or human well-being (Isibor et al., 2024). The use of new technologies, including omics and computational modelling, will facilitate the prediction of the behaviour and dangers related to nanomaterials in the environment (Isibor et al., 2024).

Conclusion

Environmental remediation is a crucial process designed to restore damaged environments, thereby protecting human health and ecological integrity. The findings from several research studies underscore the efficacy of various remediation methods, the ethical implications at play, and the necessity for continuous technical progress. It enhances microbial activity to decompose contaminants, thereby efficiently rehabilitating soil health and production. In-situ and ex-situ techniques encompass containment, separation, and decomposition approaches, which are crucial for tackling various contamination challenges. Advanced nanomaterials, including electrospun membranes, are developing as efficient alternatives for air and water filtration. Remediation has been proven to substantially mitigate health hazards, especially in children exposed to lead and chromium, hence illustrating its public health advantages. Ethical issues are crucial, as remediation strategies must address environmental justice and the equitable allocation of cleaning responsibilities.

Nanotechnology has evolved into a revolutionary method for environmental remediation, offering novel approaches to address pollution in air, water, and soil. The distinctive characteristics of nanomaterials, namely their extensive surface area and reactivity, provide effective pollutant elimination and remediation for polluted ecosystems. This chapter will examine the fundamental components of nanotechnology's contribution to environmental restoration. Carbon nanotubes excel at adsorbing heavy metals and organic molecules due to their high aspect ratio and mechanical strength. Nanogels offer precise adsorption functionalities, enabling tailored pollution control. Nanoparticles are utilised in catalysis and chemical reduction, enhancing the effectiveness of water purification systems. Nanobiosorbents and nanobiosurfactants have potential in the degradation of organic pollutants and heavy metals in agricultural soils. Nanophytoremediation utilises nanoparticles to aid plants in detoxifying contaminants, exemplifying an environmentally sustainable method. Nanoremediation frequently demonstrates greater efficiency and reduced costs compared to conventional methods. Different nanomaterials can be customised for specific pollutants, enhancing cleanup techniques.

References

Arabzadeh Nosratabad, N., Yan, Q., Cai, Z., & Wan, C. (2024a). Exploring nanomaterial-modified biochar for environmental remediation applications. *Heliyon, 10*(18), e37123. https://doi.org/10.1016/j.heliyon.2024.e37123.

Arabzadeh Nosratabad, N., Yan, Q., Cai, Z., & Wan, C. (2024b). Exploring nanomaterial-modified biochar for environmental remediation applications. *Heliyon, 10*(18), e37123. https://doi.org/10.1016/j.heliyon.2024.e37123.

Asghar, N., Hussain, A., Nguyen, D. A., Ali, S., Hussain, I., Junejo, A., & Ali, A. (2024). Advancement in nanomaterials for environmental pollutants remediation: a systematic review on bibliometrics analysis, material types, synthesis pathways, and related mechanisms. *Journal of Nanobiotechnology*, *22*(1), 26. https://doi.org/10.1186/s12951-023-02151-3.

Buehlman, M. D., Rogers, D. T., & Payne, F. C. (1998). Science plus management equals successful remediation: A case study. *Environmental Progress*, *17*(2), 111–119. https://doi.org/10.1002/ep.670170218.

Cao, J., Tan, Y., & Zhang, C. (2024). Combined Approaches for the Remediation of Cadmium- and Arsenic-Contaminated Soil: Phytoremediation and Stabilization Strategies. *Applied Sciences*, *14*(16), 7144. https://doi.org/10.3390/app14167144.

Chorsi, M. T., Tavousi, P., Mundrane, C., Gorbatyuk, V., Kazerounian, K., & Ilies, H. (2021). Kinematic Design of Functional Nanoscale Mechanisms From Molecular Primitives. *Journal of Micro and Nano-Manufacturing*, *9*(2). https://doi.org/10.1115/1.4051472.

Danish, M. S. S., Estrella, L. L., Alemaida, I. M. A., Lisin, A., Moiseev, N., Ahmadi, M., Nazari, M., Wali, M., Zaheb, H., & Senjyu, T. (2021). Photocatalytic Applications of Metal Oxides for Sustainable Environmental Remediation. *Metals*, *11*(1), 80. https://doi.org/10.3390/met11010080.

Dhanapal, A. R., Thiruvengadam, M., Vairavanathan, J., Venkidasamy, B., Easwaran, M., & Ghorbanpour, M. (2024). Nanotechnology Approaches for the Remediation of Agricultural Polluted Soils. *ACS Omega*, *9*(12), 13522–13533. https://doi.org/10.1021/acsomega.3c09776.

Dinakarkumar, Y., Ramakrishnan, G., Gujjula, K. R., Vasu, V., Balamurugan, P., & Murali, G. (2024). Fungal bioremediation: An overview of the mechanisms, applications and future perspectives. *Environmental Chemistry and Ecotoxicology*, *6*, 293–302. https://doi.org/10.1016/j.enceco.2024.07.002.

Elumalai, K., Srinivasan, S., & Shanmugam, A. (2024). Review of the efficacy of nanoparticle-based drug delivery systems for cancer treatment. *Biomedical Technology*, *5*, 109–122. https://doi.org/10.1016/j.bmt.2023.09.001.

Gu, T., Lu, Y., Li, F., Zeng, W., Shen, L., Yu, R., & Li, J. (2023). Microbial extracellular polymeric substances mitigate cadmium toxicity in rice (Oryza sativa L.) by modulating cadmium uptake, subcellular distribution, and inducing the expression of stress-related genes. *Ecotoxicology and Environmental Safety*, *257*, 114958. https://doi.org/10.1016/j.ecoenv.2023.114958.

Guerra, F. D., Attia, M. F., Whitehead, D. C., & Alexis, F. (2018). Nanotechnology for Environmental Remediation: Materials and Applications. *Molecules*, *23*(7), 1760. https://doi.org/10.3390/molecules23071760

Guo, Y., Walter, V., Vanuytsel, S., Parperis, C., Sengel, J. T., Weatherill, E. E., & Wallace, M. I. (2023). Real-Time Monitoring and Control of Nanoparticle Formation. *Journal of the American Chemical Society*, *145*(29), 15809–15815. https://doi.org/10.1021/jacs.3c02484.

Herdiana, Y., Wathoni, N., Shamsuddin, S., & Muchtaridi, M. (2022). Scale-up polymeric-based nanoparticles drug delivery systems: Development and challenges. *OpenNano*, *7*, 100048. https://doi.org/10.1016/j.onano.2022.100048.

Isibor, P. O., Kayode-Edwards, I. I., & Taiwo, O. S. (2024). Emerging Technology and Future Directions in Environmental Nanotoxicology. In *Environmental Nanotoxicology* (pp. 325–346). Springer Nature Switzerland. https://doi.org/10.1007/978-3-031-54154-4_16.

Kang, J., Tahir, A., Wang, H., & Chang, J. (2021). Applications of nanotechnology in virus detection, tracking, and infection mechanisms. *WIREs Nanomedicine and Nanobiotechnology*, *13*(4). https://doi.org/10.1002/wnan.1700.

Kazemi, S., Hosseingholian, A., Gohari, S. D., Feirahi, F., Moammeri, F., Mesbahian, G., Moghaddam, Z. S., & Ren, Q. (2023). Recent advances in green synthesised nanoparticles: from production to application. *Materials Today Sustainability*, *24*, 100500. https://doi.org/10.1016/j.mtsust.2023.100500.

Kong, X., Gao, P., Wang, J., Fang, Y., & Hwang, K. C. (2023). Advances of medical nanorobots for future cancer treatments. *Journal of Hematology & Oncology*, *16*(1), 74. https://doi.org/10.1186/s13045-023-01463-z.

Kumar, R., Kaushal, S., Verma, N., Kumar, P., Thakur, N., Kumar, A., Kumar, S., Umar, A., Almas, T., Pal, K., Baskoutas, S., Kumar, K., & Thakur, N. (2024). Nano bioaugmentation for textile dye remediation: A sustainable approach for health and environment management. *Journal of Molecular Liquids*, *415*, 126254. https://doi.org/10.1016/j.molliq.2024.126254.

Kuppan, N., Padman, M., Mahadeva, M., Srinivasan, S., & Devarajan, R. (2024). A comprehensive review of sustainable bioremediation techniques: Eco friendly solutions for waste and pollution management. *Waste Management Bulletin*, *2*(3), 154–171. https://doi.org/10.1016/j.wmb.2024.07.005.

Kuppusamy, S., Maddela, N. R., Megharaj, M., & Venkateswarlu, K. (2020). Case Studies on Remediation of Sites Contaminated with Total Petroleum Hydrocarbons. In *Total Petroleum Hydrocarbons* (pp. 225–256). Springer International Publishing. https://doi.org/10.1007/978-3-030-24035-6_9.

Lavanya, M. B., Viswanath, D. S., & Sivapullaiah, P. V. (2024). Phytoremediation: An eco-friendly approach for remediation of heavy metal-contaminated soils-A comprehensive review. *Environmental Nanotechnology, Monitoring & Management, 22*, 100975. https://doi.org/10.1016/j.enmm.2024.100975.

Li, Z., & Yu, C. (2024). Synthesis methods and paradigms of nanomaterials. In *Nanostructured Materials* (pp. 145–183). Elsevier. https://doi.org/10.1016/B978-0-443-19256-2.00030-2.

Maurya, A., Sharma, D., Partap, M., Kumar, R., & Bhargava, B. (2023). Microbially-assisted phytoremediation toward air pollutants: Current trends and future directions. *Environmental Technology & Innovation*, *31*, 103140. https://doi.org/10.1016/j.eti.2023.103140.

Mikayilov, E., Zeynalov, N., Taghiyev, D., & Taghiyev, S. (2024). Role of Computational Modeling in the Design and Development of Nanotechnology-based Drug Delivery Systems. *Chemical and Biochemical Engineering Quarterly*, *2*. https://doi.org/10.15255/CABEQ.2024.2290.

Mondal, S., & Palit, D. (2022). Prospects and implementation of nanotechnology in environmental remediation and clean up. In *Natural Resources Conservation and Advances for Sustainability* (pp. 271–287). Elsevier. https://doi.org/10.1016/B978-0-12-822976-7.00020-X.

Nguyen, T. A. H., Ngo, H. H., & Nguyen, M. K. (2022). Case studies of algae-based biomaterials for environmental remediation. In *Algae-Based Biomaterials for Sustainable Development* (pp. 291–320). Elsevier. https://doi.org/10.1016/B978-0-323-96142-4.00007-5.

Ningthoujam, R., Singh, Y. D., Babu, P. J., Tirkey, A., Pradhan, S., & Sarma, M. (2022). Nanocatalyst in remediating environmental pollutants. *Chemical Physics Impact*, *4*, 100064. https://doi.org/10.1016/j.chphi.2022.100064.

Nyabadza, A., McCarthy, É., Makhesana, M., Heidarinassab, S., Plouze, A., Vazquez, M., & Brabazon, D. (2023). A review of physical, chemical and biological synthesis methods of bimetallic nanoparticles and applications in sensing, water treatment, biomedicine, catalysis and hydrogen storage. *Advances in Colloid and Interface Science*, *321*, 103010. https://doi.org/10.1016/j.cis.2023.103010.

Osman, A. I., Zhang, Y., Farghali, M., Rashwan, A. K., Eltaweil, A. S., Abd El-Monaem, E. M., Mohamed, I. M. A., Badr, M. M., Ihara, I., Rooney, D. W., & Yap, P.-S. (2024). Synthesis of green nanoparticles for energy, biomedical, environmental, agricultural, and food applications: A review. *Environmental Chemistry Letters*, *22*(2), 841–887. https://doi.org/10.1007/s10311-023-01682-3.

Pal, D., Amyot, M., Liang, C., & Ariya, P. A. (2023). Real-time 4D tracking of airborne virus-laden droplets and aerosols. *Communications Engineering*, *2*(1), 41. https://doi.org/10.1038/s44172-023-00088-x.

Patra, J. K., Das, G., Fraceto, L. F., Campos, E. V. R., Rodriguez-Torres, M. del P., Acosta-Torres, L. S., Diaz-Torres, L. A., Grillo, R., Swamy, M. K., Sharma, S., Habtemariam, S., & Shin, H.-S. (2018). Nano based drug delivery systems: recent developments and future prospects. *Journal of Nanobiotechnology*, *16*(1), 71. https://doi.org/10.1186/s12951-018-0392-8.

Pirzada, M., & Altintas, Z. (2022). Nanomaterials for virus sensing and tracking. *Chemical Society Reviews*, *51*(14), 5805–5841. https://doi.org/10.1039/D1CS01150B.

Rathod, P. B., Singh, M. P., Taware, A. S., Deshmukh, S. U., Tagad, C. K., Kulkarni, A., & Kanagare, A. B. (2024). Comprehensive insights into water remediation: chemical, biotechnological, and nanotechnological perspectives. *Environmental Pollutants and Bioavailability*, *36*(1). https://doi.org/10.1080/26395940.2024.2329660.

Rehman, S. A., Hussain, N., Shahbaz, A., Smułek, W., & Jesionowski, T. (2024). Nanotechnology for remediation of oilfield and refineries wastewater. In *Nanotechnology for Oil-Water Separation* (pp. 83–104). Elsevier. https://doi.org/10.1016/B978-0-323-95517-1.00004-4.

Saeed, M. U., Hussain, N., Sumrin, A., Shahbaz, A., Noor, S., Bilal, M., Aleya, L., & Iqbal, H. M. N. (2022). Microbial bioremediation strategies with wastewater treatment potentialities – A review. *Science of The Total Environment*, *818*, 151754. https://doi.org/10.1016/j.scitotenv.2021.151754.

Sontakke, A. D., Mondal, P., & Purkait, M. K. (2022). *Graphene Oxide-Based Advanced Nanomaterials for Environmental Remediation Applications* (pp. 155–190). https://doi.org/10.1007/978-3-031-11996-5_6.

Tyagi, S., Kapoor, R. T., Solanki, S., Goyal, A., & Singh, R. (2024). Nanomaterial mediated wastewater treatment: a new frontier in environmental remediation. In *Microbiome-Based Decontamination of Environmental Pollutants* (pp. 31–49). Elsevier. https://doi.org/10.1016/B978-0-443-21781-4.00009-8.

Vitor, P. A. M., & Parreira, G. M. (2022). *Nanomaterials for Viral Detection* (pp. 47–69). https://doi.org/10.1007/978-3-030-86901-4_3.

Wang, L., Zhang, W., & Zhu, Z. (2024). Applications in environmental remediation. In *Full-Spectrum Responsive Photocatalytic Materials* (pp. 157–182). Elsevier. https://doi.org/10.1016/B978-0-443-13631-3.00003-3.

Wei, Q.-Y., Xu, Y.-M., & Lau, A. T. Y. (2020). Recent Progress of Nanocarrier-Based Therapy for Solid Malignancies. *Cancers*, *12*(10), 2783. https://doi.org/10.3390/cancers12102783.

Wilson, Andrew, & Conway, Amy. (2024, November 4). Soil Washing Case Studies within the Remediation Sector. *ADIPEC*. https://doi.org/10.2118/222965-MS.

Workie, A. B., Ningsih, H. S., & Shih, S.-J. (2023). A comprehensive review on the spray pyrolysis technique: Historical context, operational factors, classifications, and product applications. *Journal of Analytical and Applied Pyrolysis*, *170*, 105915. https://doi.org/10.1016/j.jaap.2023.105915.

Chapter 23

Integrative Research in Microbiology, Molecular Biology, and Nanotechnology: A Multidisciplinary Approach to Cancer Biology, Biofilm Management, and Diagnostic Sustainability

Arnab Sengupta[1]
Saikat Mazumder[1,2]
Moupriya Nag[1,*]
and Dibyajit Lahiri[1,†]

[1]Department of Biotechnology, Institute of Engineering & Management, University of Engineering and Management, Kolkata, West Bengal, India
[2]Department of Food Technology, Guru Nanak Institute of Technology, Kolkata, West Bengal, India

Abstract

With convergence in the domains of microbiology, molecular biology, and nanotechnology, significant advancements are expected in diagnostics, therapeutics, and health management. The current review encompasses the molecular basis of cancer biology, bacterial biofilm formation, and their clinical implications, as well as the green synthesis of gold and silver nanoparticles, their applications in diagnostic innovation, and their use in therapeutic strategies. Nanotechnology is becoming increasingly crucial in developing advanced diagnostics and

* Corresponding Author's Email: moupriya.nag@uem.edu.in
† Corresponding Author's Email: dibyajit.lahiri@uem.edu.in

In: Breakthrough Research and Technologies in Physics …
Editors: Koyel Ganguly, Soumyadipta Pal and Kamakhya Prasad Ghatak
ISBN: 979-8-90134-111-7

targeted therapies for solid and blood tumours, which are characterised by genetic and epigenetic alterations. At the same time, bacterial biofilms responsible for persistent infections and antibiotic resistance require alternative management strategies that integrate natural compounds and nanoparticle applications. A promising solution in this direction is the green synthesis of nanoparticles, which ensures biocompatibility and multifunctionality in biomedical applications. This review illustrates how the interdisciplinary approach to collaborative research addresses complex healthcare challenges by examining these domains. Future perspectives underscore the importance of integrating these fields with artificial intelligence and precision medicine to achieve transformative outcomes in health management.

Keywords: cancer biology, bacterial biofilms, molecular diagnostics, green nanotechnology, health management.

Introduction

Healthcare faces broadened challenges that require innovative and interdisciplinary solutions. Cancer, microbial infections, and inadequate diagnostic accuracy are factors that significantly impact health outcomes worldwide. Contemporary health issues related to cancer treatment, bacterial infection, and low diagnostic accuracy require interdisciplinary solutions. This disease still ranks as a leading cause of death globally, with more than 19.3 million new cases and 10 million deaths recorded in 2020, according to the World Health Organisation, and millions of cases diagnosed every year [1]. Its complexity arises from genetic mutations, dysregulated signalling pathways, and epigenetic modifications [2]. Similarly, bacterial biofilms complicate the clinical scenario by resisting regular treatments and facilitating chronic infections [3]. Meanwhile, the synthesis and application of green nanoparticles open up promising avenues for eco-friendly and effective healthcare solutions [4]. Simultaneously, the emergence of antibiotic-resistant bacterial infections has compounded the challenges for healthcare, with biofilm-associated infections accounting for a significant proportion of chronic infections [3]. These key findings underscore the need to incorporate ideas from microbiology, molecular biology, and nanotechnology to develop high-tech diagnostics, targeted therapies, and sustainable approaches to health management. These challenges underscore the importance of microbiology in establishing a fundamental understanding of bacterial behaviour and human-microbe interactions. Meanwhile, molecular biology and genetics will elucidate the mechanisms that govern disease progression, providing the basis for understanding microbial dynamics, including the processes of biofilm formation, microbial resistance, and host-microbe interactions. These insights are vital in addressing infections and the role of the microbiome in health and disease [5]. Indeed, in the field of molecular biology, the rapid pace of technological advances, including next-generation sequencing (NGS), CRISPR-Cas gene-editing systems, and transcriptomic analysis, has enabled a deeper understanding of the genetic alterations and regulation involved in the pathogenesis of diseases such as cancer [34]. For instance, the identification of tumour-specific mutations and epigenetic modifications has enabled the advancement of precision medicine, leading to improved outcomes for patients [6, 7].

Nanotechnology has revolutionised the biomedical field through innovative tools that provide diagnostic precision as well as therapeutic efficacy in problems associated with diagnostics and targeted therapy. Nanoparticle synthesis, particularly through green chemistry approaches, offers environmentally sustainable solutions that enhance biocompatibility and reduce toxicity [8]. Indeed, gold and silver nanoparticles have raised hopes for diagnostic applications in imaging, drug delivery, and antimicrobial treatments [9]. The integration of all these fields provides a synergistic approach to addressing healthcare issues, enabling strides that would not be possible with isolated disciplinary undertakings.

This article also examines the interconnection between microbiology, molecular biology, and nanotechnology in addressing the most pressing healthcare challenges. In detail, it focuses on three themes: (1) molecular mechanisms of solid and blood cancer development and progression; (2) the roles of bacterial biofilms in infection and resistance; and (3) green synthesis and biomedical applications of nanoparticles. Through the topics under discussion, this review will highlight the importance of interdisciplinary research in generating innovative solutions to address some of the world's current and future healthcare demands.

Molecular Biology and Cancer Research

Tumour Development: Genetic and Molecular Insights

Due to the nature of gene and epigenetic alterations, cancer causes abnormal cell signalling, allowing cells to grow beyond their limits and proliferate and metastasise at will. The proto-oncogenes, such as KRAS and EGFR, become mutated, while loss of function occurs in tumour suppressor genes, including TP53 and RB1, leading to tumour initiation and progression [10]. The abnormal pillars of the cancer hallmarks are associated with regulated pathways, such as PI3K/AKT/mTOR and Wnt/β-catenin, which are involved in sustained proliferative signalling, Immune evasion, and Angiogenesis [11].

Novel developments, such as next-generation sequencing (NGS) and single-cell RNA sequencing (scRNA-seq), have significantly advanced cancer research by enhancing the accuracy of detecting genetic mutations and the heterogeneity of tumours within them [6]. For instance, individualised therapies against mutations specific to non-small cell lung cancer, such as EGFR inhibitors, have shown considerable clinical efficacy [12]. Circulating tumour DNA (ctDNA) fluid biopsy detection was an approach to non-invasive monitoring of tumour dynamics and treatment response [13]. Epigenetics, such as DNA methylation and histone modification, is an equally important part of cancer progression. The aberrant methylation of suppressor genes, such as the BRCA1 gene, is often seen in breast and ovarian tumours [14]. The development of novel epigenetic interventions, including DNA methyltransferase inhibitors, has broadened the spectrum of cancer therapeutics [15].

Microbial Influence on Cancer

The human microbiome, particularly in its relation to cancer, plays a significant role in the initiation and progression of the disease. Although Helicobacter pylori are associated with gastric cancers, Fusobacterium nucleatum has been implicated in colorectal cancers [16]. The microorganisms that cause carcinogenesis include chronic inflammation, DNA damage, and disturbance of the immune system. For example, the reactive oxygen species produced by H. pylori cause genetic instability and tumorigenesis in gastric epithelial cells [17]. Emerging evidence also suggests that gut microbiomes can enhance the efficacy of immunotherapy against cancer. For instance, Akkermansia muciniphila and similar bacterial species have been found to improve the therapeutic effectiveness of immune checkpoint inhibitors by activating host immune pathways [16]. Such interactions may provide avenues for targeting interventions that reduce the influence of most cancers.

Bacterial Biofilms and Healthcare Challenges

Formation and Clinical Impact of Biofilms

Bacterial biofilms are composed of coalesced and attached microbes, creating communities called biofilms, which are composed of the final part of intracellular polymeric substances (EPS). This arrangement acts as a barrier against antibiotic and immune system attacks, and is a significant problem in every clinical aspect [18]. Infections caused by biofilms are chronic and often associated with medical devices, such as catheters and prosthetic implants, accounting for approximately 80% of chronic diseases [19]. The genetics of biofilm formation occur through quorum sensing (QS)-type systems, including lasR, luxS, and icaA genes [20]. The QS acyl-homoserine lactones enable bacteria to communicate and synchronise biofilm formation [21]. Biofilms are considered one of the essential contributors to antibiotic resistance. The EPS matrix blocks drug penetration, and dormant cells in biofilms exhibit a metabolism decreased by an inactive state, allowing them to become tolerant to antibiotics [22].

Strategies for Biofilm Control

Antimicrobial approaches are necessary to combat biofilm-related infections that cannot be effectively treated with conventional antibiotics. Natural compounds from plants, such as curcumin, eugenol, and flavonoids, have anti-biofilm properties due to their interference with QS signalling and EPS synthesis [23]. Curcumin inhibits biofilm formation in Pseudomonas aeruginosa, a species of bacteria, from QS-regulated genes [24]. Nanotechnology has more to offer. Green-synthesised AgNPs are potent biofilm disruptors due to the production of ROS and disruption of the bacterial membrane [25]. Nanoparticles combined with antibiotics enhance antimicrobial activity by penetrating biofilms [26].

Green Synthesis of Gold and Silver Nanoparticles

Sustainable Nanotechnology: Methods and Benefits

The greener synthesis of nanoparticles is emerging as an eco-friendly, cost-effective, and sustainable alternative to traditional chemical and physical methods of nanoparticle synthesis. In this case, biological agents such as plant extracts, microorganisms, algae, and maybe even biomolecules like proteins and enzymes, are used to synthesise gold and silver nanoparticles [25]. Green synthesis relies on natural compounds that possess intrinsic reducing and capping properties, such as polyphenols, flavonoids, and alkaloids, unlike chemical methods, which require hazardous reducing agents and stabilisers to synthesise nanoparticles. Commonly found plants suitable for nanoparticle design production include Azadirachta indica (neem), Ocimum sanctum (tulsi), and Allium sativum (garlic), which have shown great promise for antimicrobial and anticancer effects [27, 28]. Whereas microorganisms, such as Bacillus subtilis and Fusarium oxysporum, are known to produce nanoparticles by reducing metal ions from aqueous solutions [29]. The bio-reducing agents enhance the compatibility and stability of the nanoparticles, making them a candidate for biomedical applications.

Thus, green synthesis methods promote sustainability through the use of renewable source material and reduction in waste generation. It aligns with the world's efforts to minimise environmental impacts and promote the development of eco-friendly technology. Furthermore, nanoparticles synthesised via greener routes have been shown to exhibit excellent size uniformity, high stability, and improved surface functionalisation, making them ideal candidates for diagnosis, drug delivery, and antimicrobial applications [30].

Biomedical Applications

Diagnostics and therapeutics applications have been derived from the green-synthesised nanoparticles. For example, gold-nanoparticles-conjugated antibody moieties for image formation in cancer, and silver nanoparticles used as antimicrobial agents in wound dressing and coatings for medical devices [31, 32]. Biomedical scientists have extensively studied green-synthesised gold and silver nanoparticles to take advantage of their exciting physicochemical properties. Due to their outstanding optical, electronic, and thermal properties, gold nanoparticles (AuNPs) have been designed explicitly for imaging diagnosis, photothermal therapy, and biosensing applications [33]. For example, the application of advanced imaging techniques, such as surface-enhanced Raman spectroscopy (SERS), ensures that antibody-functionalised AuNPs preferentially target the tumour biomarker, making it possible for early cancer diagnosis [34]. The antimicrobial properties associated with silver nanoparticles (AgNPs) are highly potent. They exhibit the capability of producing reactive oxygen species (ROS) that disrupt the integrity of bacterial cell membranes and suppress biofilm formation. This action entails chronic infection treatment [35]. AgNPs are also applied as components included in wound dressings and surgical instruments, as well as in coatings developed for medical equipment to establish a barrier against nosocomial infections [36].

New applications of these nanoparticles are currently being explored, including their incorporation into drug delivery systems. By functionalising nanoparticles, drug solubility and stability can be improved, in addition to facilitating additive or targeted delivery and minimising side effects [47]. For example, green-synthesised AuNPs, which were loaded with doxorubicin, an anticancer drug, were found to be specific and less toxic in preclinical tests [48].

Integration of Nanotechnology in Diagnostics and Health Management Sustainability

Advanced Diagnostic Tools

Nanotechnology-based biosensors and imaging platforms enable the rapid and accurate detection of cancer markers and pathogens, revolutionising diagnostic technologies by creating platforms with unprecedented sensitivity, specificity, and speed. For example, CRISPR-Cas systems can be integrated with nanoparticles and perform accurate genetic diagnostics, therefore changing the processes of managing infectious diseases. Gold and silver nanoparticles play a significant role in the manufacture of biosensors, imaging agents, and point-of-care diagnostic devices. The specific use of mitochondria is exemplified by the development of AuNP-based biosensors in combination with CRISPR-Cas systems for detecting specific nucleic acid mutations, infectious agents, and other disease biomarkers [37]. For example, silver nanoparticles excel in colourimetric biosensing applications due to their intrinsic optical properties, which enable the visual detection of analytes without the need for complicated instruments [38]. The biosensors above also detect pathogens such as Escherichia coli and Salmonella species, as well as tumour-associated proteins [39]. They have also enabled the development of ultrasensitive diagnostic methodologies to detect cancer early and monitor its progression, thanks to their ability to enhance fluorescence and Raman signals [33].

The applications of artificial intelligence (AI) to nanoparticle-based diagnostics represent a brand-new frontier. Machine learning algorithms will allow rapid interpretation with improved accuracy of complex data.

Datasets generated by these nanodiagnostic platforms [40]. For example, AI-augmented imaging systems based on AuNP contrast agents showed more accurate detection of small tumours and metastases [41]. The following table illustrates the various techniques employed in the relevant fields (Table 1).

Applications in Health Management

Nanotechnology also significantly contributes to the improvement of health management, other than diagnosis. Nanoparticles can create an intelligent drug delivery system that delivers the drug directly to target tissues or cells, resulting in minimal off-target effects. Ligands, such as peptides or antibodies, are then directed to the disease site, where they deliver the therapeutic agent [53]. Silver nanoparticles have emerged as a potent solution against infections caused by multidrug-resistant bacteria. They diffuse into biofilms and dissolve the bacterial membrane; hence, they have proven essential against chronic diseases associated with implants and catheters [36]. Additionally, the use of nanoparticles in PPE and air filtration systems gained prevalence during the COVID-19 pandemic, where they helped enhance protection against viruses [54]. Green-synthesised nanoparticles contribute to environmental health by lessening the ecological footprints associated with conventional drug production and waste disposal. Another promising use is the application of these in water purification systems for removing pathogens and heavy metals [55]. As the breakthrough continues, the penetration of nanotechnology into healthcare management is expected to transform the delivery and accessibility of health services worldwide.

Table 1. Summarising diagnostic and laboratory techniques used in microbiology, molecular biology, nanotechnology, and cancer research, with a focus on biofilm and nanoparticle-related applications

Field	Diagnostic/Laboratory Technique	Description	Applications	Refs
Microbiology	Gram Staining	A method used to differentiate bacterial species into two groups based on the composition of their cell wall.	Bacterial identification and biofilm formation assessment.	[42]
	PCR (Polymerase Chain Reaction)	Amplifies DNA sequences to detect bacterial or viral infections.	Detection of pathogens and biofilm-related genes.	[43]
	Culturing and Plating	Growing microorganisms on selective media to identify bacterial species.	Biofilm-forming bacterial strain isolation and resistance testing.	[43]
Molecular Biology	Western Blotting	Detects specific proteins based on their size and affinity for antibodies.	Identifying biofilm-related proteins and molecular markers in cancer cells.	[44]
	RT-PCR (Reverse Transcription PCR)	Quantifies RNA to assess gene expression profiles.	Analysing gene expression in biofilm formation and cancer progression.	[45]
	Next-Generation Sequencing (NGS)	High-throughput sequencing of DNA and RNA to detect mutations and gene expression patterns.	Cancer biomarker discovery, biofilm-forming gene identification.	[17]
Nanotechnology	Dynamic Light Scattering (DLS)	Measures particle size distribution of nanoparticles.	Characterising nanoparticles for drug delivery systems and biofilm disruption.	[46]
	Atomic Force Microscopy (AFM)	Scans surface at the nanoscale to provide topographical information and mechanical properties.	Nanoparticle surface morphology and interaction with biofilms.	[9]
	Transmission Electron Microscopy (TEM)	Provides high-resolution imaging of nanoparticles and their interactions with cells and biofilms.	Imaging nanoparticles in cancer cells and biofilm-forming bacteria.	[35]

Table 1. (Continued)

Field	Diagnostic/Laboratory Technique	Description	Applications	Refs
Cancer Research	Immunohistochemistry (IHC)	Uses antibodies to detect specific antigens in tissues, commonly used in cancer diagnosis.	Detection of cancer-specific biomarkers and biofilm involvement.	[47]
	Magnetic Resonance Imaging (MRI)	Non-invasive imaging technique used for detecting tumours and tissue abnormalities.	Early cancer detection, Monitoring of biofilm infections in tissues.	[48]
	Fluorescence In Situ Hybridisation (FISH)	Uses fluorescent probes to detect specific DNA or RNA sequences in cells.	Detection of cancer-associated genes, biofilm-related pathogens.	[49]
Biofilm Studies	Confocal Laser Scanning Microscopy (CLSM)	Provides 3D imaging of biofilms and cell interactions.	Studying the structure and dynamics of bacterial biofilms	[50]
	Crystal Violet Assay	Stains biofilm biomass to quantify biofilm formation.	Quantifying biofilm formation and inhibition	[34]
	Biofilm Flow Cell Model	A system for culturing biofilms under controlled flow conditions, mimicking natural environments.	Studying biofilm growth and resistance mechanisms.	[51]
Diagnostic Techniques in Nanomedicine	Surface Plasmon Resonance (SPR)	Measures the interaction of nanoparticles with biomolecules in real-time.	Studying nanoparticle–biofilm interactions and nanoparticle-targeting.	[36]
	Colorimetric Assays	Nanoparticles (e.g., gold, silver) are used to develop assays for pathogen detection based on colour change.	Diagnostics for bacterial infections, cancer biomarkers, and disease detection.	[52]
AI & Computational Tools	Artificial Intelligence (AI) in Imaging and Diagnosis	Uses AI algorithms for image analysis to detect abnormalities in cancer cells, biofilms, and other diseases.	Enhancing diagnostic accuracy in cancer, bacterial infections, and biofilms.	[40]
	Molecular Docking	Computational method used to model interactions between molecules, such as nanoparticles and biofilm-associated proteins.	Identifying potential compounds for biofilm treatment or cancer therapy.	[37]

Discussion: Challenges and Future Perspectives

The close interaction of microbiology, molecular biology, and nanotechnology presents a new opportunity for addressing the global health challenge. Despite advances in molecular diagnostics and progress in green nanotechnology, a significant gap in translation remains in clinical practice. The interaction between microbial research and nanotechnology has produced significant achievements, including nanoparticle-based biosensors for pathogen detection and the use of nanoparticles to disrupt biofilms [35]. There are still hurdles to overcome, though, to make such green synthesis methods scalable for industrial nanoparticle production. Factors such as batch variability, scalability, and compliance must be met to ensure quality and safety [54]. The rising number of antibiotic-resistant bacteria urgently requires alternative antimicrobial strategies. Multi-target action from nanoparticles, mainly synthesised via green methods, can be a potential solution [53]. The long-term human health and environmental effects of exposure to these nanoparticles are still largely unknown, necessitating further research [54]. Advancements in molecular diagnostics, such as CRISPR-Cas systems and imaging enhanced by nanoparticles, have reinvented the way diseases are detected and treated. These technologies enable early diagnosis, targeted therapies, and real-time monitoring of treatment efficacy [37]. However, integrating these tools into low-resource settings remains a challenge due to high costs and infrastructure requirements. Collaborative efforts are needed to adapt these technologies for global accessibility. Interdisciplinary research faces challenges, including nanoparticle cytotoxicity and the scalability of green synthesis methods. Integrating these fields with artificial intelligence can enhance predictive modelling and accelerate the translation of laboratory findings into clinical practice. Future research should focus on interdisciplinary collaboration to bridge the gaps between bench and bedside. Emphasis should be placed on developing cost-effective, scalable technologies that align with sustainability goals. Additionally, exploring the use of artificial intelligence in nanomedicine and diagnostics may enhance precision, reduce costs, and streamline clinical decision-making [56].

Conclusion

This review highlights interdisciplinary research that combines microbiology, molecular biology, and nanotechnology to overcome barriers and develop transformational solutions, including those for cancer management, biofilm control, and diagnostics. Such collaboration can tackle many essential healthcare-related issues. Moreover, the intersection between artificial intelligence (AI) and nanotechnology holds excellent promise for personalised and precision medicine, making cancer treatment more effective and minimally invasive. This may be achieved through an AI-enabled diagnostics and treatment scheme for rapid and accurate cancer detection, biofilm-related disease detection, and other diseases, facilitating timely intervention. Diagnostics and therapeutic modalities that utilise AI and nanotechnology synergistically certainly have potential for revolutionising healthcare delivery. However, the reality of these technologies in the clinical environment largely relies on a strong interdisciplinary effort to develop health education and management. Thus, health education is fundamental, ensuring that both patients and healthcare providers are well-informed about their condition and its possible benefits. However, healthcare providers should continually develop and adapt to new technologies as they emerge to meet the growing demand and enhance patient care, for instance, in areas such as early cancer detection, antimicrobial resistance, and biofilm management. Health management systems should be adaptive and proactive in incorporating these new developments while addressing the issues of cost, accessibility, and infrastructure in clinical practice. Most importantly, as nanotechnology and molecular biology continue to grow, equitable access will be a vital consideration in examining these advancements.

Conflict of Interest

The author declares that there is no conflict of interest regarding the publication of this article. The research was conducted independently, and there are no financial, personal, or professional relationships that could be perceived as influencing the content presented in this manuscript.

Acknowledgments

The author would like to express sincere gratitude to the Department of Biotechnology and the PhD Cell at the University of Engineering and Management, Kolkata, for their support and guidance throughout this work.

Disclaimer

None.

References

[1] H. Sung et al., "Global Cancer Statistics 2020: GLOBOCAN Estimates of Incidence and Mortality Worldwide for 36 Cancers in 185 Countries," *CA Cancer J Clin*, vol. 71, no. 3, pp. 209–249, May 2021, doi: 10.3322/caac.21660.

[2] D. Hanahan and R. A. Weinberg, "Hallmarks of Cancer: The Next Generation," *Cell*, vol. 144, no. 5, pp. 646–674, Mar. 2011, doi: 10.1016/j.cell.2011.02.013.

[3] M. J. Mitchell, M. M. Billingsley, R. M. Haley, M. E. Wechsler, N. A. Peppas, and R. Langer, "Engineering precision nanoparticles for drug delivery," *Nat Rev Drug Discov*, vol. 20, no. 2, pp. 101–124, Feb. 2021, doi: 10.1038/s41573-020-0090-8.

[4] H. C. Flemming, J. Wingender, U. Szewzyk, P. Steinberg, S. A. Rice, and S. Kjelleberg, "Biofilms: an emergent form of bacterial life," *Nat Rev Microbiol*, vol. 14, no. 9, pp. 563–575, Sep. 2016, doi: 10.1038/nrmicro.2016.94.

[5] X. Ma et al., "Nanotechnology in healthcare, and its safety and environmental risks," *J Nanobiotechnology*, vol. 22, no. 1, p. 715, Nov. 2024, doi: 10.1186/s12951-024-02901-x.

[6] Y. Zhang, M. Li, X. Gao, Y. Chen, and T. Liu, "Nanotechnology in cancer diagnosis: progress, challenges and opportunities," *J Hematol Oncol*, vol. 12, no. 1, p. 137, Dec. 2019, doi: 10.1186/s13045-019-0833-3.

[7] Y. Xu and Z. Li, "CRISPR-Cas systems: Overview, innovations and applications in human disease research and gene therapy," *Comput Struct Biotechnol J*, vol. 18, pp. 2401–2415, 2020, doi: 10.1016/j.csbj.2020.08.031.

[8] Y. Huang et al., "Nanotechnology's frontier in combatting infectious and inflammatory diseases: prevention and treatment," *Signal Transduct Target Ther*, vol. 9, no. 1, p. 34, Feb. 2024, doi: 10.1038/s41392-024-01745-z.

[9] J. Zhang et al., "Nanoparticle-based drug delivery systems to enhance cancer immunotherapy in solid tumors," *Front Immunol*, vol. 14, Aug. 2023, doi: 10.3389/fimmu.2023.1230893.

[10] T. F. Yi, J. P. Qu, X. Lai, X. Han, H. Chang, and Y. R. Zhu, "Toward high-performance Li storage anodes: design and construction of spherical carbon-coated

CoNiO2 materials," *Mater Today Chem*, vol. 19, p. 100407, Mar. 2021, doi: 10.1016/j.mtchem.2020.100407.

[11] H. Yang, Y. Sun, R. Cai, Y. Chen, and B. Gu, "The impact of dietary fiber and probiotics in infectious diseases," *Microb Pathog*, vol. 140, p. 103931, Mar. 2020, doi: 10.1016/j.micpath.2019.103931.

[12] B. Kumar, S. Lam, M. Adam, R. Gilroy, and M. J. Pallen, "The oesophageal microbiome and cancer: hope or hype?," *Trends Microbiol*, vol. 30, no. 4, pp. 322–329, Apr. 2022, doi: 10.1016/j.tim.2021.08.007.

[13] L. Gujinović et al., "Metagenomic analysis of pioneer biofilm-forming marine bacteria with emphasis on Vibrio gigantis adhesion dynamics," *Colloids Surf B Biointerfaces*, vol. 217, p. 112619, Sep. 2022, doi: 10.1016/j.colsurfb.2022.112619.

[14] D. K. Govindarajan et al., "Biofilms and oral health: nanotechnology for biofilm control," *Discover Nano*, vol. 20, no. 1, p. 114, Jul. 2025, doi: 10.1186/s11671-025-04299-3.

[15] Y. Zhang, K. Poon, G. S. P. Masonsong, Y. Ramaswamy, and G. Singh, "Sustainable Nanomaterials for Biomedical Applications," *Pharmaceutics*, vol. 15, no. 3, p. 922, Mar. 2023, doi: 10.3390/pharmaceutics15030922.

[16] J. M. Treherne and G. R. Langley, "Converging global crises are forcing the rapid adoption of disruptive changes in drug discovery," *Drug Discov Today*, vol. 26, no. 11, pp. 2489–2495, Nov. 2021, doi: 10.1016/j.drudis.2021.05.001.

[17] H.-Y. Min and H.-Y. Lee, "Molecular targeted therapy for anticancer treatment," *Exp Mol Med*, vol. 54, no. 10, pp. 1670–1694, Oct. 2022, doi: 10.1038/s12276-022-00864-3.

[18] A. Zhao, J. Sun, and Y. Liu, "Understanding bacterial biofilms: From definition to treatment strategies," *Front Cell Infect Microbiol*, vol. 13, Apr. 2023, doi: 10.3389/fcimb.2023.1137947.

[19] S. Gaidhani et al., "Biofilm disruption activity of silver nanoparticles synthesized by Acinetobacter calcoaceticus PUCM 1005," *Mater Lett*, vol. 108, pp. 324–327, Oct. 2013, doi: 10.1016/j.matlet.2013.07.023.

[20] S. T. Asma et al., "An Overview of Biofilm Formation–Combating Strategies and Mechanisms of Action of Antibiofilm Agents," *Life*, vol. 12, no. 8, p. 1110, Jul. 2022, doi: 10.3390/life12081110.

[21] M. A. Alghamdi, A. N. Fallica, N. Virzì, P. Kesharwani, V. Pittalà, and K. Greish, "The Promise of Nanotechnology in Personalized Medicine," *J Pers Med*, vol. 12, no. 5, p. 673, Apr. 2022, doi: 10.3390/jpm12050673.

[22] B. Muñoz-Palazon et al., "Novel application of aerobic granular biofilm systems for treating nitrate-polluted groundwater at low temperature: Microbial community and performance," *J Environ Chem Eng*, vol. 10, no. 3, p. 107818, Jun. 2022, doi: 10.1016/j.jece.2022.107818.

[23] A. Lin et al., "The microbiome in cancer," *iMeta*, vol. 4, no. 5, Oct. 2025, doi: 10.1002/imt2.70070.

[24] L. Kumar et al., "Advances in Nanotechnology for Biofilm Inhibition," *ACS Omega*, vol. 8, no. 24, pp. 21391–21409, Jun. 2023, doi: 10.1021/acsomega.3c02239.

[25] V. C. Thipe, P. Bloebaum, M. Khoobchandani, A. R. Karikachery, K. K. Katti, and K. V. Katti, “Green nanotechnology,” in *Nanomycotoxicology*, Elsevier, 2020, pp. 155–188. doi: 10.1016/B978-0-12-817998-7.00007-0.
[26] Y. K. Mohanta et al., “Nanotechnology in combating biofilm: A smart and promising therapeutic strategy,” *Front Microbiol*, vol. 13, Mar. 2023, doi: 10.3389/fmicb.2022.1028086.
[27] L. A. B. Ferreira et al., “Thiol-antioxidants interfere with assessing silver nanoparticle cytotoxicity,” *Nanomedicine*, vol. 24, p. 102130, Feb. 2020, doi: 10.1016/j.nano.2019.102130.
[28] V. C. Thipe et al., “Green nanotechnology—An innovative pathway towards biocompatible and medically relevant gold nanoparticles,” *J Drug Deliv Sci Technol*, vol. 70, p. 103256, Apr. 2022, doi: 10.1016/j.jddst.2022.103256.
[29] X. Li, H. Xu, Z.-S. Chen, and G. Chen, “Biosynthesis of Nanoparticles by Microorganisms and Their Applications,” *J Nanomater*, vol. 2011, pp. 1–16, 2011, doi: 10.1155/2011/270974.
[30] M. Eghbali-Arani, S. Pourmasoud, F. Ahmadi, M. Rahimi-Nasrabadi, V. Ameri, and A. Sobhani-Nasab, “Optimization and detailed stability study on coupling of CdMoO4 into BaWO4 for enhanced photodegradation and removal of organic contaminant,” *Arabian Journal of Chemistry*, vol. 13, no. 1, pp. 2425–2438, Jan. 2020, doi: 10.1016/j.arabjc.2018.05.007.
[31] K. Gupta et al., “Green silver nanoparticles: enhanced antimicrobial and antibiofilm activity with effects on DNA replication and cell cytotoxicity,” *RSC Adv.*, vol. 4, no. 95, pp. 52845–52855, 2014, doi: 10.1039/C4RA08791G.
[32] S. Sargazi et al., “Application of Green Gold Nanoparticles in Cancer Therapy and Diagnosis,” *Nanomaterials*, vol. 12, no. 7, p. 1102, Mar. 2022, doi: 10.3390/nano12071102.
[33] L. Dykman and N. Khlebtsov, “Gold nanoparticles in biomedical applications: recent advances and perspectives,” *Chem. Soc. Rev.*, vol. 41, no. 6, pp. 2256–2282, 2012, doi: 10.1039/C1CS15166E.
[34] S. Jain, D. G. Hirst, and J. M. O’Sullivan, “Gold nanoparticles as novel agents for cancer therapy,” *Br J Radiol*, vol. 85, no. 1010, pp. 101–113, Feb. 2012, doi: 10.1259/bjr/59448833.
[35] P. R. More, S. Pandit, A. De Filippis, G. Franci, I. Mijakovic, and M. Galdiero, “Silver Nanoparticles: Bactericidal and Mechanistic Approach against Drug Resistant Pathogens,” *Microorganisms*, vol. 11, no. 2, p. 369, Feb. 2023, doi: 10.3390/microorganisms11020369.
[36] G. Franci et al., “Silver Nanoparticles as Potential Antibacterial Agents,” *Molecules*, vol. 20, no. 5, pp. 8856–8874, May 2015, doi: 10.3390/molecules20058856.
[37] Z. Huang, C. J. Lyon, J. Wang, S. Lu, and T. Y. Hu, “CRISPR Assays for Disease Diagnosis: Progress to and Barriers Remaining for Clinical Applications,” *Advanced Science*, vol. 10, no. 20, Jul. 2023, doi: 10.1002/advs.202301697.
[38] B. Wouters, S. A. Currivan, N. Abdulhussain, T. Hankemeier, and P. J. Schoenmakers, “Immobilized-enzyme reactors integrated into analytical platforms: Recent advances and challenges,” *TrAC Trends in Analytical Chemistry*, vol. 144, p. 116419, Nov. 2021, doi: 10.1016/j.trac.2021.116419.

[39] R. Kadri et al., “Role of active nanoliposomes in the surface and bulk mechanical properties of hybrid hydrogels,” *Mater Today Bio*, vol. 6, p. 100046, Mar. 2020, doi: 10.1016/j.mtbio.2020.100046.

[40] N. Kazimierczak, W. Kazimierczak, Z. Serafin, P. Nowicki, J. Nożewski, and J. Janiszewska-Olszowska, “AI in Orthodontics: Revolutionizing Diagnostics and Treatment Planning—A Comprehensive Review,” *J Clin Med*, vol. 13, no. 2, p. 344, Jan. 2024, doi: 10.3390/jcm13020344.

[41] L. Chen and Y. Wen, “The role of bacterial biofilm in persistent infections and control strategies,” *Int J Oral Sci*, vol. 3, no. 2, pp. 66–73, Apr. 2011, doi: 10.4248/IJOS11022.

[42] H. Sung et al., “Global Cancer Statistics 2020: GLOBOCAN Estimates of Incidence and Mortality Worldwide for 36 Cancers in 185 Countries,” *CA Cancer J Clin*, vol. 71, no. 3, pp. 209–249, May 2021, doi: 10.3322/caac.21660.

[43] D. T. Mitchell and R. C. Cooke, “Some effects of temperature on germination and longevity of sclerotia in Claviceps purpurea,” *Transactions of the British Mycological Society*, vol. 51, no. 5, pp. 721–729, Dec. 1968, doi: 10.1016/S0007-1536(68)80092-0.

[44] C. W. Hall and T.-F. Mah, “Molecular mechanisms of biofilm-based antibiotic resistance and tolerance in pathogenic bacteria,” *FEMS Microbiol Rev*, vol. 41, no. 3, pp. 276–301, May 2017, doi: 10.1093/femsre/fux010.

[45] J. M. Treherne and G. R. Langley, “Converging global crises are forcing the rapid adoption of disruptive changes in drug discovery,” *Drug Discov Today*, vol. 26, no. 11, pp. 2489–2495, Nov. 2021, doi: 10.1016/j.drudis.2021.05.001.

[46] V. K. Sarhadi and G. Armengol, “Molecular Biomarkers in Cancer,” *Biomolecules*, vol. 12, no. 8, p. 1021, Jul. 2022, doi: 10.3390/biom12081021.

[47] G. A. Mansoori, P. Mohazzabi, P. McCormack, and S. Jabbari, “Nanotechnology in cancer prevention, detection and treatment: bright future lies ahead,” *World Review of Science, Technology and Sustainable Development*, vol. 4, no. 2/3, p. 226, 2007, doi: 10.1504/WRSTSD.2007.013584.

[48] P. Piskunen, R. Latham, C. E. West, M. Castronovo, and V. Linko, “Integrating CRISPR/Cas systems with programmable DNA nanostructures for delivery and beyond,” *iScience*, vol. 25, no. 6, p. 104389, Jun. 2022, doi: 10.1016/j.isci.2022.104389.

[49] A. Varma et al., “The role of nanocomposites against biofilm infections in humans,” *Front Cell Infect Microbiol*, vol. 13, Feb. 2023, doi: 10.3389/fcimb.2023.1104615.

[50] N. Shreyash, S. Bajpai, Mohd. A. Khan, Y. Vijay, S. K. Tiwary, and M. Sonker, “Green Synthesis of Nanoparticles and Their Biomedical Applications: A Review,” *ACS Appl Nano Mater*, vol. 4, no. 11, pp. 11428–11457, Nov. 2021, doi: 10.1021/acsanm.1c02946.

[51] M. P. Patil, D. Ngabire, H. H. P. Thi, M.-D. Kim, and G.-D. Kim, “Eco-friendly Synthesis of Gold Nanoparticles and Evaluation of Their Cytotoxic Activity on Cancer Cells,” *J Clust Sci*, vol. 28, no. 1, pp. 119–132, Jan. 2017, doi: 10.1007/s10876-016-1051-6.

[52] M. Eghbali-Arani, S. Pourmasoud, F. Ahmadi, M. Rahimi-Nasrabadi, V. Ameri, and A. Sobhani-Nasab, “Optimization and detailed stability study on coupling of

CdMoO4 into BaWO4 for enhanced photodegradation and removal of organic contaminant," *Arabian Journal of Chemistry*, vol. 13, no. 1, pp. 2425–2438, Jan. 2020, doi: 10.1016/j.arabjc.2018.05.007.

[53] M. Ramasamy and J. Lee, "Recent Nanotechnology Approaches for Prevention and Treatment of Biofilm-Associated Infections on Medical Devices," *Biomed Res Int*, vol. 2016, pp. 1–17, 2016, doi: 10.1155/2016/1851242.

[54] L. Chen and Y. Wen, "The role of bacterial biofilm in persistent infections and control strategies," *Int J Oral Sci*, vol. 3, no. 2, pp. 66–73, Apr. 2011, doi: 10.4248/IJOS11022.

[55] S. Shah, V. Dhawan, R. Holm, M. S. Nagarsenker, and Y. Perrie, "Liposomes: Advancements and innovation in the manufacturing process," *Adv Drug Deliv Rev*, vol. 154–155, pp. 102–122, 2020, doi: 10.1016/j.addr.2020.07.002.

[56] Y. Ijjeh et al., "Quality by digital design for accelerated sustainable nanomedicine development," *European Journal of Pharmaceutical Sciences*, vol. 213, p. 107239, Oct. 2025, doi: 10.1016/j.ejps.2025.107239.

About the Editors

Professor Koyel Ganguly
Department of Basic Science and Humanities,
Institute of Engineering & Management,
Salt Lake Electronics Complex,
School of University of Engineering and Management,
Kolkata, India
Email: koyel.ganguly@iem.edu.in

Professor Koyel Ganguly completed her PhD (Physics) in the field of Gravitation and Cosmology from Jadavpur University. She has research experience of more than 12 years and teaching experience of 14 years. She has published papers in reputed international journals and book chapters as well. She is also the managing editor of International Journal of Innovative Research in Physics published by Smart Society, USA. She has edited a book titled 'Emerging trends in Science and Technology' – Proceedings of International Conference on Emerging Trends in Science and Technology published by Narosa Publishing House and has also acted as Guest editor of Special issue on 2nd International Students' Conference on 'Innovations in Science and Technology ' in International Journal of Advanced Science and Engineering. She is also the editor of the proceedings of International Conference on 'Advanced Physics for Sustainable Future: Innovations and Solutions' in European Physical Journal (EPJ) Web of conferences.

Dr. Soumyadipta Pal
Associate Professor,
Department of Basic Science and Humanities,
Institute of Engineering & Management,
Salt Lake Electronics Complex,
School of University of Engineering and Management,
Kolkata, India
Email: soumyadipta.pal@iem.edu.in

Dr. Soumyadipta Pal has completed his PhD from S N Bose National Centre for Basic Sciences in the field of Condensed Matter Physics and

Material Science. He has more than 5 years' research experience and 10 years teaching experience. He has more than 20 journal papers in SCI indexed journals. He is an editor of the International Journal of Innovative Research in Physics published by Smart Society, USA. He is also the editor of the proceedings of International Conference on 'Advanced Physics for Sustainable Future: Innovations and Solutions' in European Physical Journal (EPJ) Web of conferences.

Professor Kamakhya Prasad Ghatak
Department of Basic Science and Humanities,
Institute of Engineering & Management,
Salt Lake Electronics Complex,
School of University of Engineering and Management,
Kolkata, India
Email: kamakhyaghatakcu@gmail.com

As per the World Ranking of top 2% Scientists as prepared by Stanford University, USA, (shorturl.at/qHIJ4), Professor Ghatak is the first Dr. of Engg. Deg. awardee of Jadavpur University, India, in 1991, (h-index-37, i-10 index-201, maximum citation of a research paper-330, Total Citation. -6054, the author of 385 research papers in SCI Journals and 25 research Monographs on Nano Technology from Springer, World Scientific, De Gruyter and NOVA USA respectively) stays within less than 1% band width in the field of Applied Physics. Professor Ghatak has successfully supervised 170 PhD students in different topics of Nanotechnology and the list includes Director, Vice-Chancellor, Principal, Professor and CEO of different reputed organizations. His present research interest is Quantum Materials and from 2015-2017, in accordance with Google Scholar, his number of research publications in consecutive three years have exceeded one hundred.

Index

A

B

C

D

E

F

G

H

I

K

L

M

N

O

P

R

S

T

U

V

W